Galaxies

SCIENCES

Universe, Field Director – Fabienne Casoli

Galaxies, Subject Head – Françoise Combes

Galaxies

Formation and Evolution

Coordinated by
Françoise Combes

iSTE

WILEY

First published 2020 in Great Britain and the United States by ISTE Ltd and John Wiley & Sons, Inc.

ISTE Ltd
27-37 St George's Road
London SW19 4EU
UK

www.iste.co.uk

John Wiley & Sons, Inc.
111 River Street
Hoboken, NJ 07030
USA

www.wiley.com

Library of Congress Control Number: 2020942773

British Library Cataloguing-in-Publication Data
A CIP record for this book is available from the British Library
ISBN 978-1-78945-012-5

ERC code:
PE9 Universe Sciences
PE9_8 Formation and evolution of galaxies
PE9_9 Clusters of galaxies and large scale structures

MIX
Paper from responsible sources
FSC® C013604

Contents

Introduction

The galaxies were not identified as worlds apart, as the philosopher Immanuel Kant foresaw, until the 1920s, just over a century ago. Before then, astronomers had observed nebulae, but had not distinguished between a cloud of emitting gas, such as the Orion Nebula, and outer galaxies, such as the Andromeda Nebula. At the beginning of the 20th Century, a great debate took place to know the size of our world (the Milky Way), and the distance of the various stars in the sky. In 1924, Edwin Hubble observed the variable stars in Andromeda of the Cepheids, which Henrietta Leavitt in 1909 had shown to be a good indicator of distance. Thus, it was shown that Andromeda was a galaxy "outside" ours, located about 2 million light-years away.

Progress since then has been meteoric. We now know millions of galaxies, and determine their distance because of the expansion of the Universe and the corresponding redshift. Due to the finite speed of light, we can go back in time, observing distant galaxies in their youth. We thus observe galaxies as far as the edge of our Universe, at the limit of our horizon, which allows us to reconstruct their history.

In this book, we will first describe the various morphologies and categories of galaxies, which are essential for a better understanding of their formation and evolution. There are several classifications, depending on whether we consider the various stellar components: bulge, disc, spheroid, photometry and more or less blue colors according to the rate of star formation, or the kinematics of stars, that is to say a kinetic energy either dominated by rotation, or by the velocity of dispersion, or the fraction of gas. Disc galaxies generally contain spiral or barred structures, which are the engine of evolution. These structures give rise to resonances which will be the source of rings, or pseudo-rings, very useful to know the speed of spirals and bars.

Our Galaxy, the Milky Way, the best known and most familiar, has a barred spiral structure, yet it took a long time to be identified, because we are in its plane: it appears edge-on, obscured by dust lanes. Its structure appears more clearly in the near infrared. It consists of a thin disc, where the gas and young stars are located, and a thick disc, which dates back billions of years. It has a pseudobulge, mainly due to the bar, and its vertical resonance. There is also a more or less spherical, diffuse halo of stars, whose formation would be essentially due to the accretion of small satellite galaxies. These are destroyed by tidal interaction, and deploy in a multitude of stellar streams. We can go back to the history of the formation of the different components, by galactic archaeology, by determining the age and metallicity of stars. The GAIA astrometric satellite has recently made enormous progress in specifying the distances and proper motions of a large number of webs.

Galaxies can be divided into two main categories: disc galaxies or "late-type" and elliptical galaxies or "early-type". In our Local Universe, the majority of galaxies are disc galaxies and dominate in number, but since elliptical galaxies are more massive, they dominate in mass in the Universe. Early-type galaxies have a spheroidal morphology, and even if they sometimes have a stellar disc, it is subdominant in front of the massive bulge. Their stellar populations are old, and they have little or no gas. For a very long time, astronomers thought that the flattening of stellar spheroids was due to rotation, but when it became possible to measure their velocities, it was realized that the shapes of these spheroids were instead due to the anisotropy of the velocity dispersion: ellipticals have no or very little rotation. It is then difficult to deproject these systems, which can be triaxial. They are classified as slow or fast rotators. The origin of these differences comes from their history of formation by merger of spiral galaxies.

Late-type disc galaxies are also called spiral galaxies. Their structure of open spiral arms has long been a mystery, because the differential rotation of the disc should wind up the arms around very quickly, making them disappear. But they are actually density waves and not material arms. The arms are quite transient and several waves follow one another; on the other hand, the bars are more robust structures, which allow the galaxy to assemble its mass and evacuate the angular momentum of the accreted gas. The speed of the bar wave can be determined by the different resonances in the plane (forming rings) and perpendicular (forming peanuts or boxes). The discs of the spirals evolve essentially in a secular way, but from time to time an interaction or merger with another galaxy can increase the mass of the bulge and the mass concentration. Star formation is enriched by the accretion of gas from the cosmic filaments, which keeps the galaxy blue and active. On the other hand, the environment of a group or cluster of galaxies can dry up the gas reservoir and stop star formation. The galaxy then becomes red and passive. Among the millions of known galaxies, a clear bimodality has been observed between the red sequence and the blue cloud.

Interactions between galaxies contribute to the enrichment of galaxies with gas, which can create a starburst. Interactions are visible even at large distances because of tidal arms and morphological perturbations of galaxies. Of course, these effects are all the more visible in groups and clusters of galaxies. In the latter, galaxies are swept away from their gas by the ram pressure. Looking back in time, we can see that the number of interacting galaxies was greater in the past, in the first part of the age of the Universe. This could explain part of the history of cosmic star formation.

In today's galaxy clusters, we observe a morphological segregation: the spirals that dominate in number in the rest of the Universe are gradually disappearing to make way for lenticulars and ellipticals. Galaxies in clusters no longer form many stars. But this was not the case in the past. Already the first observers of distant clusters had noticed blue galaxies, the so-called Butcher–Oemler effect. There must even have been a time in the Universe, at the time of cluster formation, when interactions between galaxies make clusters or proto-clusters richer in star formation than elsewhere in the field. The evolution of galaxies and the history of their star formation have made a lot of progress in recent years, due in particular to several infrared satellites, which have revealed galaxies obscured by dust, where star formation is hidden in the visible, but emerges only in the infrared radiation of dust heated by young stars. The spectral distribution of the radiation from galaxies also makes it possible to distinguish the amount of star formation, and to separate the heating that is due to the energy of super-massive black holes from that due to the energy of stars. The efficiency of star formation varies over time, due to the fraction of gas in galaxies, which was much larger in the past, and to the feedback phenomena of active nuclei, corresponding to the rapid growth of black holes.

The enormous progress made in our knowledge of galaxies should not make us forget all that remains to be discovered, especially since most of the mass of a galaxy is made of exotic dark matter, the nature of which we do not know!

1

The Classification of Galaxies

Ronald BUTA
Astrophysics, University of Alabama, Tuscaloosa, USA

The classification of the forms of galaxies in a well-defined visual system is a critical step in the study of galaxies as physical objects. The Comprehensive de Vaucouleurs revised Hubble-Sandage (CVRHS) system is currently the most detailed approach that can be applied effectively to more than 95% of all galaxies. This chapter describes the different types of galaxies and the factors that may determine various morphological features.

1.1. Introduction

Galaxies are complex gravitational systems whose structure has been influenced not only by how they formed but also by the environment into which they were born. A century ago, getting a classifiable image of a single galaxy was a major effort involving long exposures of photograph plates. Today, there are classifiable images of literally millions of galaxies available through the Internet. Although it is not obvious how any galaxy arrived at its current morphological state, examination of the details of large numbers of galaxies have led to important physical insights into the roles played by both internal and external processes. It is for this reason that classical galaxy morphology and classification have survived into the modern era.

Galaxy morphology and classification began with simple descriptive terms for angular size, brightness and central concentration based on visual observations (e.g. Dreyer 1888). Although large 19th-Century reflectors did reveal through visual observation some genuine aspects of galaxy morphology, it was photography that led

Galaxies,
coordinated by Françoise COMBES.

to the classification systems of the early-to-mid 20th Century, including those of Wolf (1908), Reynolds (1920), Hubble (1926, 1936), Lundmark (1926) and Morgan (1958). On the plates of the time, which were relatively more sensitive to blue light than to red light, details of spiral arms, disks and bulges could be used as classification criteria. In spite of great observational and theoretical progress in extragalactic studies during the past century, modern galaxy classification is still basically tied to the system proposed by Hubble (1936; Figure 1.1), only now the system is applied using digital images rather than photographic plates. The main reason the Hubble system has survived for nearly a century is that the aspects Hubble focused on (degree of central concentration, pitch angle and resolution of spiral arms, amount of distinct structure) correlated with measured properties of galaxies, such as luminosities, colors, stellar populations, HI content and global star formation history. This gave his view an astrophysically relevant edge that has driven much of extragalactic research since his time.

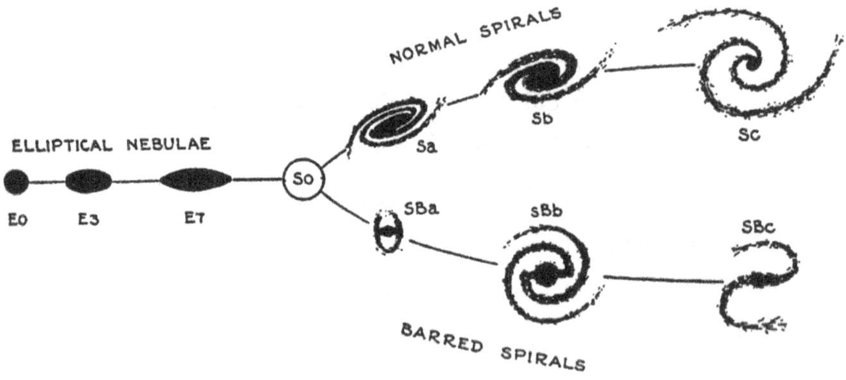

Figure 1.1. *The Hubble (1936) "tuning fork" representation of galaxy morphology*

This chapter describes the different classes of galaxies within the framework of the "Comprehensive de Vaucouleurs revised Hubble–Sandage" (CVRHS) classification system, a visual system that follows the precepts of de Vaucouleurs (1959), who proposed a personal revision of the Hubble–Sandage (HS) classification (Sandage 1961; Figure 1.2) that he believed provided a better description of galaxies without being too unwieldy. For many galaxies, the CVRHS classification is no more complicated than a de Vaucouleurs (1959) VRHS classification. However, the system has been designed to take into account more details that are of astrophysical interest today and which have become more noticeable and relevant in the era of digital astronomical imaging. These details include lenses, nuclear rings and bars, ansae bars, boxy/peanut bulges, boxy and disky elliptical galaxies, special outer rings and pseudorings, dust lanes and galactic disk warps.

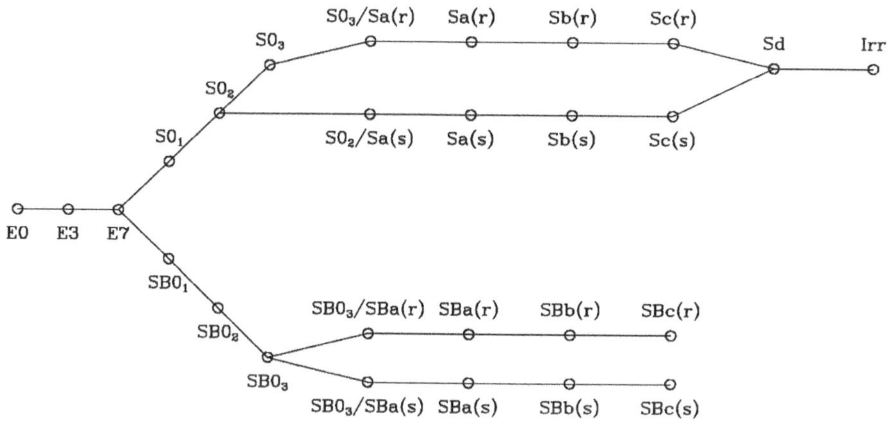

Figure 1.2. *The Hubble–Sandage (Sandage 1961) revised "tuning fork" representation of galaxy morphology, including stages of S0 galaxies*

1.2. Classes of galaxies

Hubble recognized that there are basically two classes of galaxies: disk-shaped galaxies and non-disk-shaped galaxies. In a disk galaxy, the structure is dominated by a highly flattened stellar component. Within this disk, other structures may be seen, such as spiral arms, bars, rings, a central bulge and extensive distributions of interstellar gas and dust. The way these structures are seen depends on the inclination of the plane of the disk to our line of sight. For a given galaxy, we say the inclination i is 0° when the disk is seen face-on, and 90° when the disk is seen edge-on. Disk planes are randomly oriented to the line of sight (meaning they are uniformly distributed in $\sin i$), which complicates the interpretation of highly inclined cases. Even a century ago, disk-shaped galaxies were known to be more common than non-disk galaxies to the point that the latter were considered of greater interest (Keeler 1899).

Although not all disk-shaped galaxies are spiral, the typical disk galaxy is a spiral, where luminous, outwardly winding arcs of stars and often star-forming regions form a major part of the morphology. Classic nearby spirals such as M51, M81, and M101 initially fueled the misperception that spirals are generally regular systems having only a bulge in addition to the disk and spiral arms. It was thought by Hubble that barred spirals are significantly less abundant than non-barred spirals, or what he called "normal spirals". It is now known that barred spirals are at least as abundant as non-barred spirals, and that some galaxies that appear to be non-barred in blue light can appear to be barred when imaged at infrared wavelengths (Eskridge *et al.* 2000).

The existence of non-disk-shaped (and therefore non-spiral) galaxies was at first somewhat controversial. Based on the limited plate material available in his day, Curtis (1918) believed that all galaxies were spiral, and any that did not appear to be spiral on his plates would be found to be spiral when observed with larger telescopes. Hubble, having access to better telescopes, disagreed with this conclusion and believed that genuine non-spiral galaxies existed. Hubble recognized such galaxies as elliptical galaxies, where the luminosity distribution is defined by a regular decline from a bright center to the faint outer regions. Ellipticals were thought to be characterized by no other features but their isophotal shapes, which ranged from round (E0) to a flattening approaching that of disk-shaped galaxies (E7). In fact, Hubble believed the sequence of E galaxy shapes blended smoothly into the domain of disk-shaped galaxies, which he split into two parallel sequences of normal and barred spirals characterized by the degree of central concentration, the degree of resolution of the spiral arms into what were likely to be star-forming regions and the degree of openness of the spiral arms. Hubble (1936) illustrated these features in his famous "tuning fork" of galaxy morphologies (Figure 1.1). Although this view is now obsolete, the tuning fork is still an effective way of binning galaxies into astrophysically meaningful classes.

In stellar astronomy, temporal terms are often used to describe certain kinds of stars. Stars of spectral classes O, B and A tend to be young and are known as "early-type" stars, while those of spectral classes K and M tend to be older and are called "late-type" stars. Those of spectral classes F and G are known as "intermediate-type" stars. Hubble found it convenient to use similar terms for galaxies, calling galaxies on the left part of the tuning fork early-type galaxies and those on the right part late-type galaxies. In general, E and S0 galaxies are said to be early-type galaxies, S0/a and Sa galaxies are called early-type spirals, Sc-Sm galaxies are called late-type spirals and Sab-Sbc galaxies are called intermediate-type spirals. Unlike for stars, no real temporal meaning was to be implied by these terms.

The value of the Hubble classification system, as well as its limitations, can only be appreciated by examining large numbers of galaxies and attempting to classify them within that system. The small number of astronomers who actually did this, such as Sandage, de Vaucouleurs, van den Bergh and Morgan, all saw the need for modifications or even alternative views, which led to major revisions of the system. This article uses images from the EFIGI project (Baillard *et al.* 2011; de Lapparent *et al.* 2011) to illustrate galaxies of different types within the framework of the CVRHS classification system (Buta *et al.* 2015). The classifications are from Buta (2019).

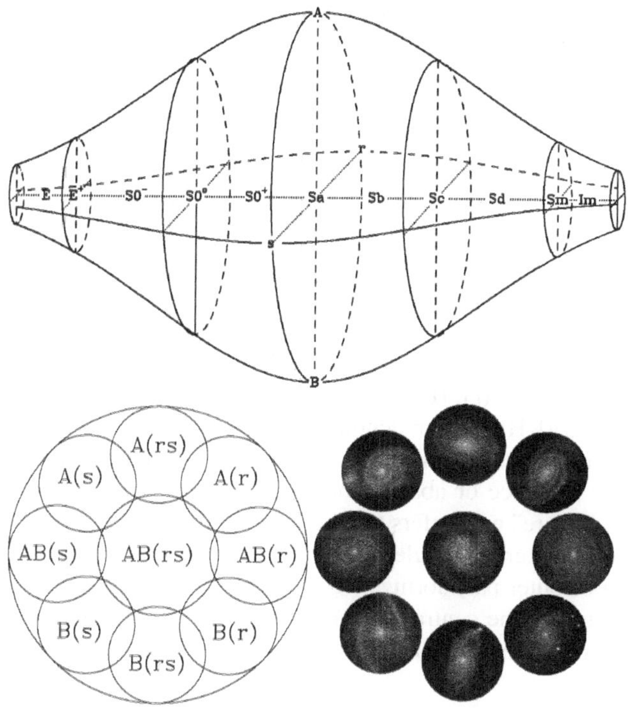

Figure 1.3. *The de Vaucouleurs (1959) revised Hubble–Sandage (VRHS) system of galaxy classification, where the Hubble "tuning fork" has become a classification volume. Top: The VRHS stage sequence. Bottom left: The cross-section of families and varieties. Bottom right: Examples of families and varieties at stage Sbc (from Buta et al. 2007)*

The basic outline of the HS system (Sandage 1961) is depicted in Figure 1.2 and that for the VRHS system is depicted in Figure 1.3. In the HS system, Sandage added the (r) and (s) subtypes, which are also used in the VRHS. The HS system still depicts galaxy morphology as a "tuning fork", but is more complicated than Hubble's original tuning fork (Figure 1.1). In the VRHS, the tuning fork is replaced with a classification volume whose long axis is the stage (E..S0..Sa, Sb..Im, etc.), and whose short axes are the family (SA, SAB, SB) and the variety [(s), (rs), (r)]. The use of "SAB" to denote galaxies having a bar of intermediate apparent strength and "(rs)" for partial inner rings (pseudorings) are major hallmarks of the VRHS and CVRHS systems. Not depicted in the volume are outer rings, large structures about twice the size of a bar in barred galaxies and nuclear rings, much smaller structures about one-tenth the size of a bar in barred galaxies. The appearance of the volume with a broad middle section and narrower ends is intended to depict how the diversity of families and varieties varies with stage. The greatest diversity is found near stage S0/a, while the diversity diminishes by stages S0 and Im either because bars become difficult to recognize or because rings are simply not present.

The CVRHS is defined by the same kind of classification volume as the VRHS, but with a more finely detailed division into various subtypes. Figure 1.4 shows a cross-section of types through any CVRHS stage from S0° to Sm. The arrangement of families and varieties is the same as in the VRHS, except that de Vaucouleurs' vision of continuity is carried further with the use of *underline notation* that he had applied mainly in his 1963 paper, "Revised Classifications of 1500 Bright Galaxies". The family sequence became SA, S$\underline{A}$B, SAB, SA$\underline{B}$, SB and the variety sequence became (s), (r$\underline{s}$), (rs), ($\underline{r}$s), (r). However, in order to recognize outer and nuclear rings in the system, the CVRHS system denotes the presence or absence of an inner feature as the "inner variety", the presence or absence of an outer feature as the "outer variety", and the presence or absence of a nuclear feature as the "nuclear variety". The term "inner feature" now refers to inner rings and lenses, not just inner rings, and the same is true for outer and nuclear features (the latter also including nuclear bars). The recognition of outer pseudorings (R′) is also one of the hallmarks of the VRHS and CVRHS systems. The central subtype in Figure 1.4, (R)SAB(rs,nr), is meant to convey the generalization of VRHS varieties to include inner, outer and nuclear varieties.

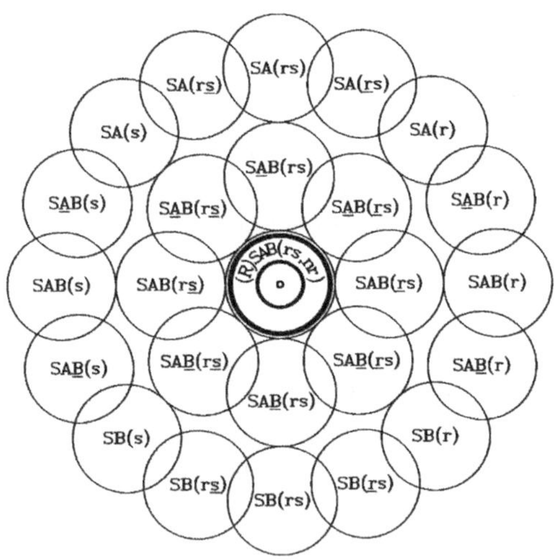

Figure 1.4. *A cross-section through the CVRHS system of families and varieties including the de Vaucouleurs (1963) underline notation. The central "cell" is meant to extend the concept of galactic varieties to outer and nuclear features, as well as additional inner features such as lenses*

de Vaucouleurs (1963) also introduced the underline notation for stages, as in SA(s)a$\underline{b}$, which means "closer to Sb than to Sa". Although this notation can be applied directly as for families and inner varieties, it has appeared mainly in classifications based on averages of multiphase (or repeat) efforts designed to check for consistency (e.g. Buta *et al.* 2015). This usage of underlines for stages, families and varieties considerably increases the number of "cells" in the CVRHS as compared to the VRHS, and one may well question whether some of these cells might not have any entrants. Buta (2019) showed that for the EFIGI sample, the cells of the CVRHS system are roughly evenly occupied among late S0s and early-type spirals, but not among later-type spirals. This means the classification volume lacks the symmetry depicted in Figure 1.3.

In the following sections, the different types of galaxies are described in more detail.

1.3. Elliptical galaxies

For more than two decades, elliptical galaxies were the subject of the most intense extragalactic research, spawning many meetings and research papers that underscored how deceptive the simple appearance of these galaxies is. When studied in detail both photometrically and spectroscopically, it becomes clear that the history of ellipticals is likely very complex. The main issues examined in detail are as follows: isophote shapes and orientations; characteristics of luminosity profiles; the presence of dust, rings, and peculiar structures; how low-luminosity ellipticals connect to high luminosity ellipticals; and the characteristic scaling relations between measured quantities such as physical size, surface brightness and central velocity dispersion.

The VRHS system as depicted in Figure 1.3 interprets E galaxies in terms of two stages: E and E^+. A normal elliptical (E) shows a smooth decline in brightness as a function of galactocentric distance, with no inflections or other apparent structure. Type E^+ galaxies are considered as "late" ellipticals showing some traces of structure. The structure can be in the form of a subtle lens, a feature that when more apparent is seen to have a shallow brightness gradient interior to a sharp edge (Sandage 1961; Kormendy 1979). In the Third Reference Catalogue of Bright Galaxies (RC3; de Vaucouleurs *et al.* 1991), the E^+ classification was also applied to brightest cluster members having a shallow brightness gradient (or an apparent extended envelope) in the outer regions. In the Morgan (1958) classification system, such objects are classified as type cD. Examples of E and E^+ galaxies are shown in Figures 1.5 and 1.6. Other aspects of E galaxy morphology include the following:

Disky versus boxy isophotes: The shapes of E galaxy isophotes carry important clues to their history. Kormendy and Bender (1996) noted how Hubble's main E galaxy classification criterion, the apparent flattening (as in, e.g. E3), had little physical significance, being determined mainly by how a galaxy is viewed. Very few ellipticals are recognized as being more flattened than E4. NGC 4623 in Figure 1.5 is as close as any real galaxy comes to being a genuine E7. Nevertheless, it shows a faint trace of a thin disk (section 1.5).

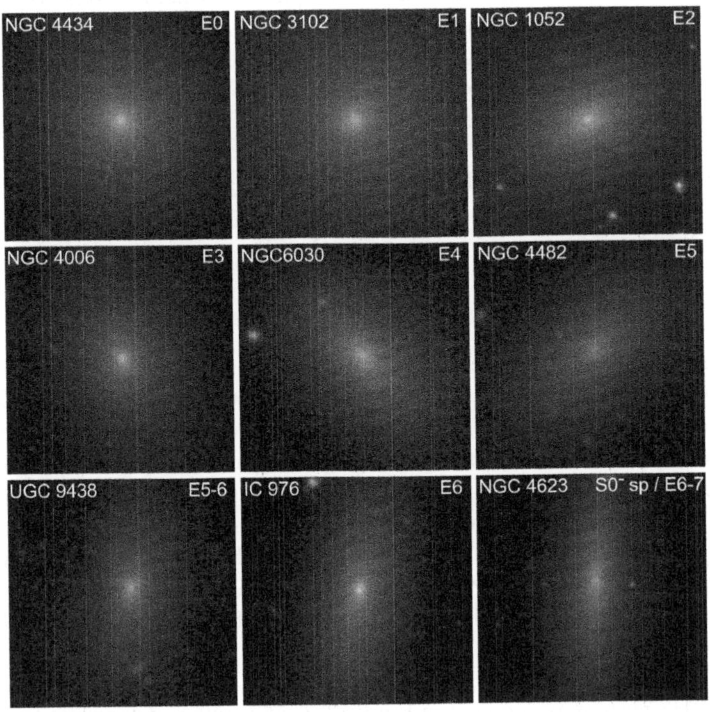

Figure 1.5. *Examples of elliptical galaxies ranging from round in apparent shape (E0) to fairly elongated (E6). Sandage and Bedke (1994) considered NGC 4623 (lower right frame) to be a possible example of a genuine E7 galaxy. Very few ellipticals are more flattened than E4*

Capaccioli (1987) noted two families of elliptical galaxies at a time when any trace of a disk component in a galaxy classified as type E was taken to mean that the galaxy was a misclassified S0 galaxy. Kormendy and Bender (1996) proposed a reclassification of ellipticals into disky and boxy categories based on the sign of the $\cos 4\theta$ relative Fourier deviation from perfectly elliptical isophotes. If the relative amplitude of this term is positive, the isophotes are said to be cuspy (or "disky"), but if negative, the isophotes are boxy. Cuspy isophotes are interpreted as indicating the presence of a subtle disk component, likely to be made of accreted material. Boxy isophotes are also thought to be indicative of interactive history. If an E galaxy is disky, it is classified as type E(d), while if an E galaxy is boxy, the type is E(b). The distinction of these subcategories can depend on how we are viewing the structures. For example, if the subtle disk in an E galaxy is oriented more nearly face-on rather than edge-on, then the isophotes will not necessarily show a cuspy shape. Both boxiness and diskiness are favored to be seen in the edge-on view. Several examples are shown in Figure 1.7. The CVRHS adopts the E(b) and E(d) terminology of Kormendy and Bender (1996) when these distinctions are obvious in a visual inspection of images.

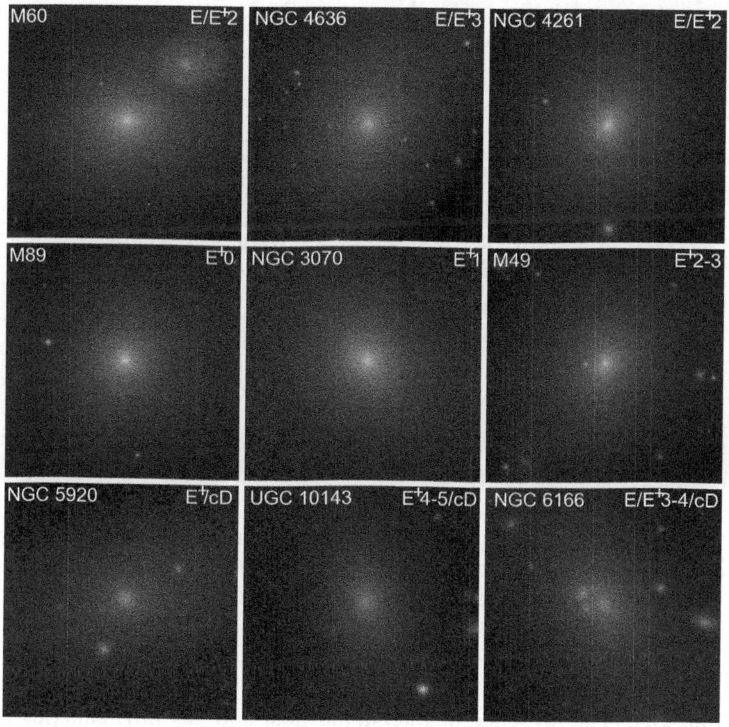

Figure 1.6. *E^+ galaxies are generally elliptical galaxies with very subtle traces of structure, usually in the form of a lens, a feature with a shallow brightness gradient interior to a sharp edge. Lenses are more prominent in S0 galaxies. The bottom row shows three brightest cluster members that are best interpreted as Morgan cD (supergiant) galaxies but which have also been considered E^+ galaxies by de Vaucouleurs*

Isophote twisting: Isophote twisting, where the ellipticity and position angle of the major axis of the isophotes of an E galaxy change systematically with increasing radius, is thought to be the principal evidence favoring triaxial intrinsic shapes (e.g. Fasano and Bonoli 1989). In a triaxial galaxy, there are three principal axes, each having a different radial scale-length. This highlights how the *observed* ellipticity of an E galaxy cannot be interpreted as easily as that of a disk-shaped galaxy. There is also the possibility, as noted by Nieto (1988), that some isophote twisting in E galaxies is due to an incipient bar structure in a very "early" SB0 galaxy. Tsatsi *et al.* (2017) detected evidence of prolate rotation in eight early-type galaxies, again supporting the idea of triaxial shapes.

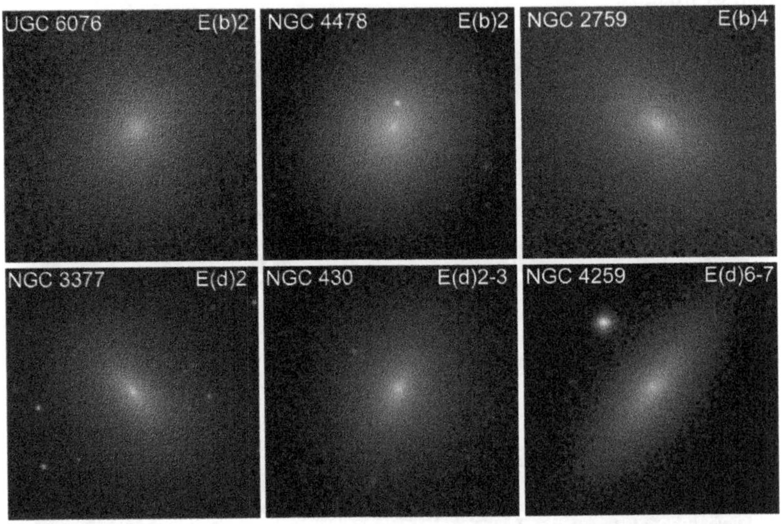

Figure 1.7. *Examples of disky and boxy elliptical galaxies. Generally, the disky/boxy nature of E galaxy isophotes can only be determined with detailed isophotal ellipse fits. The examples illustrated here are more obvious cases that can be distinguished in color displays of the images*

Dust and gas: The presence of interstellar dust and gas in E galaxies is a likely indicator of interaction/merger history (Schweizer 1987). The orientations of planar dust lanes may be tied to the likely triaxial structure of ellipticals. This follows from the presence of *minor axis* dust lanes, where the dust appears along the short apparent axis of the isophotes (Bertola and Galletta 1978). In disk-shaped galaxies, the planar dust lane always appears parallel to the apparent major axis. In general, minor, major and intermediate axis dust lanes are found in E and S0 galaxies. Examples are shown in Figure 1.8. In the CVRHS system, dust-lane Es are specified as E (dust-lane).

Figure 1.8 also includes the peculiar example of NGC 4459, where a small dusty ring, here recognized as a nuclear ring, is found. Although the presence of such a feature would normally warrant a classification of SA(r)0^+ (Buta *et al.* 2007), the ring is clearly a small disk embedded in a much larger elliptical galaxy.

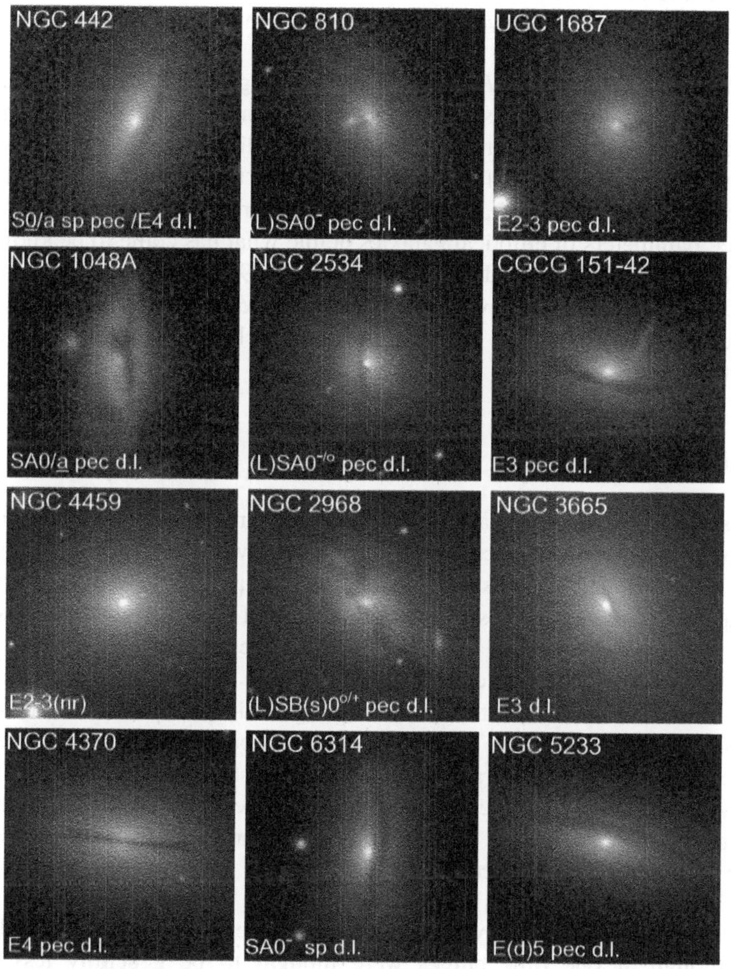

Figure 1.8. *Twelve examples of dust-lane (d.l.) early-type galaxies, including ellipticals and S0s. Major axis (e.g. NGC 442, 4370 and 6314) and minor axis (e.g. NGC 810) cases are included*

Luminosity profiles: de Vaucouleurs (1948) first recognized the well-known "$r^{\frac{1}{4}}$ law" for ellipticals, and for a while it was generally thought that this characterized the luminosity distributions of all E galaxies. When the Sérsic (1968) $r^{\frac{1}{n}}$ approach was used (e.g. Huang *et al.* 2013 and references therein), it became clear that $n = 4$

only characterized the more luminous ellipticals (see review by Graham 2013). Lower luminosity ellipticals have lower values of n. Other aspects of luminosity profiles include the distinction between "core Es", which have a flat central profile, and "power-law Es", which have a central cusp (Kormendy and Bender 1996).

The fundamental plane of E galaxies: Elliptical galaxies are characterized by a well-defined interconnectedness of their physical parameters. In the three-dimensional space defined by the parameters R_e, the effective radius that transmits half the total luminosity; I_e, the average surface brightness within this radius; and σ_o, the central velocity dispersion, an important characteristic emerges. The correlations between these three parameters of elliptical galaxies define what has come to be known as the "fundamental plane" (Kormendy and Djorgovski 1989). The plane tells us that larger E galaxies tend to have lower average surface brightness than smaller E galaxies, and that more luminous E galaxies have higher central velocity dispersion than do lower luminosity E galaxies.

1.4. Spiral galaxies

The spiral structure of galaxies was discovered more than 170 years ago. The subtle patterns were first detected in 1845 with the world's largest telescope at the time, the "Leviathan of Parsonstown" located in central Ireland. William Parsons, the Third Earl of Rosse, visually saw the spiral arms of the "Whirlpool Galaxy" M51 with his newly built 72-inch speculum metal reflector. In the parlance of 19th Century astronomy, M51 was called a "nebula", not a galaxy, although the general view at the time was that most or all nebulae were distant systems of stars like the Milky Way ("Island Universe" hypothesis). Parsons built the Leviathan partly to test this idea. The discovery of spiral structure added mystique to the nebulae, and led to alternative ideas as to what the nebulae actually were. It would be nearly a century after Parsons' discovery that any serious understanding of the nature of spiral structure would be achieved (section 1.11).

Spiral galaxies are generally two-component systems consisting of a bulge and a disk. Although at one time bulges were thought to be generally less flattened components than disks, it is now clear that bulges include a mix of dissimilar structures, such as spheroidal "classical" bulges, highly flattened "pseudobulges" (Kormendy and Kennicutt 2004) and "boxy/peanut" bulges, the latter thought to be due to edge-on views of bars (e.g. Lutticke and Dettmar 1999).

The classification of spirals is generally based on a rough correlation between the degree of central concentration and the character of the spiral arms. Hubble had noticed that galaxies with tightly wrapped, relatively smooth arms tended to have bright central bulges, while galaxies having open, relatively patchy spiral arms tended to have very small bulges. Hubble (1926) called the former cases Sa galaxies and the latter cases Sc galaxies, with Sb galaxies being intermediate between the two types. This observation set the stage for galaxy classification for nearly a century. However, the correlation works best for non-barred galaxies. It is poorer for barred galaxies that, even in cases with smooth, tightly wrapped arms, can have very small bulges. An example is NGC 3351, type SBb, which has a very small bulge in the center of a bright nuclear ring (Buta *et al.* 2007). Because of such inconsistencies, Sandage (1961) advocated basing spiral stage classifications (i.e. Sa, Sb, etc.) mainly on the appearance of the spiral arms.

Figure 1.9 shows a full CVRHS stage sequence for non-barred galaxies from stage S0° to stage Sm, that is, from the intermediate S0 stage (Figure 1.3) to the latest stage on the spiral sequence. The spiral sequence begins with the stage S0/a, which is considered to be a transition type between S0s and spiral galaxies. Type S0/a is a legitimate type in the sense that it is easily recognized and the continuity that de Vaucouleurs envisioned seems well represented by the type. Nevertheless, this apparent continuity does not necessarily imply that all S0s are correctly placed in the Hubble (1936) "tuning fork". The stage generally begins with pseudorings made of tightly wrapped spiral structure as in NGC 809. In the CVRHS classification, stage S$\underline{0}$/a is closer to S0 than to Sa, while S0/$\underline{a}$ is closer to Sa than to S0.

The sequence for non-barred galaxies in Figure 1.9 shows the rough correlation between central concentration and stage. Bulges are most prominent at stages Sbc and earlier, and are least prominent at stages Sc and later. The sequence shows well how arms are smooth at stage Sa and knotty, well-resolved, more open features at stage Sc. Intermediate stages are as well defined as regular stages: Sab galaxies often resemble Sa galaxies but with a greater degree of resolution into star-forming regions; Sbc galaxies typically have the bulge of an Sb galaxy in a disk with Sc arms; Scd is recognized as an Sc galaxy with only a trace of central concentration; and Sdm galaxies are typically bulge-less asymmetric systems with an offset bar and one spiral arm longer than the other. Similar underline stages (e.g. S$\underline{a}$b. Sc$\underline{d}$) are used throughout the CVRHS sequence (de Vaucouleurs 1963).

Figure 1.10 shows the same kind of stage sequence for barred galaxies. Initially, Hubble (1926) believed that non-barred galaxies were the "normal" form of spirals, with perhaps maybe 20% of the spirals being barred. He nevertheless envisioned barred spirals as falling on a sequence parallel to that of non-barred spirals. In the Hubble Atlas of Galaxies, Sandage (1961) smoothly connects non-barred and barred S0s with the non-barred and barred spiral prongs, respectively, which is also true for the VRHS and the CVRHS systems. The same kinds of types are recognizable among barred spirals as among non-barred spirals. However, Figure 1.10 shows the small bulge effect in early-type barred spirals, an example being NGC 5610 whose smooth arms wrap into an outer pseudoring but whose bulge is no more prominent than that in an Sc galaxy.

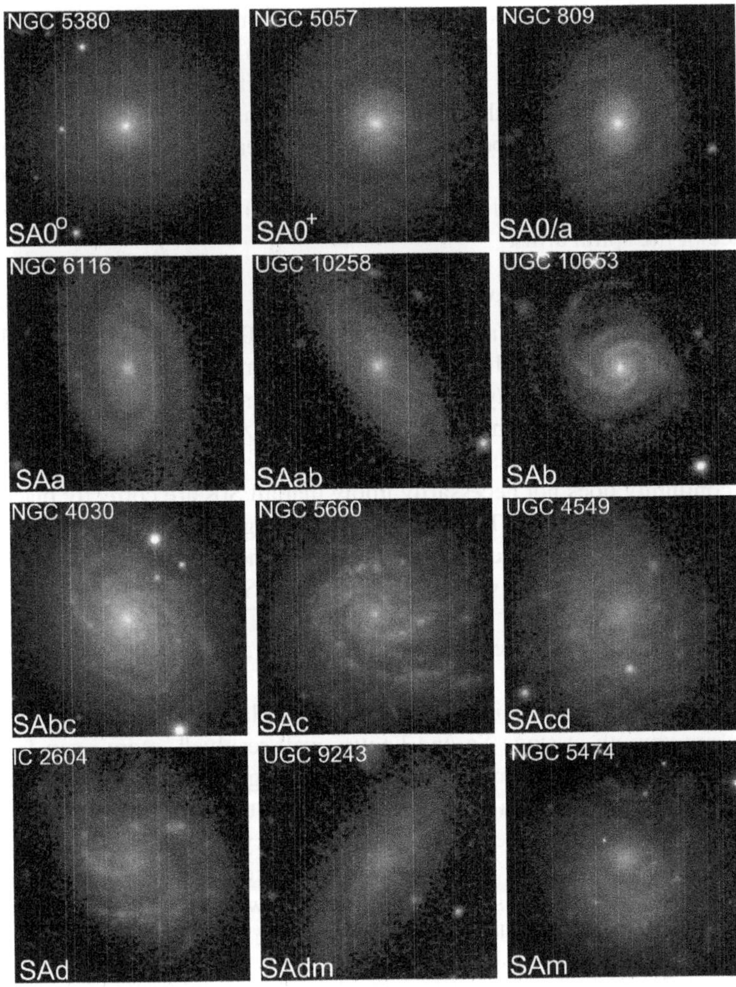

Figure 1.9. *A sequence of stages for non-barred galaxies in the VRHS/CVRHS system*

The bars of spiral galaxies: As already noted in section 1.2, the CVRHS classification of bars utilizes five categories, SA, S$\underline{A}$B, SAB, SA$\underline{B}$ and SB (de Vaucouleurs 1963), in a sequence of increasing visual bar strength (Figure 1.11). The classification is based on the length, contrast and axis ratio of the bar or bar-like feature. Although leading dust lanes are not a classification criterion for bars, such lanes are often seen in barred spirals of types S0/a to Sbc and could impact the apparent bar strength. Based on CVRHS classifications, the bar fraction is about 50%

for SAB, SA$\underline{\text{B}}$ and SB cases, but increases to 67% if S$\underline{\text{A}}$B is included (Buta *et al.* 2015; Buta 2019). The cosmological significance of the bar fraction is discussed by Sheth *et al.* (2008).

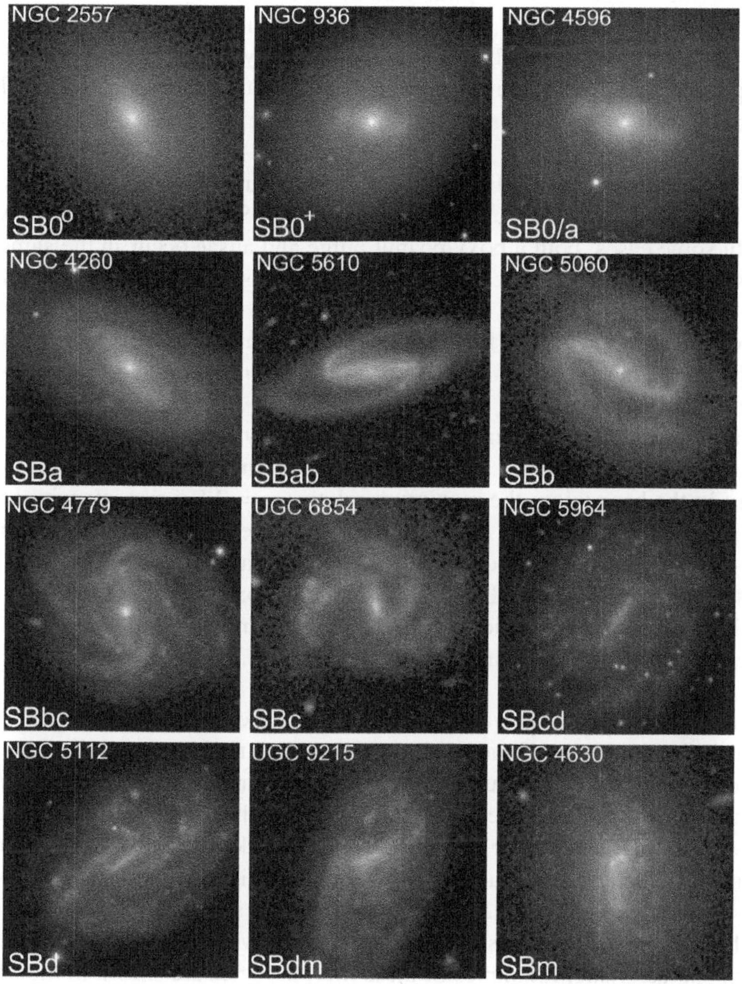

Figure 1.10. *A sequence of stages for barred galaxies in the VRHS/CVRHS system*

There are additional aspects of bars recognized in CVRHS classifications. Most bars are regular bars, the kinds of features seen in classic barred spirals like NGC 1300 or NGC 1365. Others are "ansae-type" bars, where the bar appears to have "handles" or enhanced spots at the ends (Martinez-Valpuesta *et al.* 2007). Several examples are shown in Figure 1.12, which are classified using the symbols SB_a or

SAB_a. In some cases, the appearance of ansae-type bars suggests a regular bar in the process of actual dissolution. On the other hand, Athanassoula *et al.* (2016) have recently used numerical simulations to show that ansae could form in the disk-shaped remnant of the merger of two spiral galaxies. Another aspect of bar classification comes from box/peanut bulge galaxies. A box/peanut bulge galaxy is generally an edge-on disk-shaped galaxy where the bulge has boxy isophotes and has the look of an "X" pattern crossing the center. One of these is shown in Figure 1.12 and several others are shown in Figure 1.13 (bottom row). The appearance of a boxy/peanut bulge can depend on whether the bar is viewed end-on or broadside-on. The X is believed to be a result of the line of sight view through vertical resonant bar orbits (e.g. Abbott *et al.* 2017). The classifications SB_x and SAB_x are used for boxy/peanut bulges. In some cases, both a boxy/peanut bulge and ansae occur in the same system (as in NGC 5445; Figure 1.12). In such cases, the classification is SB_{ax} or SB_{xa}.

Although the CVRHS bar family classification can be consistently applied, it is still a visual judgment and is not the most effective way of quantifying bar strength. It is also technically based on blue light images (the historical waveband of galaxy classification) where the appearance of the bar may be affected by dust and star formation. More quantitative approaches to bar strength include the maximum ellipticity in the bar region, the maximum relative $m = 2$ Fourier intensity amplitude $A_2 = (I_2/I_0)_{\max}$ and the ratio of the maximum tangential force to the mean radial force in the bar region, all based on near-infrared images (e.g. Combes and Sanders 1981; Buta and Block 2001; Buta 2012). Garcia-Gómez *et al.* (2017) describe the application of a two-dimensional Fourier transform technique to more reliably characterize the strengths of bars in disk galaxies.

Inner varieties: The inner variety of any disk-shaped galaxy refers to the presence or absence of an inner ring (Sandage 1961). If an inner ring is present, the inner variety is (r). In a spiral, the spiral structure breaks from near the location of the inner ring. The inner variety is (s) if there is no inner ring and the spiral structure either winds all the way to near the center of the galaxy or breaks directly from the ends of a bar. In many galaxies, a partial inner ring made of tightly wrapped spiral structure is seen. As noted in section 1.2, such "pseudorings" are recognized by the symbols (rs) in the variety sequence: (s), (r$\underline{\text{s}}$), (rs), ($\underline{\text{r}}$s), (r), where the underlines denote the dominant characteristic. Examples of these morphologies are shown in the upper row in Figure 1.14.

Inner rings are most common in barred galaxies, but also appear in non-barred galaxies. Some non-barred galaxies with rings could be evolved remnants of an earlier barred phase, owing to the possibility that bars may dissolve in much less than a Hubble time due to a buildup of the central mass concentration (Norman *et al.* 1996).

Figure 1.11. *A sequence of increasing apparent bar strength*

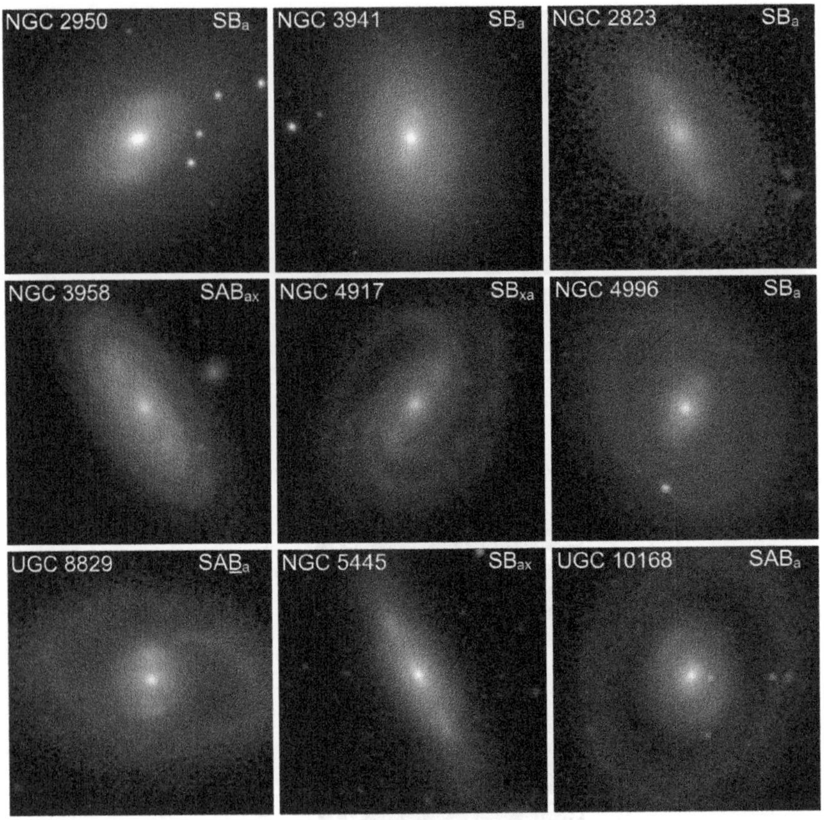

Figure 1.12. *Bars showing enhanced "handles", or ansae. The features appear in spot, linear or curved form*

Figure 1.13. *Three edge-on galaxies showing boxy/peanut-type bulges*

A different inner variety sequence is sometimes applicable. As noted in section 1.3, early-type galaxies often show inner lenses, which are features located in the same place where an inner ring would be seen. If a bar is present, the bar usually fills the inner lens in one dimension (Kormendy 1979). The symbol for an inner lens is (l) and that for an inner ring-lens is (rl). These are used in the sequence: (r), (r̲l), (rl), (rl̲), (l). Examples of these morphologies are shown in the bottom row of Figure 1.14. In some cases, the inner variety is r′l, meaning an inner pseudoring-lens. In an actual classification, the inner variety is in parentheses between the family and the stage [as in, e.g. SB(r)b, SAB(rs̲)cd, SA(l)0/a, etc.].

The relation between inner rings and inner lenses is unclear. One possibility is that an inner lens is a highly evolved inner ring. This might account for the existence of inner ring-lenses (rl), which appear to be low contrast inner rings. However, Kormendy (1979; see also Bournaud and Combes 2002; Gao *et al.* 2018) proposed another interpretation: that inner lenses represent dissolved bars. Bar dissolution is possible because the presence of a bar not only heats the disk component, but also causes resonance effects that force stars onto orbits that do not support the bar. An example of the latter is the formation of a *nuclear bar,* which is a small secondary bar that forms inside a primary bar. Such features are recognized with the symbol (nb) in the CVRHS classification system, and are often significantly misaligned with a primary bar if present.

An interesting aspect of inner rings and lenses is that the former are most common in barred galaxies, but the latter are most common in non-barred galaxies. A possible reason for this is that bar dissolution could leave behind a lens that was formerly the inner part of the bar, called a *barlens* (Laurikainen *et al.* 2013). A barlens [symbolized by (bl)] is generally the roundish, inner component of a bar that often is mistaken for a classical bulge. Examples are shown in Figure 1.18. Athanassoula (2016) interprets barlenses as the three-dimensional inner sections of bars that appear as boxy/peanut bulges in the edge-on view. The ends of the bar are much flatter than this inner section. In general, the boxy character of these inner sections is not very evident in the near face-on view. However, in some bars, an inner boxy zone is seen even in a lower inclination view (examples: NGC 7020, IC 4290, IC 5240; Buta *et al.* 2007).

Another interesting aspect of inner rings is that these features have a wide range of intrinsic shapes (deprojected minor-to-major axis ratio 0.5 to 1.0; Buta 2019) and are often regions of intense star formation. The distribution of star formation in inner rings is sensitive to this range: the more elongated the ring, the greater the concentration of HII regions around the major axis points (Crocker *et al.* 1996; Grouchy *et al.* 2010). The effect is especially evident in cuspy-shaped inner rings, of which NGC 6782 is the best example (Lin *et al.* 2008). It is also seen in NGC 3081 (Buta and Purcell 1998).

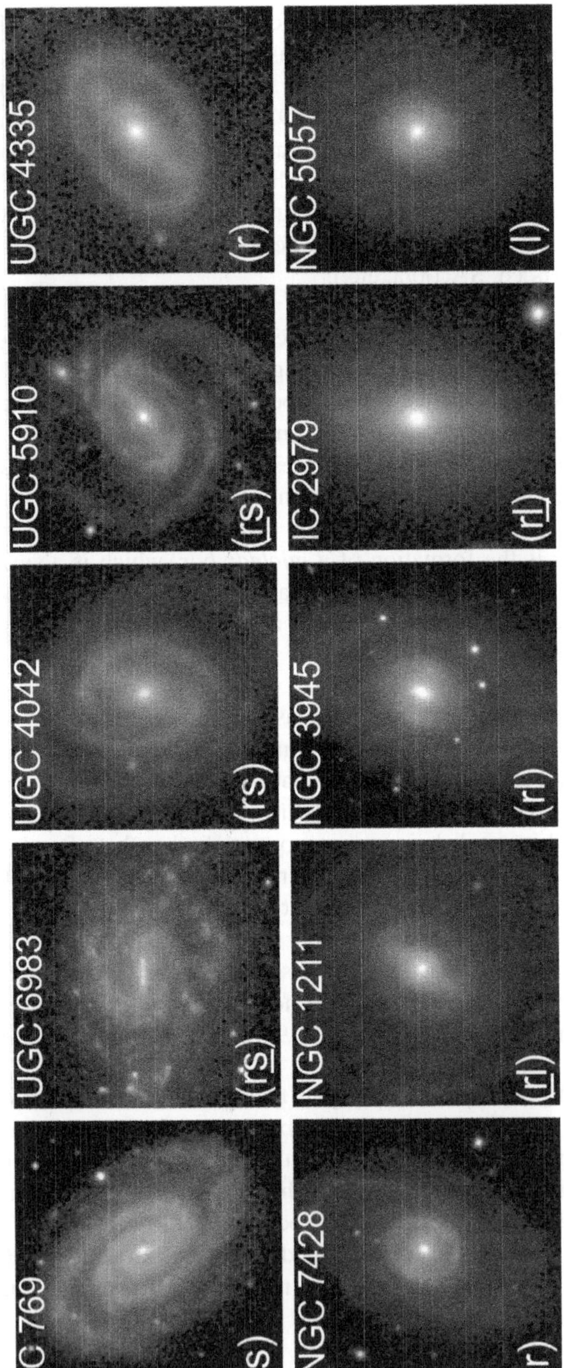

Figure 1.14. *Examples of different inner varieties*

Nuclear varieties: The nuclear variety of a disk-shaped galaxy refers to the presence of nuclear structure, usually in the form of a nuclear ring (nr), nuclear pseudoring (nr′), nuclear spiral (ns), nuclear bar (nb), nuclear lens (nl) or nuclear ring-lens (nrl). The features tend to be small and therefore are recognizable mainly in nearby galaxies. The features also have a wide range of linear diameters, from a few hundred pc to nearly 5 kpc (Comerón *et al.* 2010). In some cases, a nuclear ring is crossed by a nuclear bar. Figure 1.15 shows several examples of spiral galaxies having a nuclear ring. In a CVRHS classification, the nuclear variety appears with the inner variety. For example, NGC 3081 is classified as (R_1R_2')SAB(r,nr,nb)0/a, where the inner/nuclear variety is (r,nr,nb). Knapen (2005) determined that 21±5% of spiral galaxies have nuclear rings or related features.

Figure 1.15. *Examples of spiral galaxies having a nuclear ring*

Outer varieties: Just as inner varieties refer to inner rings and lenses mainly, outer varieties refer to outer rings and lenses. While a typical inner ring or pseudoring in a barred galaxy has about the same size as the bar, the typical outer ring is about twice the size of the bar. Outer features are typically more diffuse than inner features, and of lower surface brightness. The main types of outer features include outer rings (R), outer pseudorings (R′), outer lenses (L), outer ring-lenses (RL) and outer pseudoring-lenses (R′L). An example of each of these, including a doubled outer ring case (RR), is shown in Figure 1.16. In all cases, the outer feature classification is positioned ahead of the family classification, as in (RL)SA(l)0° or (R′)SB($\underline{r}$s)ab.

In spiral galaxies, outer features tend to be pseudorings or pseudoring lenses. The best-defined outer rings tend to be found in $S0^+$ or S0/a cases. The galaxies NGC 2859 (Figure 1.16) and 3945 (Figure 1.14) are two of the best examples of outer rings. Both are classified by Buta (2019) as (R)SA$\underline{B}_a$(rl,bl,nb)0^+, meaning they are "late" S0s. A typical outer pseudoring is seen in NGC 5610, type (R′)SB(s)ab (Figure 1.10). Outer pseudorings are much more common than outer rings.

Among outer features are the resonant subclasses, that is, features that strongly resemble the kinds of resonance rings that form in numerical simulations of barred galaxies, specifically the models of Schwarz (1981) and Byrd *et al.* (1994). Figure 1.17

shows examples of the R_1, R_1', R_1R_2' and R_2' subclasses of outer rings and pseudorings. The way to recognize these features is described by Buta (2017b). The outer resonant subclasses constitute only a small fraction of outer features. Most disk-shaped galaxies do not have an outer feature, but among the ones that do, outer pseudorings of type R' (excluding the resonant subclasses) are most abundant. The formation of rings has been reviewed in Buta and Combes (1996) and is discussed further in section 1.12.

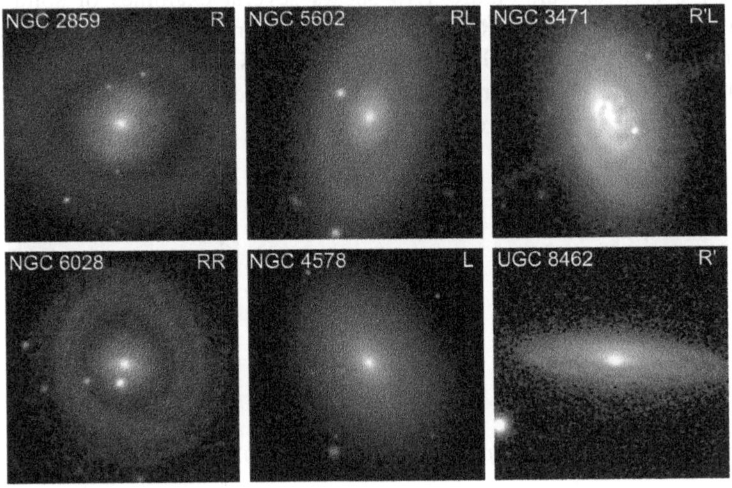

Figure 1.16. *Examples of different types of outer features. The feature type is in the upper right of each frame*

1.5. S0 galaxies

S0 galaxies are generally viewed as "armless" disk galaxies. Once referred to as "lenticulars" because of their double convex lens-like appearance in the edge-on view, S0s first appeared as a galaxy type in Hubble's 1936 book, *The Realm of the Nebulae*, where he presents his "tuning fork" depiction of his system (Figure 1.1). Although ellipticals, normal spirals and barred spirals were the main types recognized in Hubble's original 1926 classification system, the possibility of S0 galaxies was raised even then. Something was puzzling about the original system in that Sa galaxies usually had fully developed arms while SBa galaxies did not. To account for this, Hubble hypothesized the existence of a class of disk-shaped galaxies lacking spiral arms. He placed these at the juncture of the "tuning fork", where the most flattened-looking E galaxies link with the two parallel sequences of spiral galaxies. As Hubble increased his database of photographic images of "nebulae", real examples of the hypothesized group of armless disk galaxies were found. Hubble was working on a revision to his system but died before publishing it. Sandage (1961) used Hubble's notes to make the revision: S0s were interpreted in terms of two sequences of three "normal" stages: $S0_1$, $S0_2$ and $S0_3$, and three barred stages $SB0_1$, $SB0_2$ and $SB0_3$ (Figure 1.2). The subdivision of non-barred S0s was based on the development of structure such as the appearance of dust lanes, while the subdivision of barred S0s was based on the development of the bar from something very subtle to a more prominent feature.

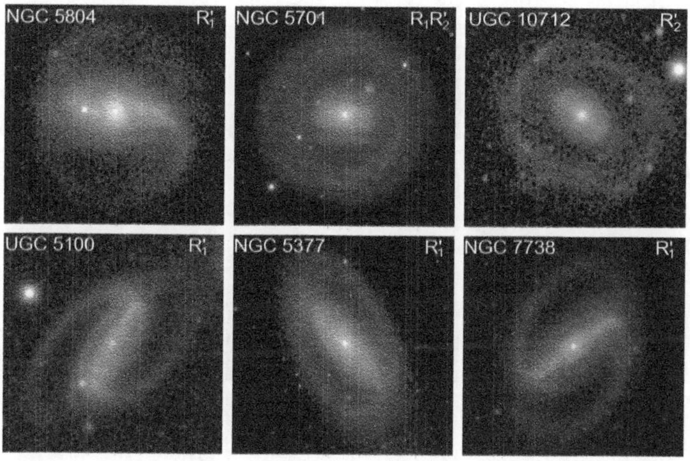

Figure 1.17. *Outer resonant features. The feature type is in the upper right of each frame*

The tuning fork arrangement of galaxy types with S0s included began to be questioned when it was realized that S0s had lower luminosities than Sa or SBa galaxies and thus do not really link smoothly to those parts of the Hubble sequence. This led to the "parallel-sequence" interpretation of S0s, whereby S0s form a sequence parallel to the spirals (not to be confused with the normal/barred spiral prongs of the tuning fork). This interpretation is based on a nurture view that S0s are former spiral galaxies that have been stripped of their interstellar gas and dust through environmental interactions (van den Bergh 1976). On a sequence between spirals and S0s, van den Bergh placed what he called "anemic spirals" or galaxies in an intermediate state of stripping. Real examples of galaxies in such an intermediate state have been found in the Virgo Cluster (Koopmann and Kenney 2004) and the Coma Cluster (Yagi *et al.* (2010).

Figure 1.19 shows several S0 galaxies classified in the CVRHS system. The CVRHS system treats S0s differently from the tuning fork and parallel sequence ideas. Like spirals, S0s are divided into stages, families and varieties (Figure 1.3). The stage for S0s has three subdivisions: $S0^-$, $S0^\circ$ and $S0^+$, a sequence of increasing structure (but not necessarily decreasing bulge to total luminosity ratio, which determines stages along the spiral sequences). Intermediate CVRHS stages from average multiphase classifications (as in Buta *et al.* 2015 and Buta 2019) are $S0^{-/\circ}$ and $S0^{\circ/+}$.

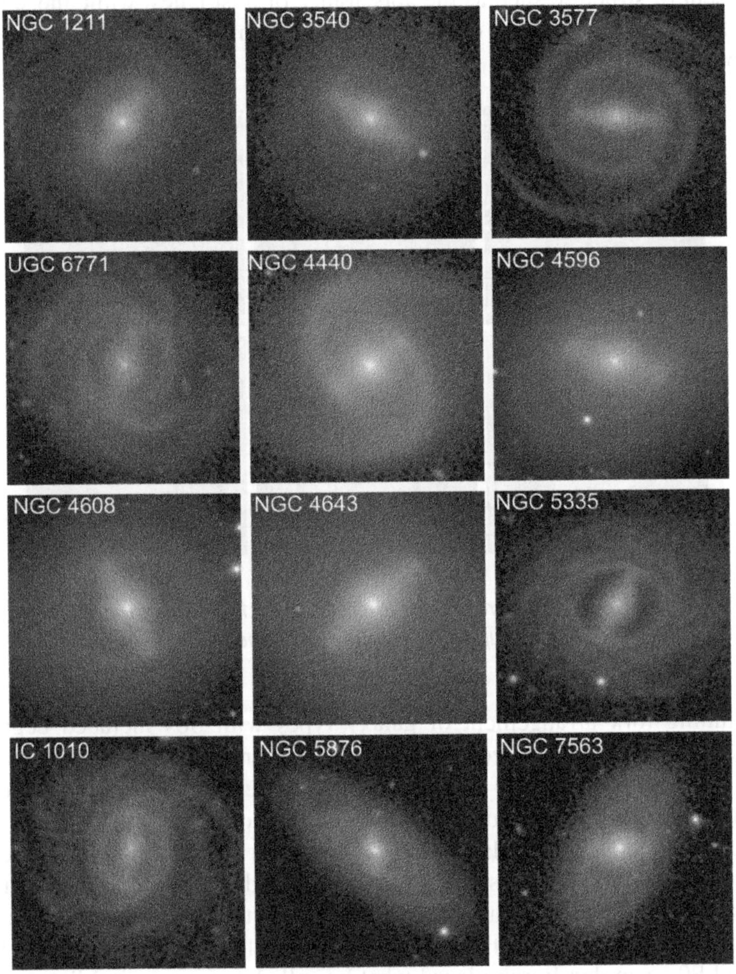

Figure 1.18. *Twelve examples of barred galaxies having a "barlens", which refers to the roundish section inside the bar. Several of these show light deficits around the bar, especially NGC 5335 (Buta 2017b; see also Gadotti and de Souza 2003b; Kim et al. 2016)*

Lenses are common in early S0 galaxies. For example, typical $S0^-$ and $S0^{-/\circ}$ galaxies have mostly featureless extended envelopes that often have a subtle outer edge. Such features are classified as outer lenses (L). Faint traces of inner lenses (l) are also seen in such galaxies. As noted in section 1.4, the origin of lenses is uncertain.

Some or most inner lenses are thought to be due to the dissolution of bars (Kormendy 1979), but the existence of outer and nuclear lenses suggests a possible link to the evolution of rings.

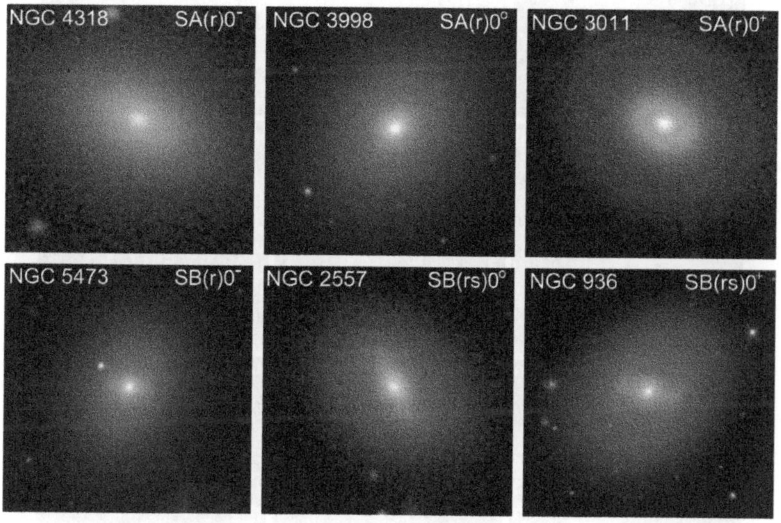

Figure 1.19. *Stage sequences for both barred and non-barred S0 galaxies*

1.6. Magellanic spiral and irregular galaxies

In distance-limited samples of galaxies, Magellanic spirals and irregulars are the most common types of galaxies. Encompassing a wide range of luminosities from dwarfs to giants, Magellanic spirals and irregulars are generally HI rich systems with considerable active star formation. Several examples are shown in Figure 1.20.

The Magellanic Clouds are considered to be examples of SB(s)m galaxies, i.e. extreme late-type galaxies having clear bars but very subtle spiral structure showing a characteristic asymmetry with a single main arm and weaker features (de Vaucouleurs and Freeman 1972). This asymmetry is recognizable even in the nearly edge-on view (Buta *et al.* 2015). The transition stage from Sd to Sm, or Sdm, is also well defined in showing two arms with one much longer than the other. As for Scd and Sd galaxies, Sdm, Sm and Im galaxies are often barred, except for low luminosity dwarfs that may lack the mass required to show such a feature. Non-barred giant late-type galaxies like NGC 5474 show a similar asymmetry to SBm galaxies and can be classified as type SA(s)m.

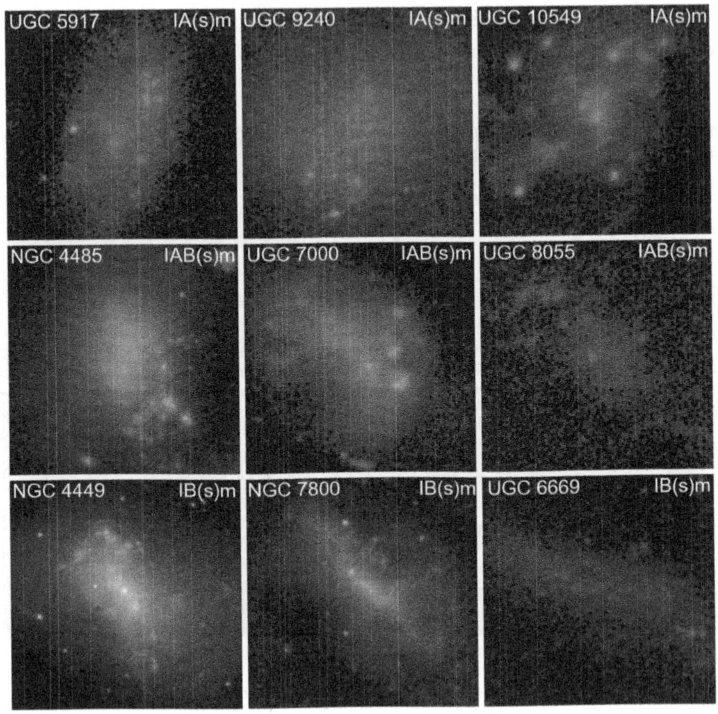

Figure 1.20. *Magellanic irregular galaxies*

1.7. Dwarf elliptical, S0, and spheroidal galaxies

In the environment of the Virgo Cluster, the most common type of galaxy is not an irregular, but what Sandage and Binggeli (1984) called a dwarf elliptical (dE) type. Of nearly 2,000 catalogued galaxies in the Virgo Cluster, Sandage and Binggeli (1984) found that 80% were of this type. A dE galaxy is a low luminosity system with a relatively smooth luminosity distribution that resembles a conventional elliptical galaxy but has a low surface brightness and a luminosity distribution closer to an exponential than to the $r^{\frac{1}{4}}$ law. In some cases, there is subtle structure similar to what is seen in conventional S0 galaxies, and the system is referred to as a dwarf S0, or dS0, galaxy. In either case, the object may be *nucleated*, and if so, the classification is dE,N or dS0,N. In contrast, dwarf spirals, or dS types, are extremely rare.

The exact placement of the dE and dS0 systems within the Hubble tuning fork can be examined by studying the luminosity distributions and investigating scaling relations among elliptical galaxies of all luminosities. Based on such studies,

Kormendy and Bender (2012) concluded that dE and dS0 galaxies are not merely the low luminosity extensions of classical E galaxies, but are *environmentally modified* former irregular and dwarf irregular galaxies. Kormendy and Bender (2012) suggested that all such galaxies be referred to as "Spheroidal" galaxies, and that their existence supports the idea of the van den Bergh (1976) parallel-sequence classification.

In the CVRHS system, the dE and dS0 classification symbols of Sandage and Binggeli (1984) are adopted in the form dE/dS0/Sph or dE/S0,N/Sph. Ann *et al.* (2015) show numerous examples.

1.8. Edge-on galaxies

Edge-on disk galaxies can be more difficult to classify than face-on galaxies, owing to the foreshortening of features such as spiral arms, bars and rings. Nevertheless, the degree of central concentration is often still detectable in spite of heavy planar extinction, and the degree of resolution in the disk can still be used, so that the stage is still discernible (Figure 1.21). Features like inner varieties have to be left out of the classification, leaving types such as Sb$\underline{c}$ sp, Sd spw, etc. The characteristic "sp" (meaning "spindle-shaped") is only used to single-out edge-on or nearly edge-on disks, and does not imply prolate rotation. A galaxy classified as "spw" is one that shows evidence of disk warping in the edge-on view. Some extreme cases of disk warping are shown in Buta *et al.* (2015).

Embedded disks and thick disks: An embedded disk system is a case where a disk galaxy is embedded within a clear 3D component, usually an E galaxy of low flattening. In a disky elliptical galaxy, classified as type E(d) in the notation of Kormendy and Bender (1996), the presence of a faint disk is inferred from subtle distortions of isophotes. As noted in section 1.3, disky isophotes are detected through positive values of the cos(4θ) relative Fourier component of deviations from a perfectly elliptical shape. Embedded disk systems are similar to disky ellipticals, only the disk is much more obvious.

The 3D systems the disks are embedded in may be considered part of the bulge, and the disk may appear to underextend, fill or overextend the isophotes of that system (Buta *et al.* 2015). While disky E systems are best detected when the disk is close to edge-on, more prominent disks need not be so inclined. Figure 1.22 shows several examples. Embedded disks are recognized in CVRHS classification by a notation such as, for example, S0$^-$ sp/E(d)2-3 where the "S0$^-$ sp" part is the spindle-shaped edge-on disk component and the "E(d)2-3" refers to a low flattening E galaxy the disk is embedded within. Obvious embedded disks, where the background spheroidal system has a low flattening, are not common. In general, the flattening of the spheroidal part of an embedded disk system, and the character of the isophotes as boxy or disky, are estimated using a color display of the galaxy's isophotes. In rare cases, the extended component is boxy rather than cuspy, as in the case of NGC 4638 where the CVRHS type is S0$^-$ sp/E(b)3.

Figure 1.21. *Edge-on galaxies from stage Sa to stage Im. The "sp" stands for "spindle" and is used to characterize the edge-on orientation. If "spw", it means the edge-on disk shows signs of warping*

Most spiral and S0 galaxies do not have dominant classical bulges, but many have a two-component disk: the *thin disk*, which has a vertical scale height of a few hundred pc, and the *thick disk*, which shares the same plane with the thin disk but has a vertical scaleheight of about 1 kpc. Especially in spirals, the two disks can have very different stellar populations, with population I material occupying the thin disk and population II material occupying the thick disk. As for embedded disks, thin and thick disks are recognized in CVRHS classifications by combining a normal stage classification with an E(d) or E(b) classification as in, for example, S0$^-$ sp/E(d)7. In this case, the "S0$^-$ sp" part refers to the thin disk and determines the stage index T of the galaxy. The "E(d)7" part refers to the thick disk, whose apparent flattening and isophotal character (cuspy or boxy) are judged from a color display. Note that thick disks may have cuspy, boxy or perfectly elliptical isophotes, the latter requiring no (d) or (b) added to the classification.

Figure 1.22. *Embedded disks and thick/thin disks. The main embedded disk cases shown are NGC 4660, NGC 5382, NGC 3675 and NGC 4474, and the thick/thin disk examples include NGC 4452, NGC 4527, NGC 4710 and NGC 4544*

1.9. Morphology of interacting and merging galaxies

Interacting and merging galaxies make up about 2–4% of nearby galaxies (Knapen and James 2009) and are often unclassifiable in any standard Hubble-based morphological bins. Figure 1.23 shows several examples, all of which are classified as Pec (merger) in the CVRHS system. Distorted shapes, tidal tails, sharp edges ("shells" or "ripples"; e.g. Schweizer and Seitzer 1988), multiple nuclei, and in some cases enhanced star formation characterize these objects. Although major mergers of two spiral galaxies have been thought to generally lead to an elliptical galaxy as a remnant (e.g. Schweizer 1982), more recent numerical studies have shown that a disk galaxy can also be a remnant of such mergers (Athanassoula *et al.* 2016).

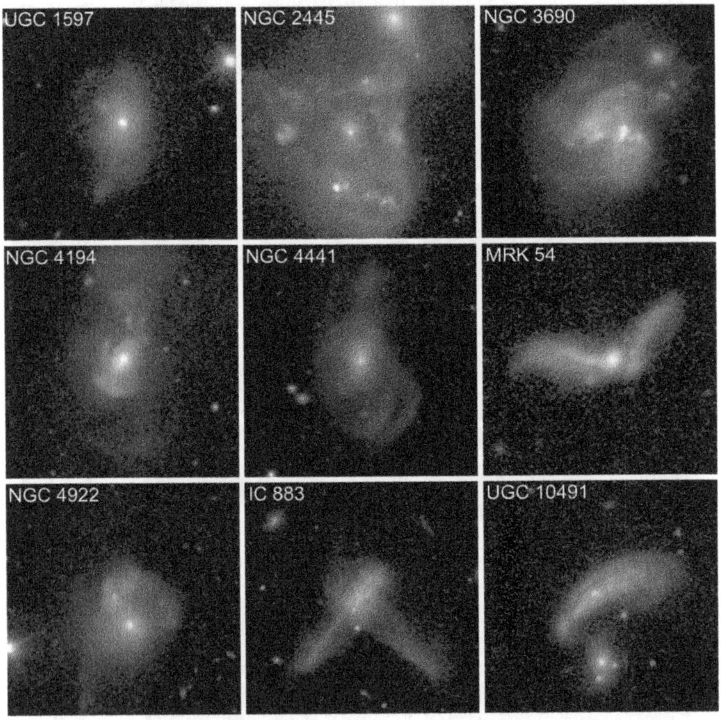

Figure 1.23. *Examples of interacting and merging galaxies*

The types of rings in normal galaxies that define inner, outer and nuclear varieties are likely features that result from internal dynamics (section 1.12). Figure 1.24 shows examples of the extremely rare types of ring-like features that likely result from a cataclysm, such as a galaxy interaction or encounter. Collisional ring galaxies (RGs; upper frames in Figure 1.24) are one type of cataclysmic ring; these are thought to result from a head-on collision between a disk galaxy and a companion (Appleton and

Struck-Marcell 1996). Polar ring galaxies (PRGs; lower frames in Figure 1.24) are a second type; these are thought to originate from the disruption of a small satellite galaxy along a polar orbit around a more massive disk galaxy (Schweizer *et al.* 1983). None of the examples of this type of feature shown in Figure 1.24 are certain cases. The possible PR feature in NGC 2918 is very faint but favorably oriented nearly edge-on (Whitmore *et al.* 1990). The possible case in NGC 4247 may be nearly face-on and less clearcut.

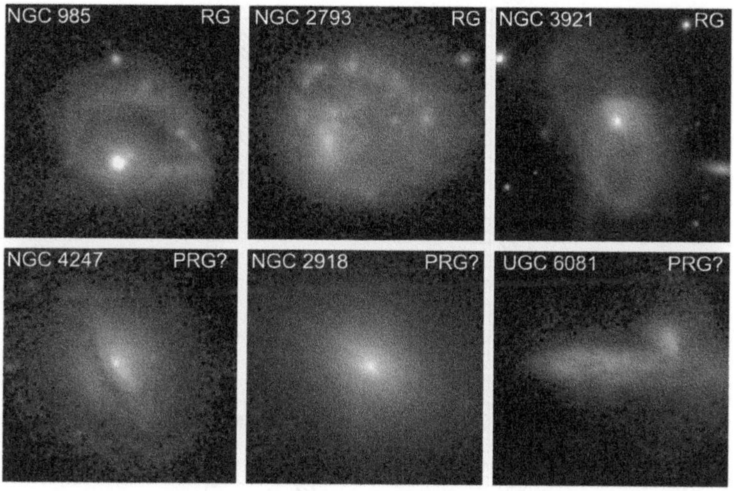

Figure 1.24. *Possible examples of cataclymic ring galaxies. Top row: collisional ring galaxies. Bottom row: polar ring or inclined ring galaxies*

1.10. General properties along the CVRHS sequence

1.10.1. *Morphological systematics*

The classification of a large sample of galaxies in the CVRHS system allows investigation of the systematics of different morphological types and classes, i.e. how aspects of morphology might vary with stage and family. Figure 1.25 first shows the distribution of stages for three galaxy samples: (a) a sample of 1240 galaxies from the Spitzer Survey of Stellar Structure in Galaxies (S^4G; Buta *et al.* 2015); (b) a sample of 441 galaxies from the Analysis of the Interstellar Medium in Isolated Galaxies (AMIGA; Buta *et al.* 2019; Verdes-Montenegro *et al.* 2005); and (c) a sample of 2,628 galaxies from the Extraction of the Forms of Galaxies from Images survey (French acronym: EFIGI; Buta 2019; Baillard *et al.* 2011). Each plot is restricted to galaxies inclined at $\leq 60°$. The three graphs show how the distribution of stages depends sensitively on the selection criteria for a given sample. The S^4G sample is distance-limited and used 21 cm radial velocities to judge distances; this accounts for the significant emphasis of this sample on extreme late-type (Sd-Sm) spirals and Magellanic irregulars (Im). The EFIGI sample was selected largely on the basis of angular diameter and emphasizes mainly Sb-Sc spirals. The AMIGA sample of isolated spirals is similar in showing an emphasis on Sb-Sc spirals.

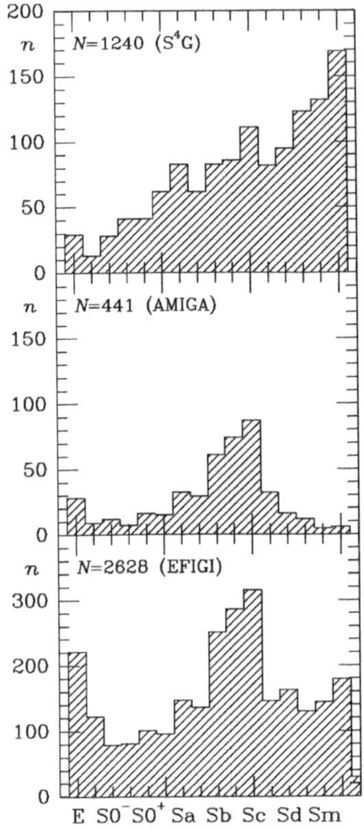

Figure 1.25. *Distribution of CVRHS stages from (a) Buta (2019; EFIGI), (b) Buta et al. (2019; AMIGA) and (c) Buta et al. (2015; S^4G). The histograms are each restricted to relatively face-on galaxies ($i \leq 60°$)*

Figure 1.26 shows the bar, inner variety and outer variety fractions (f_{bar}, f_{IV}, and f_{OV}, respectively) as a function of stage for subsets of EFIGI galaxies inclined $\leq 66°$. The numbers are N = (a) 2,440, (b) 749, (c) 1,309, (d) 260 and (e) 2,642 galaxies. Figure 1.26(a) shows that the bar fraction has a minimum near stage Sc, with the highest bar fractions for extreme late-type spirals; Figure 1.26(b) shows that inner rings and well-defined inner pseudorings have the highest fraction around stage Sab; Figure 1.26(c) shows that (s) variety is uncommon for types Sa<u>b</u> and earlier, but jumps to $\geq 80\%$ for stages Sc and later; Figure 1.26(d) shows that lenses, ring-lenses and pseudoring-lenses are found mainly among stages Sb and earlier; and Figure 1.26(e) shows that outer features (rings, pseudorings, lenses and ring lenses) are most common between stages S0$^+$ and Sb. Although outer features are rare for types later than Sb, Figure 1.26 shows a secondary peak among extreme late-type spirals.

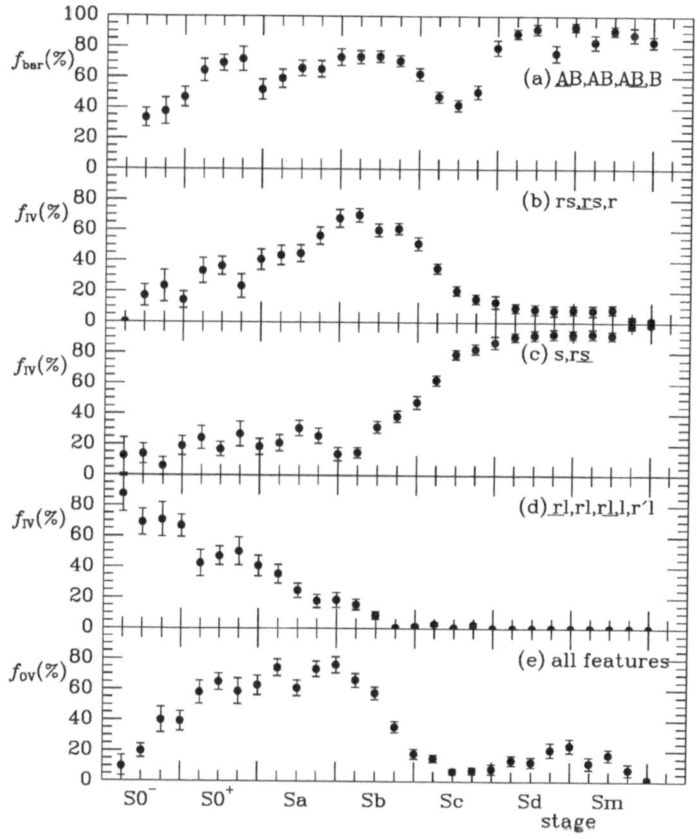

Figure 1.26. *Systematics of CVRHS classifications from Buta (2019; EFIGI). The samples used for these graphs were restricted to inclinations ≤66°*

1.10.2. *Astrophysical systematics*

The CVRHS sequence of galaxy types from ellipticals to irregulars has astrophysical significance. This is because some measured properties of galaxies correlate with position along the sequence. For example, the integrated color index, $(B-V)_T^o$, ranges from ≈0.9 for E galaxies to ≈0.4 for magellanic irregular galaxies. Over the same range, the color index $(U-B)_T^o$ varies from ≈0.5 to −0.2. Color images of galaxies of different types available through the Sloan Digital Sky Survey show these differences directly. Types earlier than Sa tend to be yellowish, indicating the predominance of an old stellar population, while types Sc and later tend to be mostly bluish, indicating the predominance of a younger stellar population. Intermediate types (Sa to Sbc) show the presence of both populations, with bars and bulges usually yellowish and spiral arms generally bluish. As shown by Buta *et al.* (1994), the color indices of galaxies have a dependence on type that is well approximated by an integral Gaussian, showing little variation from type E to type $S0^+$, followed by a more rapid variation from type S0/a to types Sd and Sdm. This strong dependence on type, which carries into other colors such as $(V-R)_T^o$, $(R-I)_T^o$ and $(B-H)_T^o$ (Buta and Williams 1995; Buta 1995a), has been interpreted in terms of a star formation history where the star formation rate declines exponentially with different decay rates (e.g. Kennicutt *et al.* 1994; Gusev *et al.* 2015).

Other galactic properties that vary along the VRHS sequence include the average surface brightness and HI mass-to-blue luminosity ratio. The average surface brightness ranges from $\approx$12.4 mag arcmin^{-2} for stages E to $S0^+$ to 14.9 mag arcmin^{-2} for stage Im, a factor of 10 drop (Buta *et al.* 1994). This correlation is directly related to one of Hubble's classification criteria for spirals: that Sa galaxies have more prominent central concentrations than do Sc galaxies. The latest stages (Sd-Im) have virtually no central concentration and the lowest average surface brightnesses. Similarly, there is a relatively smooth variation in HI mass-to-blue light ratio across the spiral sequence, ranging from 0.08 at stage S0/a to 0.5 at stage Im, a factor of 6.5 change (Buta *et al.* 1994).

1.11. Other approaches to galaxy classification

The CVRHS system is only one approach to galaxy classification, but it has several advantages: (1) a high focus on features that are likely intimately connected to dynamics and evolution, such as bars, rings and spirals; (2) correlation with star formation history; and (3) the broadest perspective on galaxy morphology without being too unwieldy. However, in the era of large imaging surveys like the Sloan Digital Sky Survey (SDSS; Gunn *et al.* 1998, 2006; York *et al.* 2000), the number of classifiable images of galaxies available is literally in the millions. CVRHS classification by a single individual (for example, Buta 2019) will likely be impractical for samples much larger than 20,000 objects. Another approach to galaxy classification is needed.

Nair and Abraham (2010) provide classifications for 14,034 SDSS galaxies in the redshift range $0.01 < z < 0.1$ in a system similar to the CVRHS. In this case, a single professional astronomer classified each galaxy twice to check for consistency. In the studies of Fukugita *et al.* (2007), Baillard *et al.* (2011) and Ann *et al.* (2015), 3–10 professional astronomers classified either the whole or part of samples of 2253, 4458 and 5836 SDSS galaxies, respectively. In each case, the final classification is an average of the multiple astronomers, usually after intercomparing results and taking into account "personal equations" of each classifier. One of the largest applications of the multiclassifier approach was made by Kartaltepe *et al.* (2016), who had the input of 65 professional astronomers.

In spite of these efforts, the number of galaxies that have been classified by experts is still relatively small. This led some professional astronomers to try "crowd-sourcing" galaxy classification by engaging with the general public and enlisting the help of "citizen scientists". The first major crowd-sourcing effort is described by Lintott *et al.* (2008) and is called "Galaxy Zoo 1" (GZ1). In GZ1, a web interface was created to display SDSS color images of several hundreds of thousands of galaxies accompanied by a set of "buttons" for selecting rudimentary elements of morphology, such as whether a galaxy is disk-shaped or not. This was followed by "Galaxy Zoo 2" (Willett *et al.* 2013), which allowed classifiers to specify more detail, such as bars, rings and spiral arm character. The classifications in GZ2 were combined using a vote scheme, and an attempt was made to correct the classifications for biases due to resolution and distance. GZ2 provided classifications for nearly 300,000 SDSS galaxies based on about 80,000 participants. Willett *et al.* (2013) provide short abbreviations of these classifications that are compared with the CVRHS classifications of Buta (2019) and Buta *et al.* (2019) in Table 1.12. In these abbreviations, E galaxies are smooth while S galaxies are disk shaped with structure. The other codes are described in Appendix A of Willett *et al.* (2013). Four of the galaxies in Table 1.12 are shown in Figure 1.27.

1.12. Interpretations of morphology

We have seen that galaxy morphology includes a bewildering array of complex structures, and that the evolutionary path that any galaxy took to reach its current morphological state is not obvious. Nevertheless, some reasonable judgments can be made. This section summarizes some aspects of morphology that are well understood, and others that are still largely uncertain.

Galaxy formation: The best current theory of galaxy formation is the Λ Cold Dark Matter model (White and Rees 1978). In this model, galaxy clusters begin as tiny fluctuations in the temperature of the cosmic microwave background radiation that are expanded during the inflationary period to much larger sizes. The fluctuations become "seeds" of cold dark matter within which baryonic matter collects. The formation of collapsing proto-galactic clouds is enhanced by the extra gravity of the dark matter. Whether a galaxy becomes a disk galaxy or a non-disk galaxy depends on how much angular momentum the collapsing cloud has from tidally induced torques due to neighboring clouds (Ryden 1988) and on how effectively it converts its baryonic material into stars. If this conversion is largely complete within a billion or so years after collapse, then the object could become an elliptical galaxy. If instead the rate of conversion is slower, the proto-galactic cloud may have time to form a disk. Multiple mergers of small objects with this disk-shaped object could lead to the build up of what is generally known as a "classical bulge". This formation scenario is not expected to form a bulge-less, pure disk galaxy, and it is also well-established that some ellipticals have likely formed from mergers of disk-shaped galaxies (Schweizer 1982).

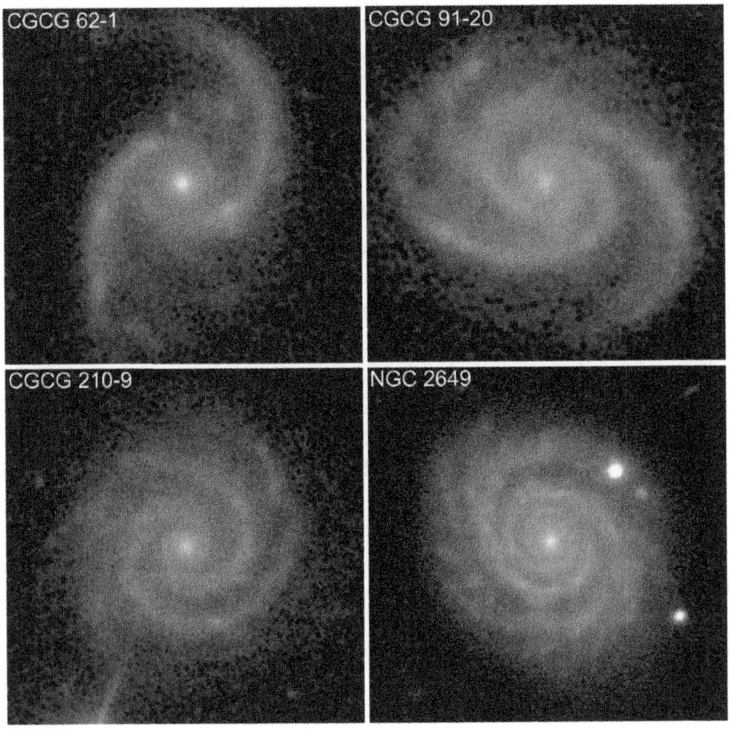

Figure 1.27. *Images of four relatively isolated grand-design spirals (Buta et al. 2019)*

Although the "formative" phase of a galaxy never really ends (i.e. a galaxy may continue to accrete material long after the formation period largely ended, just as planets still accrete interplanetary debris even now), at some point the accretion rate slows down and galactic changes occur very much more slowly. This is the time when *secular evolution* takes over as the dominant mechanism of change (Kormendy and Kennicutt 2004). Secular evolution can be driven entirely by internal processes or by external processes. The idea is that the formative phase of galaxy evolution is fairly rapid compared to secular evolution. To interpret the morphology of some nearby galaxies, we should appeal to secular evolutionary processes.

The origin of S0 galaxies: Dressler (1980) describes the morphology–density relation in rich galaxy clusters, where the most common types of galaxies found are E and S0. This relation, which was first described by Hubble and Humason (1935), led to the general idea, originally proposed by Spitzer and Baade (1951), and elaborated upon by van den Bergh (1976), that S0 galaxies are former spirals that have been stripped of their interstellar gas and dust. The stripping shut down ("quenched") star formation and essentially "killed" the spiral arms. In this scenario, as already noted in section 1.5, S0s cannot be transition stages between ellipticals and spirals, but must form a sequence of decreasing bulge to total luminosity ratio parallel to spirals. This idea is strongly supported by kinematic studies of early-type galaxies (Cappellari *et al.* 2011), photometric studies of dE and dS0 galaxies (Kormendy and Bender 2012) and by multicomponent analysis of a large sample of S0s (Laurikainen *et al.* 2011).

Galaxy 1	CVRHS type 2	Willett *et al.* 2013 Type 3
NGC 5057	(RL)SA(l)0^+	E(r)r
NGC 6116	SA(s)a	Sb2m
UGC 10258	(R′)SA(s)ab	Sb2t
NGC 2649	SA(rs)bc	Sc2t
CGCG 62-1	SA(rs)bc	Sc2l
CGCG 91-20	SA(s)bc	Sc2t
CGCG 210-9	SA(s)b$\underline{c}$	Sc2t
UGC 4549	SA(rs)cd	Sc?t
IC 2604	SA(s)d	Sc3m
iUGC 9243	SA(s)dm	Sc?m
NGC 4260	SB_x(r$\underline{s}$)a	SBc2t
NGC 5610	(R′)SB(s)ab	SBc2m
UGC 6854	SB(s)c pec	SBc2m(i)
NGC 5964	SB(s)cd	SBc2m
NGC 5112	SB(s)d	SBc?m
UGC 9215	SB(s)dm	Sd?l(i)

Table 1.1. *Comparison of classifications*

Recent studies of S0 bulges (e.g. Laurikainen *et al.* 2013; Gao *et al.* 2018) have shown that while pseudobulges are present in the class, most S0 bulges are classical in nature. As noted by Gao *et al.* (2018), this strongly rules out any connection between S0s and later-type spiral galaxies (whose bulges are often pseudobulges).

The VRHS actually anticipates a strong relation between S0s and spirals. This is because in the VRHS, inner and outer varieties are carried across the spiral sequence into the S0 domain. The S0 stage sequence, $S0^-$, $S0^\circ$ and $S0^+$, is not the same as the spiral sequence, but is based on the appearance of decaying patterns, often in the form of rings and lenses, and sometimes as extremely subtle spiral patterns. This suggests a link between S0s and some galaxies that once were spirals. The main problem is that the S0 stage sequence and the spiral stage sequence are not compatible and cannot be considered "parallel".

The sophistication of modern numerical simulations has led to an alternative idea for the origin of some S0s. Rather than simply being stripped spirals, many could be the result of mergers of spirals. This is shown by Eliche-Moral *et al.* (2018), who compare detailed images of S0 galaxies with the results of the GalMer *n*-body merger simulations (Chilingarian *et al.* 2010). Eliche-Moral *et al.* (2018) argue that GalMer major mergers show features that are commonly seen in S0 galaxies, such as ovals and lenses, and also predict a very low bar fraction for S0s. That S0s do indeed have fewer strong bars than spirals was shown by Buta *et al.* (2010) based on the relative bar torque parameter Q_g.

The origin of spiral structure: The most popular idea, known as "density wave" theory, views spiral structure as a rotating pattern through which a galaxy's stars and interstellar gas clouds must pass (Lin and Shu 1964; Toomre 1977; for a more recent review, see Dobbs and Baba 2014). The arms are not material objects, always made of the same stars, but are waves of density where stars and gas clouds experience a temporary "slowdown". It is thought that gas compression in the arms could lead to the enhanced star formation often seen in spiral arms.

Spiral patterns can be leading or trailing in their sense of winding. A leading spiral opens into the direction of rotation, while a trailing spiral opens opposite the direction of rotation. Studies of extinction in tilted spirals shows that trailing arms are the rule (e.g. de Vaucouleurs 1958); leading spirals are very rare but do exist (e.g. Buta *et al.* 2004). A mechanism in density wave theory known as "swing amplification" was identified by Toomre (1981) as a possible way of explaining the long-lived nature of some spirals. The idea uses a combination of differential rotation, epicyclic theory and self-gravity to "swing" a weak leading spiral instability into a much stronger trailing spiral, and explains the preponderance of trailing spiral patterns.

Although density wave theory could account for the velocity perturbations across spiral arms (e.g. Visser 1980), and also could be used to determine spiral pattern speeds (Roberts *et al.* 1975), the theory left open the question of the origin of spiral structure and the impact of a spiral on the evolution of a galaxy. Attention turned to a search for driving mechanisms, which led to a focus on bars and companions. Bars especially became primary features for affecting the secular evolution of galaxies, because a bar can drive gas into the center of a galaxy and build up what is known as a "pseudobulge", or bulge made of disk material. Based on passive orbit studies, it was concluded that secular evolution primarily affects the interstellar gas distribution, and likely can only change a galaxy of intermediate type by at most one Hubble stage interval (e.g. stage Sbc to Sb) during a Hubble time.

Kormendy (1979) and Kormendy and Norman (1979) brought attention to the role of bars in driving galactic secular evolution. Kormendy and Norman (1979) identified bars as the likely driving mechanism for what they referred to as "global" spiral patterns, i.e. well-defined spirals that extend across an entire galactic disk. If not a bar, then the perturbing effects of a close companion may also drive a density wave spiral. However, there are relatively isolated non-barred galaxies with strong global spiral patterns; Figure 1.27 shows four examples from Buta *et al.* (2019).

Although global spirals are the best-known patterns in galaxies, many galaxies show spiral structure that cannot be called "global". This is because their patterns are mainly in the form of disconnected pieces or ill-defined spirals, called flocculent (Elmegreen 1981). Such spirals have been interpreted in terms of star-forming complexes sheared by differential rotation into short spiral arcs (Gerola and Seiden 1978). Dobbs *et al.* (2018) compare numerical simulations with the ill-defined arms of M33, and conclude that the arms form as a result of gravitational instabilities in the stellar and gaseous distributions, coupled with a high amount of stellar feedback.

Dobbs and Baba (2014) summarize the general view of spiral structure today: "With the possible exception of barred galaxies, spiral arms are transient, recurrent, and initiated by swing-amplified instabilities in the disk".

The origin of bars in spiral galaxies: The general idea has been that bars form through a natural process known as the "bar instability". This is the tendency for an initially axisymmetric but cold stellar disk to naturally form a bar (Hohl 1971). Numerical simulations showed that bars develop in systems where the bulk of the kinetic energy is in the form of ordered motions, i.e. in a rotating disk-shaped galaxy (e.g. Miller and Smith 1979).

Gadotti (2004) argues that while this mechanism seems likely for spirals, it may not be the origin of bars in S0 galaxies because these tend to be found in a cluster environment and have a higher relative abundance of classical bulges (as opposed to pseudobulges). The light-deficient regions around some bars suggest a mechanism where a bar forms at the expense of the disk (Gadotti and de Souza 2003a; Kim *et al.* 2016).

The origin of galactic rings: The general view is that the vast majority of galaxy rings are associated with orbital resonances in galactic disks. A resonance ring is a feature that can, on the basis of morphology and other characteristics, such as star formation, be linked to a specific orbital resonance with a bar, oval or spiral density wave. The idea of a resonance ring first appeared in discussions of density wave theory (for example, Lin 1970), but it was the modeling of barred galaxies that led to a better understanding of these features (Schwarz 1981, 1984).

Resonances occur where $m(\Omega - \Omega_p) = n\kappa$, where Ω is the circular angular velocity, Ω_p is the pattern speed (uniform angular rotation rate of the spiral pattern), κ is the radial (epicyclic) frequency, and m/n is the number of radial oscillations per single orbit in the reference frame rotating at Ω_p. Both Ω and κ are functions of galactocentric radius. The main ring-forming resonances are the outer and inner Lindblad resonances (OLR and ILR, respectively), for which $m = 2$ $n = \pm 1$, and the outer and inner 4:1 resonances (O4R and I4R, respectively), for which $m = 4$ and $n = \pm 1$. In any normal galaxy, there will likely be two ILRs, with the innermost one called the inner ILR (or IILR) and the outermost one called the outer ILR (or OILR). Another important (but not necessarily ring-forming) resonance is the corotation resonance (CR), where $\Omega_p = \Omega$. The locations of all of these resonances can be determined by plotting the curves Ω and $\Omega \pm \kappa/m$ versus galactocentric radius and reading the radii where these functions equal Ω_p.

Support for the resonance idea comes principally from (a) the morphology of galactic rings (Buta and Crocker 1991; Buta 1995b); (2) statistical studies of intrinsic shapes and orientations of galactic rings with respect to bars (Buta 1995b; Comerón *et al.* 2014); and (3) detailed studies and numerical modeling of individual cases such as NGC 3081 (Buta and Purcell 1998) and NGC 6782 (Lin *et al.* 2008). Rings are relatively narrow features, most frequently associated with gas. In particular, nuclear rings are often conspicuous sites of star formation. In general, rings form when gas is accumulating in the bar resonance locations, since the gravity torques from the bar are then canceling out. The angular momentum transport is discussed in detail in Buta and Combes (1996).

Figure 1.28 shows three galaxies from the Catalogue of Southern Ringed Galaxies (Buta 1995b) having well-defined outer resonant features. The white circles in the images show the estimated location of the CR in each galaxy based on the "gap method", a way of determining the radius of the CR using dark gaps lying along the line perpendicular to the bar axis (Buta 2017b). By using this method, and assuming the rotation curves for the galaxies are approximately flat, the schematics show the locations of the other major resonances, I4R, O4R and OLR. The results from these galaxies and 47 others examined by Buta (2017b) provide a consistent picture of systems where R_2' rings lie slightly outside the OLR, R_1 and R_1' rings lie near but inside the OLR, and inner rings and bars lie close to the I4R. The ILR was once thought to play a major role in nuclear ring formation, but this view was challenged by Regan and Teuben (2004), who argued that the most reliable interpretation of these features is in terms of an orbit transition region inside the bar.

Ringed galaxies and invariant manifolds: Romero-Gomez *et al.* (2006) proposed an alternative to the resonance interpretation of galactic rings. In this idea, the R_1, R_1', R_2' and R_1R_2' subclasses of outer rings and pseudorings in barred galaxies can all be interpreted in terms of invariant manifolds that emanate from the unstable L_1 and L_2 Lagrangian points near the ends of a bar. In this interpretation, rings are not necessarily resonant features, but are features mapped by the manifolds as tube-guided homoclinic, heteroclinic and escaping chaotic orbits. In homoclinic orbits, a spiral emanates from L_1 or L_2 and closes at L_2 or L_1, respectively, after winding 180°; this is a morphology consistent with R_1 and R_1' rings. In heteroclinic orbits, a spiral emanates from L_1 or L_2 and closes at L_1 or L_2, respectively, after winding 360°; this is a morphology consistent with the double outer ring/pseudoring morphology R_1R_2'. Finally, in an escaping orbit, a spiral emanates from L_1 or L_2 and winds outward, never to intersect the opposing Lagrangian point; these can account for the morphology of R_2' outer pseudorings.

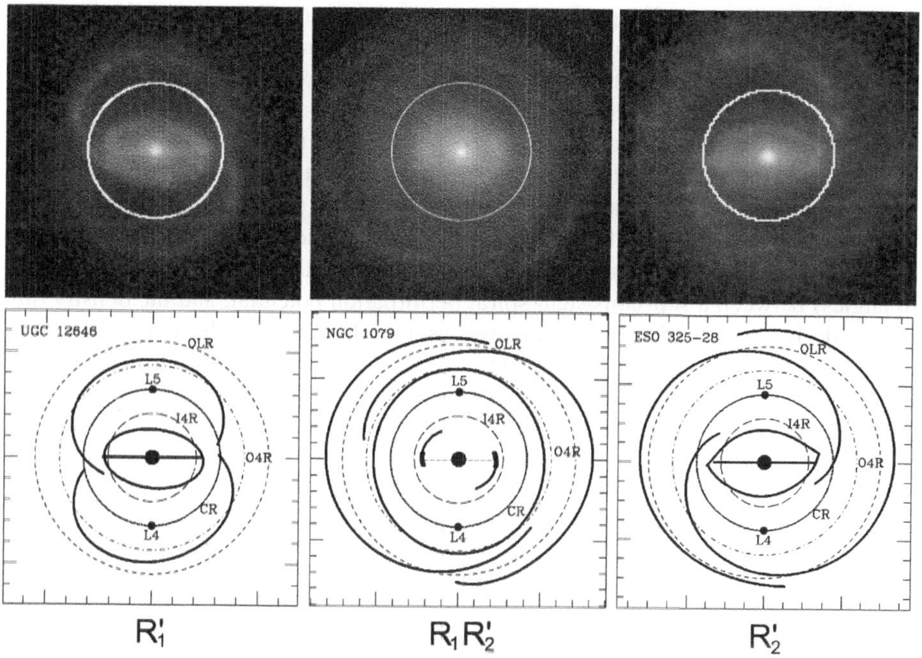

Figure 1.28. *Three galaxies showing outer resonant subclass features. The white circle in each image is the estimated location of the corotation resonance (CR) based on the gap method of Buta (2017b). The schematics show the locations of other resonances relative to visually mapped ring and pseudoring features, assuming a flat rotation curve. The method assumes that light deficits on the line perpendicular to the bar axis trace the location of unstable Lagrangian points L_4 and L_5 (indicated as L4 and L5 each plot). These points are assumed to trace the location of the CR. The other resonances highlighted are the inner 4:1 resonance (I4R), the outer 4:1 resonance (O4R) and the outer Lindblad resonance (OLR)*

The potential-density phase shift and galactic secular evolution: Models of barred galaxies like those of Schwarz (1981, 1984) and Rautiainen and Salo (2000) suggest that rings are products of the evolution of spiral patterns near resonances. That is, a ring begins as a pseudoring and then evolves into a more closed feature. However, very few galaxies show the influence of resonances as strongly as do cases like NGC 3081 or any galaxy showing outer resonant subclass features. This is because the typical inner variety for spirals is (s), the typical nuclear variety is no nuclear feature and the typical outer variety is no outer feature. The question to ask, then, is not why some galaxies have rings, but why so many do not.

The theoretical work of Zhang (2018 and references therein) challenges previous views on galaxy dynamics and provides a real mechanism for the secular evolution of the stellar distribution of a spiral galaxy, not just the interstellar gas distribution. The galaxy-disk mass distribution, contributed mostly by stellar mass, is part of the so-called "basic state" used to calculate spiral density wave perturbations. In addition to the axisymmetric (i.e. azimuthally averaged) mass distribution, determined observationally from disk surface brightness and color, the basic state specification includes also the axisymmetric rotation speed (with contributions from the halo and bulge, in addition to the contribution from the disk) and velocity dispersion, with all three usually specified as a function of galactocentric radius. The key factor for the evolution of the basic-state mass distribution is an azimuthal phase shift (phase offset in angles) between a self-consistent spiral perturbation potential, and the density this potential gives rise to. The spiral can arise as a mode in an initially featureless disk, and if it can achieve a quasi-steady state, angular momentum can be taken away from the inner disk, transferred outward and deposited onto the outer disk, leading to the slow buildup of a central mass concentration and an extended outer disk.

The main departure of Zhang's work from previous studies of galactic dynamics is the elevated role of collective effects on the dynamical evolution of galaxies. This refers to the correlated small-angle scatterings stars would experience as they move across the spiral density wave crest. The density wave itself was shown in Zhang's work to be collisionless shocks related to "dissipative structures". Another important aspect of the work is that the phase shift is not simply an abstract concept from potential theory, but is manifestly measurable from any image which approximates the stellar mass distribution. From such images, one can derive a graph of the phase shift as a function of radius, and in such a graph a single mode appears as a single positive bump followed by a single negative bump (e.g. Buta and Zhang 2009). The point where the phase shift changes from positive to negative is the location of the mode's CR, which is generally considered the most important resonance for any model pattern in a galaxy. This location is also the divide between radial mass inflow and outflow patterns.

1.13. Artificial galaxies and the future of galaxy classification

Galaxy classification will continue to have value in extragalactic studies for several reasons: (1) the continuing availability and also planned production of high-quality, multifilter, ground-based and space-based, digital imaging surveys; (2) the usefulness of the Internet for "crowd sourcing" galaxy classification through the Galaxy Zoo project and its offshoots; (3) the sophistication of modern numerical simulation programs that can make model galaxies that are so realistic that they can be classified like real galaxies; and (4) the advances in artificial neural networks that have made automatic galaxy classification feasible on a scale not previously achieved.

The SDSS will continue to play a role in all of this through its many data releases, the latest of which, DR 16, will occur in December 2019. In general, SDSS images are of very high quality for both qualitative and quantitative studies of galaxies. However, the SDSS footprint has largely excluded the southern sky. The Panoramic Survey Telescope and Rapid Response System (Pann-STARRS; Kaiser *et al.* 2010) cover the part of the sky north of declination $-30°$. Although many of the Pan-STARRS images have peculiar defects, many are still very useful. Sophisticated databases in progress, or planned, include the Dark Energy Camera Legacy Survey (DECaLS; Blum *et al.* 2016), the Kilo-Degree Survey (KiDS; de Jong *et al.* 2013), the Hyper Suprime-Cam Subaru Strategy Program (HSC-SSP; Aihara *et al.* 2018), the Large Synoptic Survey Telescope (LSST; Ivezic̀ *et al.* 2019), the European Space Agency Euclid mission (Laureijs *et al.* 2012) and the Wide Field Infrared Survey Telescope (WFIRST; Gehrels *et al.* 2015). Although the goals of many of these new surveys are mainly cosmology related (with focus on dark energy and dark matter), the public availability of these imaging databases will facilitate deeper, higher resolution studies of both nearby and very distant galaxies.

The Illustris project (Vogelsberger *et al.* 2014) is one of the most comprehensive simulational surveys of galaxies ever made. The project takes advantage of advances in simulation programming, storage space and computing power, and includes as many physical processes as relevant, such as feedback due to AGN, supernovae, and supermassive black holes, mass and internal dynamics of dark and baryonic matter and the efficacies of star formation, to produce synthetic galaxies at such a realistic level that, once converted into observational units, can match the morphology of many SDSS galaxies. An expert morphologist would be able to examine these artificial galaxies and classify many within the Hubble tuning fork or in some other manner. Snyder *et al.* (2015) show using non-parametric measures of galaxy morphology that Illustris synthetic galaxies approximate $z \approx 0$ galaxies well enough to match the correlation between star formation and galaxy morphology. A great advantage of the project is the ability of the survey to show how different synthetic galaxies evolve under the conditions of the models. For example, followed to high redshifts, the Illustris models reproduce the observed characteristic that galaxies become more irregular at higher redshifts than at lower redshifts (Genel *et al.* 2014).

Like the SDSS, many of the new digital imaging surveys will provide detailed information on the morphologies of literally *millions of galaxies* over a wide range of redshift. The classifications of these galaxies are critical to cosmological studies of galaxy formation and evolution. While significant subsets of galaxies are and will continue to be classified visually by single or multiple experts, the vast majority of galaxies will have to be classified in other ways. The Galaxy Zoo project (Lintott *et al.* 2008) recognized that there will never be enough professional experts to classify hundreds of thousands of galaxies, and instead enlisted the volunteer assistance of tens of thousands of non-expert "citizen scientists" to classify galaxies in enough detail to be cosmologically useful. The Galaxy Zoo Team did not require that volunteers use the CVRHS system to classify galaxies, but instead allowed volunteers to select a few basic characteristics through the use of a small number of buttons in a web interface. How GZ2 classifications compare with some CVRHS classifications was shown in section 1.11. However, as effective as Galaxy Zoo has been for large-scale galaxy classification, there will still be hundreds of thousands or millions of galaxies that will need to be classified.

Dieleman *et al.* (2015) describe an effective method for automatic classification that uses deep learning with convolutional neural networks. The method inputs multifilter SDSS images of a large sample of galaxies having known morphological classifications. Domínguez Sánchez *et al.* (2019) use the crowd-sourced GZ2 classifications and the T-classifications of Nair and Abraham (2010) to obtain automatic classifications of 670,000 galaxies.

Many other aspects of galaxy morphology are worth examining; these are described in more detail in the review articles by Buta (2012, 2013).

1.14. References

Abbott, C.G., Valluri, M., Shen, J., *et al.* (2017). *MNRAS*, 470, 1526.

Aihara, H., *et al.* (2018). *PASJ*, 70, 84.

Ann, H.B., Seo, M., Ha, D.K. (2015). *ApJS*, 217, 27.

Appleton, P., Struck-Marcell, C. (1996). *FCPh*, 16, 111.

Athanassoula, E. (2016) In *Galactic Bulges,* Laurikainen, E., Peletier, R., Gadotti, D. (eds). ASSL, vol. 418. Springer International Publishing, Switzerland, p. 391.

Athanassoula, E., Rodonov, S.A., Peschke, N., *et al.* (2016). *ApJ*, 821, 90.

Baillard, A., Bertin, E., de Lapparent, V., *et al.* (2011). *A&A*, 532, 74.

Bertola, F., Galletta, G. (1978). *ApJ*, 226, L115.

Blum, R.D., *et al.* (2016). *BAAS*, 228, 317.

Bournaud, F., Combes, F. (2002). *A & A*, 392, 83.

Buta, R. (1995a). *ApLC*, 31, 109.

Buta, R. (1995b). *ApJS*, 96, 39.

Buta, R. (2012). Galaxy morphology. In *Secular Evolution of Galaxies*, Falcón-Barroso, J., Knapen, J.H. (eds). Cambridge University Press, Cambridge, 155.

Buta, R. (2013). Galaxy morphology. In *Planets, Stars, & Stellar Systems, Vol. 6*, Oswalt, T.D., Keel, W. C. (eds). Springer, Dordrecht, p. 1.

Buta, R. (2019). *MNRAS*, 488, 590.

Buta, R.,Verdes-Montenegro, L., Damas-Segovia, A., *et al.* (2019). *MNRAS*, 488, 2175.

Buta, R., Block, D.L. (2001). *ApJ*, 550, 243.

Buta, R., Combes, F. (1996). *FCPh*, 17, 95.

Buta, R., Crocker, D.A. (1991). *AJ*, 102, 1715.

Buta, R., Purcell, G.B. (1998). *AJ*, 115, 484.

Buta R., Williams K.L. (1995). *AJ*, 109, 543.

Buta, R., Zhang, X. (2009). *ApJS*, 182, 559.

Buta, R., Byrd, G.G., Freeman, T. (2004). *AJ*, 127, 1982.

Buta, R., Corwin, H.G., Odewahn, S.C. (2007). *The de Vaucouleurs Atlas of Galaxies*. Cambridge University Press, Cambridge.

Buta, R., Laurikainen, E., Salo, H., *et al.* (2010). *ApJ*, 721, 259.

Buta, R., Mitra, S., de Vaucouleurs, G., *et al.* (1994). *AJ*, 107, 118.

Buta, R., *et al.* (2015). *ApJS*, 217, 32.

Byrd, G.G., Rautiainen, P., Salo, H., *et al.* (1994). *AJ*, 108, 476.

Capaccioli, M. (1987). In *The Structure and Dynamics of Elliptical Galaxies*, de Zeeuw, T. (ed.). *IAU Symposium* 127, Reidel, Dordrecht, p. 47.

Cappellari, M., *et al.* (2011). *MNRAS*, 416, 1680.

Chilingarian, I.V., Di Matteo, P., Combes, F., *et al.* (2010). *A&A*, 518, A61.

Combes, F., Sanders, R.H. (1981). *A&A*, 96, 164.

Comerón, S., Knapen, J.H., Beckman, J.E., *et al.* (2010). *MNRAS*, 402, 2462.

Comerón, S., *et al.* (2014). *A&A*, 562, 121.

Crocker, D.A., Baugus, P.D., Buta, R. (1996). *ApJS*, 105, 353.

Curtis, H.D. (1918). *Publ. Lick Obs.*, 13, 9.

de Jong, J.T.A., *et al.* (2013). *The Messenger*, 154, 44.

de Lapparent, V., Baillard, A., Bertin, E. (2011). *A&A*, 532, 74.

de Vaucouleurs, G. (1948). *Annales d'Astrophysique*, 11, 247.

de Vaucouleurs, G. (1958). *ApJ*, 127, 487.

de Vaucouleurs, G. (1959). *Handbuch der Physik*, 53, 275.

de Vaucouleurs, G. (1963). *ApJS*, 8, 31.

de Vaucouleurs, G., Freeman, K.C. (1972). *Vistas in Astronomy*, 14, 163.

de Vaucouleurs, G., de Vaucouleurs, A., Corwin, H.G., *et al.* (1991). *Third Reference Catalogue of Bright Galaxies*. Springer, New York.

Dieleman, S., Willett, K.W., Dambre, J. (2015). *MNRAS*, 450, 1441.

Dobbs, C., Baba, J. (2014). *PASA*, 31, e035.

Dobbs, C.L., Pettitt, A.R., Corbelli, E., *et al.* (2018). *MNRAS*, 478, 3793.

Domínguez Sánchez, H., Huertas-Company, M., Bernardi, M., *et al.* (2018). *MNRAS*, 476, 3661.

Domínguez Sánchez, H., *et al.* (2019). *MNRAS*, 484, 93.

Dressler, A. (1980). *ApJ*, 236. 351.

Dreyer, J.L.E. (1888). A new general catalogue of nebulae and clusters of stars. *Mem. Roy. Astr. Soc.*, 49, 1.

Eliche-Moral, M.C., Rodríguez-Pérez, C., Borlaff, A., *et al.* (2018). *A&A*, 617, A113.

Elmegreen, D.M. (1981). *ApJS*, 47, 229.

Eskridge, P., *et al.* (2000). *AJ*, 119, 536.

Fasano, G., Bonoli, C. (1989). *A&AS*, 79, 291.

Fukugita, M., *et al.* (2007). *AJ*, 134, 579.

Gadotti, D. (2004). *PASP*, 116, 591.

Gadotti, D., de Souza, R. (2003a). *Ap&SS*, 284, 527.

Gadotti, D., de Souza, R. (2003b). *ApJ*, 583, L75.

Gao, H., Ho, L.C., Barth, A.J., *et al.* (2018). *ApJ*, 867, 100.

Garcia-Gómez, C., Athanassoula, E., Barberà, C., *et al.* (2017). *A&A*, 601, A132.

Gehrels, N., *et al.* (2015). *J. Phys., Conf. Ser.* 610, 012007.

Genel, S. *et al.* (2014). *MNRAS*, 445, 175.

Gerola, H., Seiden, P.E. (1978). *ApJ*, 223, 129.

Graham, A.W. (2013). In *Planets, Stars, and Stellar Systems, Vol. 6*, Oswalt, T.D., Keel, W. C. (eds). Springer, Dordrecht, p. 91.

Grouchy, R.D., Buta, R., Salo, H., *et al.* (2010). *AJ*, 139, 2465.

Gunn, J.E., Carr, M., Rockosi, C., *et al.* (1998). *AJ*, 116, 3040.

Gunn, J.E., *et al.* (2006). *AJ*, 131, 233.

Gusev, A.S., *et al.* (2015). *A&A*, 59, 899.

Hohl, F. (1971). *ApJ*, 168, 343.

Hubble, E. (1926). *ApJ*, 64, 321.

Hubble, E. (1936). *The Realm of the Nebulae*. Yale University, New Haven, CT.

Huang, S., Ho, L.C., Peng, V.Y., *et al.* (2013). *ApJ*, 766, 47.

Hubble, E., Humason, M.L. (1931). *ApJ*, 74, 43.

Ivezi, Z., *et al.* (2019). *ApJ*, 873, 111.

Kaiser, N., *et al.* (2010). In *Proceedings of SPIE*, Vol. 7733, Ground-based and Airborne Telescopes III, 77330E.

Kartaltepe, J.S., *et al.* (2016). *ApJS*, 221, 11.

Keeler, J.E. (1899). *MNRAS*, 60, 128.

Kennicutt, R.C., Tambln, P., Congdon, C.W. (1994). *ApJ*, 435, 22.

Kim, T., Gadotti, D., Athanassoula, E., *et al.* (2016). *MNRAS*, 462, 3430.

Knapen, J., James, P.A. (2009). *ApJ*, 698, 1437.

Koopmann, R., Kenney, J.D.P. (2004). *ApJ*, 613, 866.

Kormendy, J. (1979). *ApJ*, 227, 714.

Kormendy, J., Bender, R. (1996). *ApJ*, 464, 119.

Kormendy, J., Bender, R. (2012). *ApJS*, 198, 2.

Kormendy, J., Djorgovski, S. (1989). *ARAA*, 27, 235.

Kormendy, J., Kennicutt, R. C. (2004). *ARA&A*, 42, 603.

Kormendy, J., Norman, C. (1979). *ApJ*, 233, 539.

Laureijs, R., *et al.* (2012). *SPIE Proceedings,* Space Telescopes and Instrumentation: Optical, Infrared, and Millimeter Wave, 8442, OT.

Laurikainen, E., Salo, H., Athanassoula, E., *et al.* (2013). *MNRAS*, 430, 3489.

Laurikainen, E., Salo, H., Buta, R., *et al.* (2011). *MNRAS*, 418, 1452.

Lin, C.C. (1970). In *The Spiral Structure of Our Galaxy*, Becker, W., Kontopoulos, G.I. (eds). *IAU Symposium 38*, Reidel, Dordrecht, p. 377.

Lin, C.C., Shu, F.H. (1964). *ApJ*, 140, 646.

Lin, L.-H., Yuan, C., Buta, R. (2008). *ApJ*, 684, 1048.

Lintott, C., *et al.* (2008). *MNRAS*, 389, 1179.

Lundmark, K. (1926). *Archive for Mathematics, Astronomy, and Physics*, 19(8), 1.

Lutticke, R., Dettmar, R.-J. (1999). *Ap&SS*, 265, 393.

Martinez-Valpuesta, I., Knapen, J.H., Buta, R. (2007). *AJ*, 134, 1863.

Miller, R. H., Smith, B.F. (1979). *ApJ*, 227, 785.

Morgan, W.W. (1958). *PASP*, 70, 364.

Nair, P., Abraham, R.G. (2010). *ApJ*, 714, L260.

Nieto, J.L. (1988). *Bol. Acad. Nac. Cienc. Cordoba, Argentina*, 58(3-4), 239.

Norman, C.A., Sellwood, J., Hasan, H. (1996). *ApJ*, 462, 114.

Rautiainen, P., Salo, H. (2000). *A&A*, 362, 465.

Regan, M., Teuben, P. (2004). *ApJ*, 600, 595.

Reynolds, J.H. (1920). *MNRAS*, 80, 746.

Roberts, W.W., Roberts, M.S., Shu, F.H. (1975). *ApJ*, 196, 381.

Romero-Gómez, M., Masdemont, J.J., Athanassoula, E., *et al.* (2006). *A&A*, 453, 39.

Ryden, B.S. (1988). *ApJ*, 329, 589.

Sandage, A. (1961). *The Hubble Atlas of Galaxies*, Carnegie Inst. of Wash., Washington, D.C., Publ. no. 618.

Sandage, A., Binggeli, B. (1984). *AJ*, 89, 919.

Schwarz, M.P. (1981). *ApJ*, 247, 77.

Schwarz, M.P. (1984). *MNRAS*, 209, 93.

Schweizer, F. (1982). *ApJ*, 252, 455.

Schweizer, F. (1987). In *The Structure and Dynamics of Elliptical Galaxies*, IAU Symposium 127, de Zeeuw, T. (ed). Reidel, Dordrecht, p. 109.

Schweizer, F., Seitzer, P. (1988). *ApJ*, 328, 88.

Schweizer, F., Whitmore B.C., Rubin V. (1983). *AJ*, 88, 909.

S'ersic, J.L. (1968). *Atlas de Galaxias Australes*. Observatorio Astronomico, Cordoba, Argentina.

Sheth, K., *et al.* (2008). *ApJ*, 675, 1141.

Snyder, G.F., *et al.* (2015). *MNRAS*, 454, 1886.

Spitzer, L., Baade, W. (1951). *ApJ*, 113, 413.

Toomre, A. (1977). *ARA&A*, 15, 437.

Toomre, A. (1981). In *The Structure and Evolution of Normal Galaxies*, Fall, S.M., Lynden-Bell, D. (eds), Cambridge University Press, Cambridge, p. 111

Tsatsi, A., Lyubenova, M., van de Ven, G., *et al.* (2017). *A&A*, 606, A62.

van den Bergh, S. (1976). *ApJ*, 206, 883.

Verdes-Montenegro, L., Sulentic, J., Lisenfeld, U., *et al.* (2005). *A&A*, 436, 443.

Visser, H. (1980). *A&A*, 88, 159.

Vogelsberger, M., *et al.* (2014). *MNRAS*, 444, 1518.

White, S.D.M., Rees, M.J. (1978). *MNRAS*, 183, 341.

Whitmore, B.C., Lucas, R.A., McElroy, D.B., *et al.* (1990). *AJ*, 100, 1489.

Willett, K.W., *et al.* (2013). *MNRAS*, 435, 2835.

Wolf, M. (1908). Publ. of the Astrophysical Institute, Koenigstuhl-Heidelberg, 3, 109.

Yagi, M., Yoshida, M., Komiyama, Y., *et al.* (2010). *AJ*, 140, 1814.

York, D.G., Adelman, J., Anderson, J.E., *et al.* (2000). *AJ*, 120, 1579.

Zhang X. (2018). *Dynamical Evolution of Galaxies*. de Gruyter GmbH, Berlin/Boston.

2

Our Galaxy, the Milky Way

Paola Di MATTEO
Observatoire de Paris (GEPI), Meudon, France

2.1. Introduction

Our Galaxy is a barred, spiral disc galaxy, with a Sb-Sbc morphology in the Hubble classification. Its total stellar mass is $M_* = 5 \pm 1 \times 10^{10} M_\odot$, its baryonic mass (stars plus cold gas) is $M_* = 6.3 \pm 0.5 \times 10^{10} M_\odot$ and its total mass (dark matter included) is $M_{TOT} = 1.3 \pm 0.3 \times 10^{12} M_\odot$ inside the Galactic virial radius $r_{vir} = 282 \pm 30$ kpc (Bland-Hawthorn and Gerhard 2016). Its current star formation is of $1 - 2\ M_\odot/\text{year}$, which places it in the "green valley", that is halfway between the blue cloud of actively star forming galaxies, and the red passive sequence (Licquia *et al.* 2015). Given its actual rate of star formation, and its total gas content, the Galaxy should stop forming stars in less than 5 Gyr, unless a new supply of gas will come to revitalize its star formation activity.

There has been a long debate in the astronomical community about the exceptionality or normality of our Galaxy when compared to galaxies of similar mass or living in similar environments: the Milky Way appears exceptional in the Tully–Fisher relation (see, Hammer *et al.* 2007, but also Licquia *et al.* (2016)) but rather normal when its star formation history is compared to galaxies of similar type, at different redshifts (Snaith *et al.* 2014). The two major difficulties encountered when comparing the Galaxy to other disc galaxies is, first, to have a robust description of what the Milky Way really is, in terms of structure, mass content and

For color versions of the figures in this chapter, see www.iste.co.uk/combes/galaxies.zip.

Galaxies,
coordinated by Françoise COMBES.

stellar populations, and, second, to define a sample of galaxies (observed or simulated) representative of the Milky Way properties and environment, in order to guarantee an appropriate and well-grounded comparison. Exceptional or normal, it appears anyway that our Galaxy has a stellar mass at the peak of the distribution of galaxy masses in the Local Universe. Understanding its characteristics, and its evolutionary history is thus fundamental not only for Galactic studies, but also for the possibility to make use of our understanding of the Milky Way to interpret the properties of other galaxies of similar mass.

This chapter aims at introducing the reader to the main properties of our Galaxy, and to explain how the Galactic community is making use of these properties to trace a tentative scheme of the Milky Way formation and evolution. We caution the reader that the field of Galactic studies is experiencing a very rapid and deep revision, because of the extraordinary data, in terms of quality and quantity, delivered by the ESA astrometric mission, *Gaia*[1] (see, Gaia Collaboration *et al.* 2016, 2018), and complementary spectroscopic surveys. In this respect, the reader needs to be conscious that what we present in the following pages is a description of our Galaxy as we understand it today. This description may become outdated very rapidly, as the analysis of the *Gaia* data will be pushed further, and next *Gaia* releases will be available, together with new spectroscopic surveys. The following pages should constitute, in all cases, an introduction to our current understanding of the Galaxy, and also the basis which the scientific community will build on in its future vision.

Because of our position in the Galaxy, at a few tens of parsecs from the disc midplane, we miss a global view of our Galaxy, and of its grand design components. The understanding of its global structure is relatively recent, and it is still ongoing. It requires to make use of observations at different wavelengths to trace the distribution of gas and stars from the outskirts of the disc and halo up to the very dense central regions, and also of the comparison to models, which help to deproject the observations, and decipher the Galaxy 3D structure.

Schematically, our Galaxy is made of a thin disc of dust (constituting less than 1% of its mass), gas (about 10% of its mass), and stars (90% of its mass), whose distribution is neither smooth nor radially symmetric, but concentrates in some main structures, like a central bar and a number (2 or 4) of spiral arms, where most of the star formation currently takes place. The borders of the stellar disc are not well known, but possibly extend up to about 20–30 kpc from the Galactic center. The Sun, whose distance from the center of our Galaxy is estimated at about 8.2 ± 0.1 kpc (Bland-Hawthorn and Gerhard 2016), is halfway between the center and the periphery of our Galaxy. The mean thickness of this disc, which the Sun belongs to, and which is usually referred to as the Galactic "thin disc", is about 300 pc. Young stars (ages less

1 https://sci.esa.int/gaia.

than 6-7 billions years) are mostly confined there, while intermediate age (7–10 billion years) and most of the old stars (ages greater than 10 billion years) are redistributed in a thicker, disc-like configuration, known as the Galactic "thick disc".

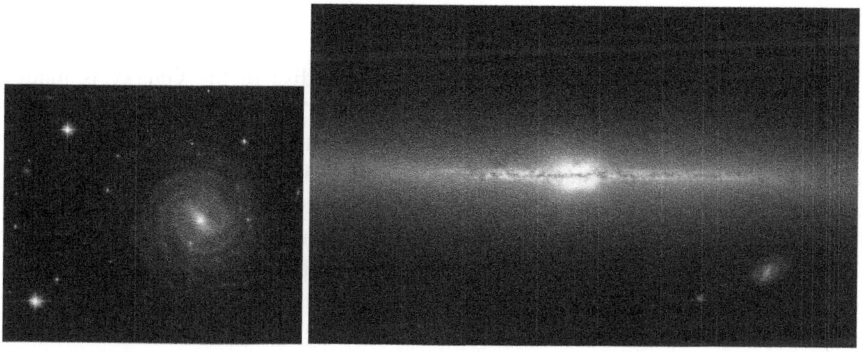

Figure 2.1. Left panel: *The Sb-type barred spiral galaxy UGC 12158, often considered a Milky Way analogue for its appearance (Credit: ESA/Hubble & NASA).* Right panel: *An infrared image of the Milky Way galaxy, from the 2 Micron All Sky Survey (Credit: 2MASS/J. Carpenter, M. Skrutskie, R. Hurt). The two galaxies on the bottom-right part of the image are the Large and Small Magellanic Clouds*

In the inner regions, the Galaxy is dominated in mass and light by the so-called "bulge", that is a structure rounder than the disc, that extends for about 2 kpc in radius from the Galactic center, and for at least 1 kpc in height above and below the plane. Differently from the thin disc, that is the site of ongoing star formation, the bulge is mostly old and metal poor.

Finally, all these components (bulge, thin and thick disc) are surrounded by a diffuse and light (a few percent of the total stellar mass) distribution, which has a spherical shape at first approximation, and which extends for tens of kpc from the Galactic center. This component, called the Galactic "stellar halo", is made of the oldest and most metal-poor stars in the Galaxy, and as such it is considered to trace the first epochs of the Galaxy formation. The halo is also the region of the Galaxy where about 70% of the current population of Galactic globular clusters are found, the remaining fraction being rather associated with the bulge and the thick disc.

This chapter is organized as follows. In sections 2.2–2.4, we present the global structure of our Galaxy, its gaseous and stellar discs, bulge and stellar halo. In section 2.5, we discuss the distribution of the dark matter content of the Galaxy and its shape. In section 2.6, we revise the stellar components presented in the previous sections, in view of their kinematics, chemistry and ages. Section 2.7 proposes a possible evolutionary path for our Galaxy based on the properties described in the previous sections. Finally, section 2.8 opens to the future of Galactic astronomy and archeology, which will be undoubtedly deeply redesigned in the next decade.

2.2. Baryonic discs and their spiral structure

The large majority of the baryonic mass in the disc of the Galaxy is made up of stars (80–90%), while the remaining 10–20% is in the form of interstellar matter, which is prevalently made up of gas (99%) and dust (1%). In the following, we will present the main characteristics of the disc gaseous and stellar components, cautioning the reader that a supplementary hot (temperatures greater than 10^5 K) and extended gaseous halo, or corona, should surround the Milky Way disc. Its mass may be as large as $2.5 \pm 1 \times 10^{10} M_{\odot}$, and it should extend for several hundreds kiloparsecs from the Galactic center (Bland-Hawthorn and Gerhard 2016). The presence of the Galactic corona may explain why all the closest satellites of the Milky Way, with the exception of the Magellanic Clouds, are gas depleted (Tolstoy *et al.* 2009): the absence of gas in these systems would be indeed the result of ram-pressure stripping by the hot Galactic medium.

2.2.1. *Neutral, ionized and molecular gas*

Despite it constitutes only a small fraction of the total baryonic mass budget, the study of the gas distribution in the Galactic disc, initiated in the second half of the last century, has probed to be essential to understand the Milky Way large structure. The gas is mostly made up of hydrogen (74%), either in atomic or molecular form, helium (24%), and only for a small percentage of its mass is heavier elements, generically referred to as "metals". It is mostly by studying the hydrogen distribution in the Galactic disc that it has been possible to measure the first Galactic rotation curves, and to evidence the large scale, spiral structure of our Galaxy (see, Kwee *et al.* 1954; van de Hulst *et al.* 1954; Oort 1958, for pioneering works on these fields). In the following, we briefly recall the main forms of hydrogen that can be found in the disc of the Milky Way – neutral, ionized and molecular (see Figure 2.2 for their distributions) – and how, by its study, it is possible to reconstruct the large-scale structure of the disc.

Atomic neutral hydrogen, referred to as HI, is the most conspicuous gaseous component in the Galaxy, representing about $M_{HI} = 8 \times 10^9 M_{\odot}$ inside 60 kpc from the Galactic center (Kalberla and Kerp 2009), that is about 65% of all hydrogen in the Galaxy. The HI gas has temperatures between 100 K and about 3,000 K. It does not emit detectable visible light, but it can be traced by its emission at the 21-cm line wavelength. Because of the transparency of the Galaxy to radio wavelengths, and because of its ubiquity, since the discovery of the 21-cm line in the early 1950s of the last century, radio observations have been essential to study the interstellar medium, the Galaxy large-scale structure and its rotation curve. While a robust knowledge of the HI distribution inside the solar circle is still lacking, because of ambiguities in estimating the distances of HI clouds (the so-called kinematic-distance ambiguity that affects also molecular gas clouds, see (Roman-Duval *et al.* 2009)), outside the solar radius the HI distribution can be approximated by an exponential disc with a scale length $R_{d,HI} = 3.75$ kpc, and an average thickness h_z that increases with the distance R from the Galactic center (a feature referred to as the "flaring" of the Galactic disc, and that also affects the stellar distribution). This flaring can be quantified by approximating it through an exponential relation of the form $h_z = h_0 e^{(R-R_\odot)/R_0}$, with $h_0 = 0.15$ kpc, and $R_0 = 9.8$ kpc for $5 \leq R \leq 35$ kpc. Besides the overall flaring, the HI disc is also strongly warped, that is the average HI distribution deviates from planarity outside the solar circle, for $R \geq 9$ kpc, and the warp extends at least up to $R \sim 40$ kpc, reaching a height of 5 kpc at $R \sim 20$ kpc.

Atomic ionized hydrogen, referred to as HII, is interstellar atomic hydrogen located around massive young stars (O and B spectral type) that is ionized by their far ultraviolet radiation. Its temperature is several thousand Kelvin. It constitutes about 16% of all Galactic hydrogen inside 60 kpc, for a total mass of about $M_{HII} = 2 \times 10^9 M_\odot$ (Kalberla and Kerp 2009). Because O and B stars have typical lifetimes of a few million years, HII regions are among the best tracers of recent massive star formation. They are the most luminous objects in the Galaxy at wavelengths from the mid-infrared to the radio and, as such, they can be seen across the entire Galactic disc, thus tracing its large-scale structure[2]. As for the HI, also the distribution of the HII regions shows a warping and a flaring outside the solar circle (Anderson *et al.* 2019).

Molecular hydrogen, referred to as H_2, represents the third state in which hydrogen can be found in the Galaxy. It represents about 20% of all the Galactic hydrogen inside 60 kpc, for a total mass of about $M_{H_2} = 2.5 \times 10^9 M_\odot$ (Kalberla and Kerp 2009). Because this molecule does not radiate at radio wavelengths, the CO molecule is often used as a tracer of H_2 in the Galaxy. CO is indeed expected to be associated with H_2, and because it is ubiquitous in the Galaxy, it is considered as a fundamental tracer of the molecular gas. Differently from HI, for which most of the mass is outside the solar radius, molecular gas appears concentrated in the innermost Galaxy, with most of the CO emission coming from a region located between 4 and 6 kpc from the Galactic center, the so-called "molecular ring" (see, Stecker *et al.* 1975; Dame *et al.* 2001), even if its physical association with a true ring feature is questioned these days (Rodriguez-Fernandez and Combes 2008; Dobbs and Burkert 2012). As for HI and HII, also the distribution of molecular gas is found to flare and warp in the outer disc, even if the effects are milder than those found for the neutral hydrogen distribution.

2 HII regions can also be detected at optical wavelengths, but in this case their detection is limited at a few kpc from the Sun, because of the dust extinction.

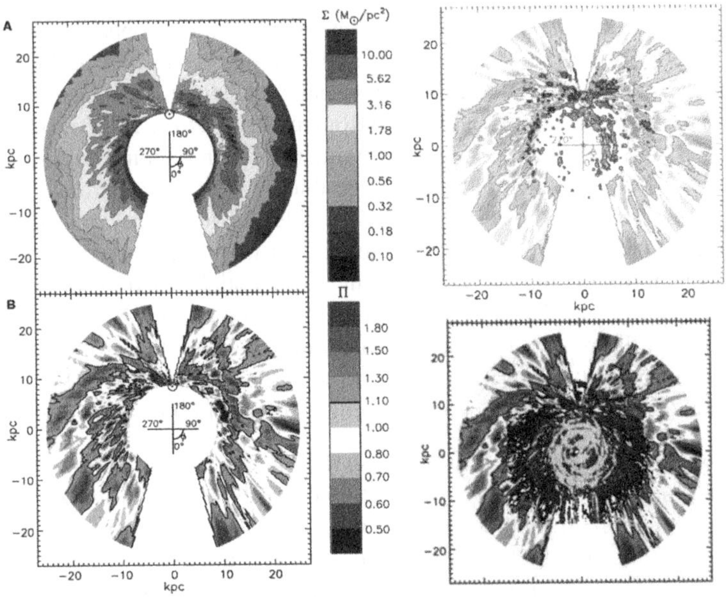

Figure 2.2. Left column, top panel: *Map of the surface density of HI in the Galactic disc. The location of the Sun is marked at (0, 8.5 kpc) by the solar symbol. Two regions in the anticenter and Galactic center direction are excluded because of unreliable distances. From (Levine* et al. *2006)*. Left column, bottom panel: *Map of the strength of the surface density perturbations in the HI distribution, defined as the ratio of the surface density map shown above divided by its median value. From (Levine* et al. *2006)*. Right column, top panel: *HII regions in the Galactic disc, superimposed to the HI map from Levine* et al. *(2006). From Hou* et al. *(2009)*. Right column, bottom panel: *Map of the molecular gas distribution in the inner galaxy from Pohl* et al. *(2008), superimposed to the HI map from Levine* et al. *(2006). The Sun is indicated by the yellow dot. Adapted from Englmaier* et al. *(2011)*

2.2.2. *Thin and thick stellar discs*

The large majority of stars in the Galaxy are redistributed in a disc. Despite being the most massive component of the Milky Way, understanding the structure of the stellar disc (or stellar discs, as we will see in the following) is not an easy task. As for the gaseous component, the place of the Sun, close to the Galaxy midplane, prevents an immediate view of the stellar disc large-scale structure that can be reconstructed only by deriving distances for a large number of stars. Differently from the gaseous component, mostly observed at radio and submillimeter wavelengths, not affected by extinction, the optical and near-infrared observations of stars in the disc are severely impacted by the presence of dust, which in some directions severely limits the maximal distance from the Sun that can be reached by observations at these wavelengths.

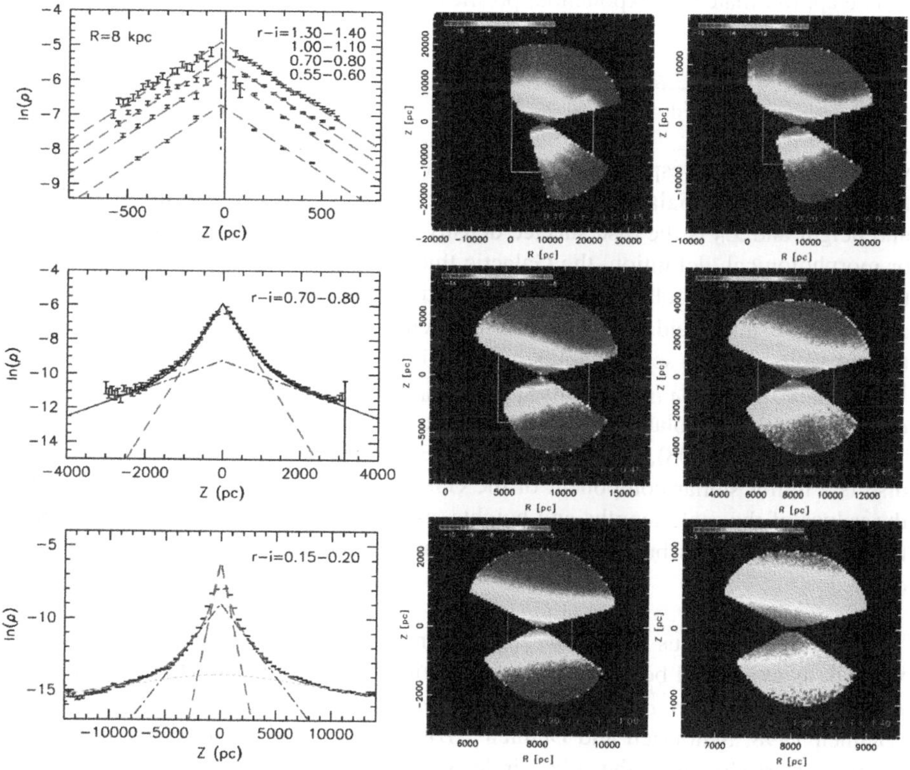

Figure 2.3. Left column: *Vertical distribution of SDSS stellar counts at the solar radius and for different $r - i$ color bins. The bluer the stars, the more distant they are, on average, from the Sun. The lines are exponential models fitted to the points. The dashed lines in the top panel correspond to a fit with a single, exponential disc with a scale height of 0.27 kpc. The dashed line in the middle panel corresponds to a sum of two discs with scale heights of 0.27 kpc (thin disc) and 1.2 kpc (thick disc). This latter becomes important for heights $|z| > 1$ kpc. The dotted line in the bottom panel shows the contribution of the stellar halo, which appears necessary to include when stars at large distances from the Sun are considered.* Right column: *Stellar number density as a function of Galactic cylindrical coordinates R, the in-plane distance from the Galactic center, and z, the height above the Galaxy midplane, for different $r - i$ color bins, as indicated. The panels on the top contain stars with blue $r - i$ colors, and which are, on average, the most distant from the Sun. All plots in this figure are adapted from* Jurić et al. *(2008)*

Overall, the Galactic stellar disc can be described as made up of two main components, a thin and a thick disc, each having volume density distributions that can be approximated by exponential profiles, of the form

$$\rho_{\rm d}({\rm R}, {\rm z}) = \frac{\Sigma_0}{2{\rm z_d}} \exp(-\frac{|{\rm z}|}{\rm z_d} - \frac{\rm R}{\rm R_d}) \qquad [2.1]$$

where R and z are, respectively, the in-plane distance from the Galactic center and the height from the Galactic plane; R_d and z_d are, respectively, the disc scale length and height and Σ_0 is the disc surface density at $R = 0$ and $z = 0$. In this geometric, or morphological, definition, the Galactic thick disc appears as the excess of stars in the vertical direction relative to that predicted by a single exponential (see Figure 2.3 and Gilmore and Reid (1983)). When based on this geometrical decomposition, typical values of the scale lengths and heights for the thin and the thick discs are $R_{d,thin} = 2.6$ kpc, $R_{d,thick} = 3.6$ kpc, $z_{d,thin} = 0.3$ kpc and $z_{d,thick} = 1$ kpc (Jurić *et al.* 2008). At the solar vicinity, the thick-to-thin disc volume density ratio is about 10% (Jurić *et al.* 2008), which has been often interpreted as evidence that the thick disc is a minor stellar component of the Galaxy. As we will see in section 2.6, both the values of the scale lengths and heights, as well as the weight of the thick disc to the total stellar mass budget, strongly depend on the way thin and thick discs are defined.

As for the gaseous disc, also the stellar disc, in its outer parts, is flared and warped, as evidenced both from the distribution of stellar tracers (López-Corredoira *et al.* 2002; Reylé *et al.* 2009; Skowron *et al.* 2019) and from their kinematics (Dehnen 1998; Schönrich and Dehnen 2018; Poggio *et al.* 2018). The stellar warp seems to start already at the solar circle (see Figure 2.4), and it is coherent, both in amplitude and orientation, with the HI warp[3] (Momany *et al.* 2006). The maximum of the amplitude of the stellar warp in the southern hemisphere (negative latitudes) seems to coincide with the locus of an overdensity, called Canis Major, initially interpreted as the remnant of a dwarf galaxy recently cannibalized by the Milky Way (Martin *et al.* 2004), but later suggested to be rather related to the outer disc structure (Momany *et al.* 2006). This overdensity seems also linked to a stellar ring feature, called the Monoceros Ring (Newberg *et al.* 2002), confined close to the Galactic plane. As for Canis Major, also the origin of the Monoceros Ring is unclear. It could be related to the flare of the Galactic disc, or constitute a stellar stream leftover of a recent accretion. Whatever the origin of Canis Major and Monoceros, it is clear that the outer Galactic disc presents a complex and perturbed structure, as often observed in many other disc galaxies.

3 Note that also the dust distribution is warped and coherent in amplitude and orientation both with the stellar and the HI warp

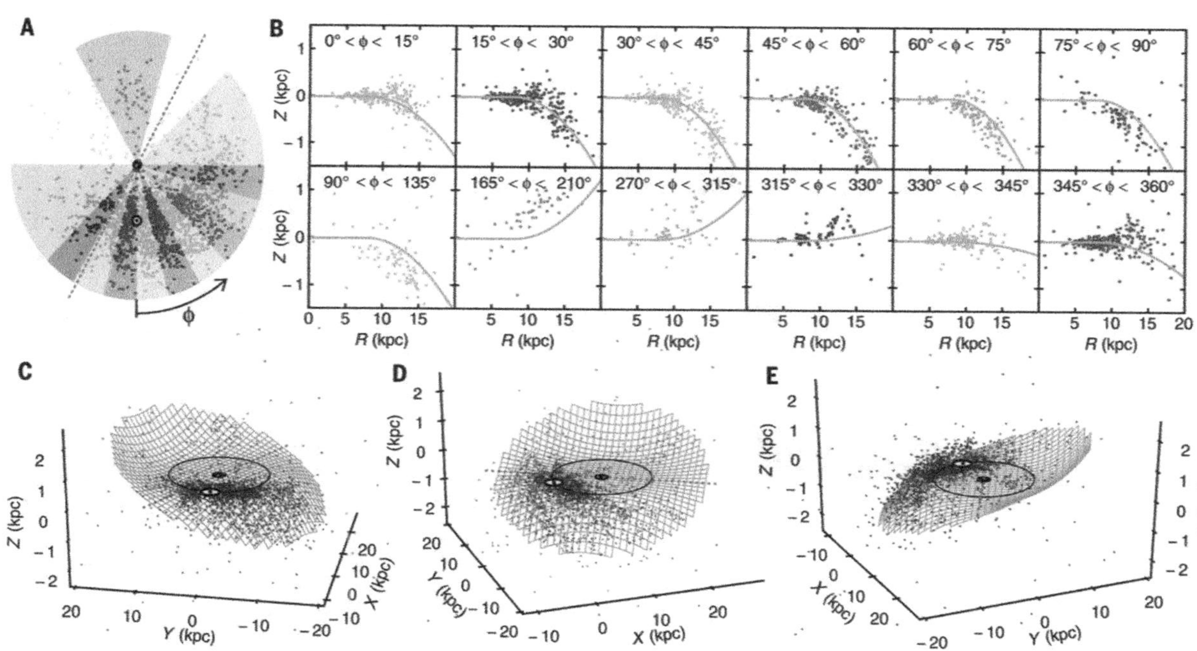

Figure 2.4. *The Galactic warp as traced by classical Cepheids (Skowron* et al. *2019, from). The Cepheids distribution in the* (R, z) *plane is shown in plot (B), for different intervals in azimuth, as defined by the sectors defined in plot (A). The gray line in each panel corresponds to the warp model shown in plots (C), (D) and (E), as projected in the given azimuthal sector. Plots (C), (D) and (E) show a model of the Milky Way warp from three viewing angles. As in plot (A), the Sun is marked by the yellow dot. The line of nodes, separating parts of the Galaxy warped toward negative and positive* z*, is marked by the gray dotted line in plot (A)*

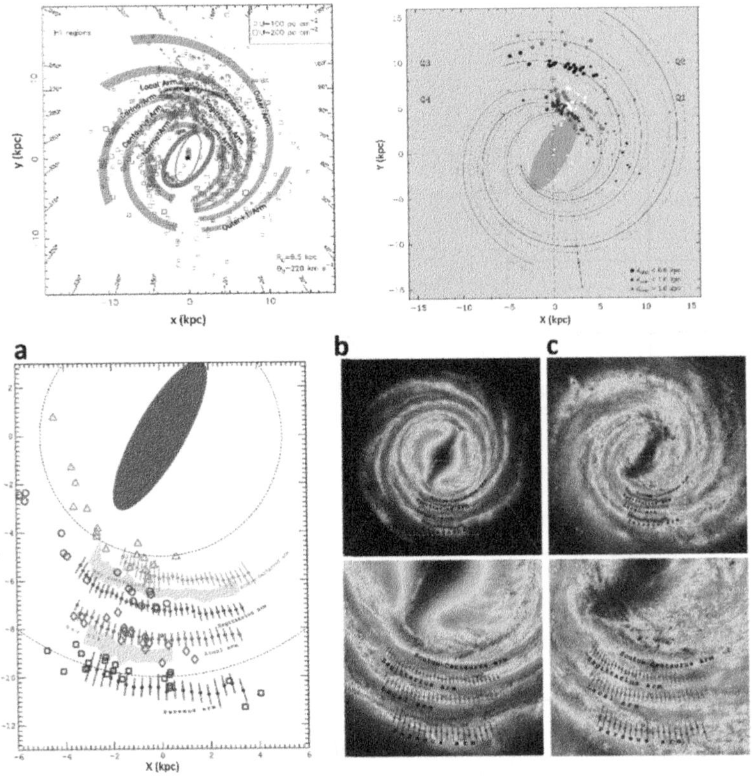

Figure 2.5. Top-left panel: *The distribution of HII regions overlaid with a possible spiral arm model for the Milky Way. The Sun is located at x = 0.0 kpc, y = 8.5 kpc. Adapted from Hou and Han (2014).* Top-right panel: *Distribution of high-mass star-forming regions with measured trigonometric parallaxes. The Sun is at (0,8.15) kpc. Different spiral arms are identified, with (from the inner to the outer disc) the 3-kpc arm, the Norma–Outer arm, the Centaurus-Scutum arm, the Sagittarius-Carina arm, the Local arm and the Perseus arm. Adapted from Reid* et al. *(2019).* Bottom-left panel*: Milky Way spiral arms (colored symbols with error bars) in the stellar distribution as revealed by Gaia DR2. The two gray-shaded areas show the location of stars possibly associated with resonances of the Milky Way bar. Large symbols without error bars show the distribution of high-mass star-forming regions from Reid* et al. *(2014). The gray oval represents the orientation of the Milky Way bar, and the solar position is marked by a yellow star.* Bottom-right panels*: Two new artists views of the Milky Way including a possible four-armed spiral arm model, in agreement with the spiral structure revealed by Gaia DR2 in the stellar distribution. These images are modified from the original Milky Way galaxy rendering by R. Hurt and N. Risinger. Adapted from (Khoperskov* et al. *2019)*

2.2.3. *Spiral structure from gaseous and stellar tracers*

When gas flows into spiral arms, it is expected to shock, cool and compress to form dense clouds where star formation occurs. It is by tracing the spatial distribution of gas, in its neutral, atomic and ionized phases, as well as of massive, young star forming regions, and young stars, that we can try to reconstruct the large-scale spiral structure of the Galaxy (see, Vallée 2017, for a recent review). The Sun appears to belong to the so-called "Local Arm" that is in between the "Sagittarius-Carina" and the "Perseus" arms (see Figure 2.5, top-left panel). Moving toward the Galactic center, other structures appear, as the Centaurus–Scutum arm and the Norma Arm. In the outer disc, beyond the Perseus arm, gaseous tracers and young massive star forming regions suggest the presence of one, and possibly two, more structures: the "Outer" and "Outer+1" Arm. All these arms are not necessarily independent, and some of them can indeed be part of the same large-scale structure. However, deriving from these arms, or portion of them, a global picture of the spiral structure of the Galaxy is complicated by at least two factors: (1) different tracers do not necessarily overlap, and thus can predict slightly different patterns and connections between them; (2) in the inner disc, we still lack enough tracers with sufficiently good distance estimates to robustly reconstruct the spiral arm structure and its link to the Galactic bar. An example of these difficulties is shown in Figure 2.5 (top panels): the Local Arm appears as a connecting arm between Perseus and Sagittarius-Carina, when HII tracers are used, but looks as an independent arm when massive, young star forming regions are used as tracers. Despite these difficulties, several works now converge to suggest that the Milky Way, as traced by the gas and massive star forming regions, appears as a four-arm spiral galaxy, with the Sun belonging to a local, possibly not independent, arm. We caution the reader, however, that a simpler, mainly two-arm spiral model is still not excluded (see, Dame and Thaddeus 2011), and that spiral arms may appear stronger in some tracers (gas and star forming regions), than in others (stars).

While the search for reconstructing the spiral structure of the Galactic disc has mainly focused on young stellar tracers and gas, all disc stars are expected to cross spiral arms and hence contribute to their overdensities, even the oldest. In other words, the entire disc should keep the imprint of its spiral structure. However, the search for spiral structures among stars of different ages has been historically very complicated. The limiting magnitude of Galactic surveys, and the effect of extinction both naturally have the effect of limiting the search for stellar spiral arms to a limited region around the Sun, and hence to prevent a large-scale view. Moreover, there is a second effect, which is related to the kinematics of stars: the older the stars, the larger their radial excursions around their orbital guiding radius is (see, Binney and Tremaine 1987, an effect known as "epicyclic motion"). As a result of these epicyclic oscillations, all morphological features in the stellar disc, as spiral arms, tend to be blurred and weakened, the older the population under examination is. A methodology to reveal the spiral arm structure of the Galactic disc by making use of stellar tracers of all ages, not restricted only to young stars, has been only recently proposed (Khoperskov *et al.* 2019). Even if these studies are preliminary, they have the potential to allow us to explore the spiral arm structure of the Galactic disc over a large range of tracers (see Figure 2.5, bottom panels). It is only by comparing all of them (gaseous components, star forming regions, young and old stars, etc.) that we can understand the nature of the spiral structure and its link to other Galactic components, like the bar/bulge, for example.

2.3. The central kiloparsecs: the bar and the bulge

The study of the structure of the central few kiloparsecs of the Milky Way is a very successful example of how observations and models can proceed together to build a coherent description of the Galaxy, or of some of its main components. Until the beginning of the 1990s, the bulge of the Milky Way – that is, the central stellar overdensity, visible at $|l| \lesssim 15$ and $|b| \lesssim 10$ – was thought to be a spheroidal component, that is "an elliptical that happens to have a prominent disc around it" (Renzini 1999), and as such possibly formed by a series of merging events at very high redshift.

In the early 1990s, however, two concomitant results emerged. On the one hand, N-body simulations started to investigate the way to form box or peanut shapes in the center of galaxies, because these shapes were found to exist in a number of external disc galaxies (see, for example, de Souza and Dos Anjos 1987). On the other hand, it was becoming increasingly clear that some intrinsic difference between bulges of disc galaxies and elliptical galaxies existed: boxy shapes were found to be much more pronounced in bulges than in elliptical galaxies, and none elliptical galaxy was known to have a peanut structure; box and peanut-shaped bulges of disc galaxies were found to be characterized by cylindrical rotation – that is, the rotation of stars in these bulges is independent on their height from the disc (see section 2.6) – while elliptical galaxies were not (see, Combes *et al.* 1990). Moreover, at least half of all spiral galaxies were found to have a stellar bar in their central regions (de Vaucouleurs 1963), and hence the link between stellar discs, bars and boxy-peanut shape bulges naturally started to be investigated. By analyzing these N-body simulations, it appeared soon clear that box and peanut-shapes bulges in disc galaxies were the manifestation of the same phenomena: a stellar bar thickened because of dynamical instabilities (Raha *et al.* 1991) and because of the existence of a resonance between the vertical oscillations of stars and the bar pattern speed, with stars trapped at this resonance constituting the backbone of the box/peanut-shape structure (Combes *et al.* 1990). Both scenarios were able to show that the "boxiness" of "peanutiness" was an effect of the bar viewing angle (Combes and Sanders 1981), the bulge appearing increasingly peanut-shaped the more the bar projection is close to edge-one (see Figure 2.6). At the same time, near-infrared and infrared observations of the central regions of our Galaxy, less affected by extinction than those at optical wavelengths, were revealing its boxy shape (see, Weiland *et al.* 1994; Dwek *et al.* 1995, and Figure 2.7), thus supporting earlier suggestions that the Galaxy was barred based on earlier studies of non-circular gas motions (Liszt and Burton 1980; Binney *et al.* 1991). Because of several bulge surveys conducted since then, we now have definite evidence that the Milky Way bulge is a boxy-peanut shaped bulge, making it the closest example to study this class of objects. Because a boxy-peanut bulge is a thickened stellar bar, and a stellar bar forms out of disc material, these findings naturally imply that the Milky Way bulge is mostly made of stars formed in the Galactic disc, and then trapped into it. A few decades of research on this field have thus allowed us to drastically change our vision and understanding of this component: the Galactic bulge is not the by-product of galaxy mergers, but rather the result of the galactic evolution occurring on much longer time scales (several billion years).

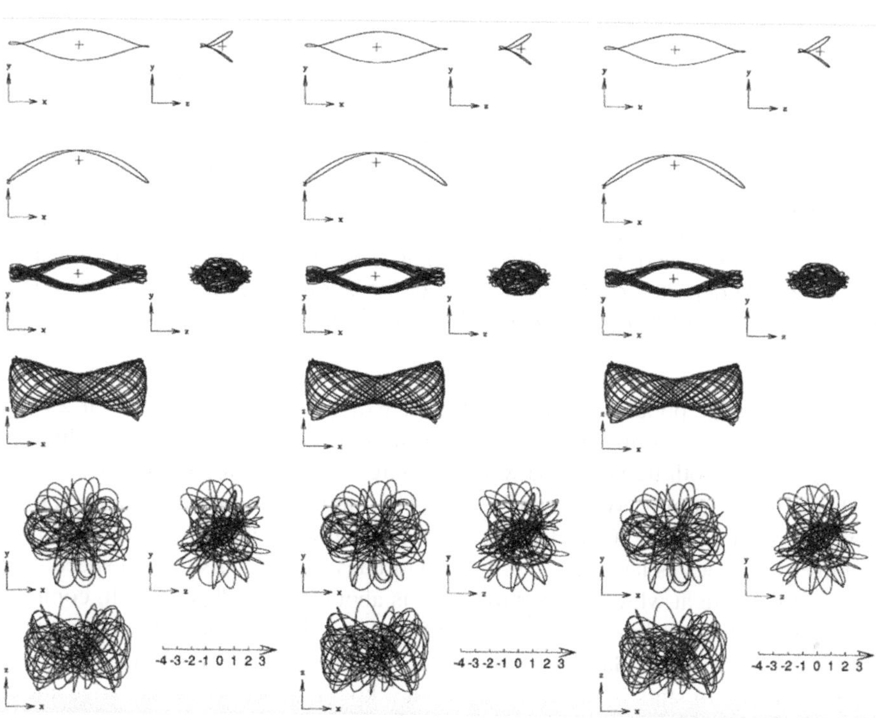

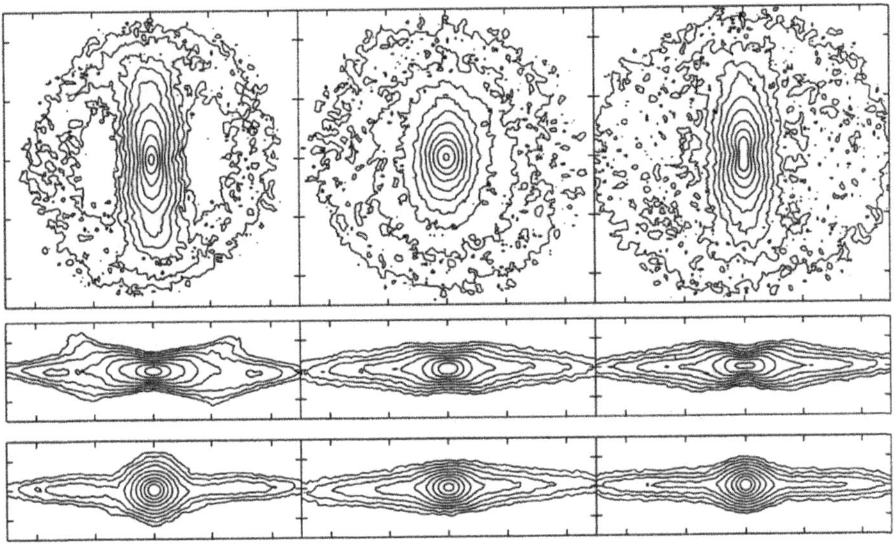

Figure 2.6. Top panels: *Regular orbits supporting a boxy/peanut-shaped bulge (left and middle panels), and a chaotic orbit in the same spatial region (right panel). For each orbit, the three Cartesian projections are shown. Adapted from Combes* et al. *(1990).* Bottom panels: *Face-on, side-on and end-on views of a bar and boxy/peanut-shaped bulge, in three different N-body models. The shape of the bulge (more peanut or boxy-shape) depends on the orientation of the bar relative to the observer line-of-sight. Adapted from Athanassoula and Misiriotis (2002)*

Besides stellar population studies, which will be discussed in section 2.6, the structural link between the bulge and the disc is also confirmed by the finding that there exists a smooth transition between the bulge scale height and that of the stellar bar that extends from it (Wegg *et al.* 2015). This stellar bar, the thin extension of the boxy/peanut shape in the Galactic disc, appears to have a half-length of about 5 kpc, reaching longitudes as large as $|l| \sim 30°$ at latitudes $|b| < 5°$. Its pattern speed – that is the angular rate at which the bar rotates – is about 40 km/s/kpc, which, combined with the Galactic rotation curve (section 2.5), places the bar corotation radius – that is the distance from the Galactic center where the bar pattern speed equals the mean angular velocity of stars in the disc – at about 6 kpc. It appears that the bulge and the bar, taken together, contain the majority of the mass of the inner Galactic disc (see, about $2 \times 10^{10} M_{\odot}$, Portail *et al.* 2017). Given their mass and size, they represent the engines of the Galactic disc dynamics.

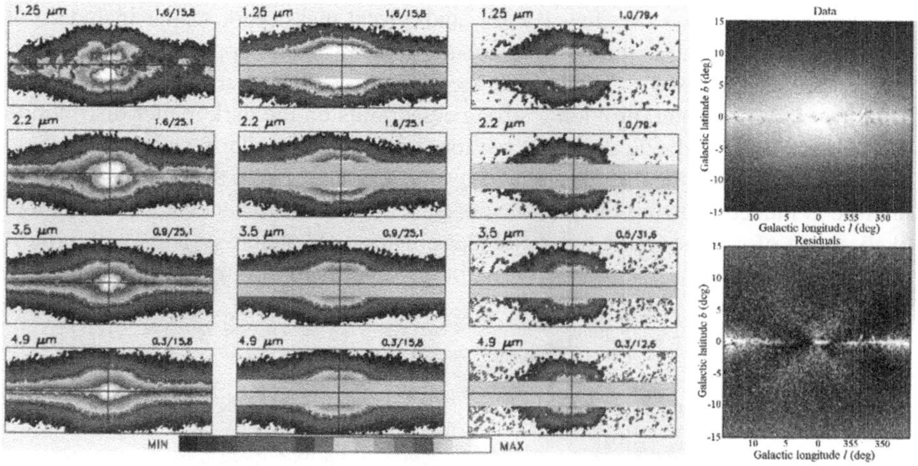

Figure 2.7. Left panels*: The Galactic bulge as observed in the near-infrared by the Diffuse Infrared Background Experiment (DIRBE). Each row in the figure corresponds to observations at a different wavelength, as reported in the top-left corner of each panel. Different columns correspond to: full DIRBE observations (left); extinction-corrected observations (middle); extinction-corrected observations after subtraction of the disc (right). Adapted from Weiland* et al. *(1994).* Right panels*: A modern view of the Galactic bulge and of its peanut shape as revealed by the Wide-Field Infrared Survey Explorer (WISE). The top panel shows the WISE data, the bottom panel the residuals, after subtraction of an exponential model for the disc. Adapted from Ness and Lang (2016)*

2.4. The stellar halo

While there is mounting evidence that the inner regions of the Galaxy are for the very large majority made of stars formed in the Galaxy itself over time, it is in the stellar halo that we need to look to find traces of past and ongoing satellite accretion events. Numerical simulations indeed suggest that the stellar halos of disc galaxies like the Milky Way may contain hundreds of stellar streams, relics of the merging of other galactic systems, with masses comparable to or significantly smaller than our own Galaxy at the time of their accretion. Some of these streams have indeed been found in the recent years. The most prominent and famous of them is the Sagittarius stream (see Figure 2.8), a stellar band of stars that wraps around the Galaxy on a polar-like orbit (Newberg *et al.* 2002; Majewski *et al.* 2003). This stream is made up of stars that were once associated with the Sagittarius dwarf spheroidal galaxy (see, Ibata *et al.* 1994, the closest Milky Way companion, at about 24 kpc from the Sun) and that have been stripped from it, because of the tidal effects exerted by the Galactic gravitational field. The luminosity of the detected portion of the Sagittarius stream should correspond to about 70% of the original luminosity the progenitor galaxy, which should initially have had a stellar mass of a few $10^9 M_\odot$ and a total mass (dark matter included) of a few $10^{10} M_\odot$. Because stars in the streams follow, at first approximation, similar orbits, they can be used to reconstruct the Galactic potential at tens of kiloparsecs from the Galactic center, providing strong constraints on the dark matter distribution even far from the Galactic disc (see section 2.5). Not all stellar streams in the halo have, however, an extragalactic origin. The analysis of data from the *Gaia* astrometric satellite is indeed showing the presence of tens of very narrow stellar streams, more likely made of stars stripped from globular clusters, old stellar systems with typical masses of a few $10^5 M_\odot$, mostly orbiting in the inner halo of our Galaxy (Ibata *et al.* 2019).

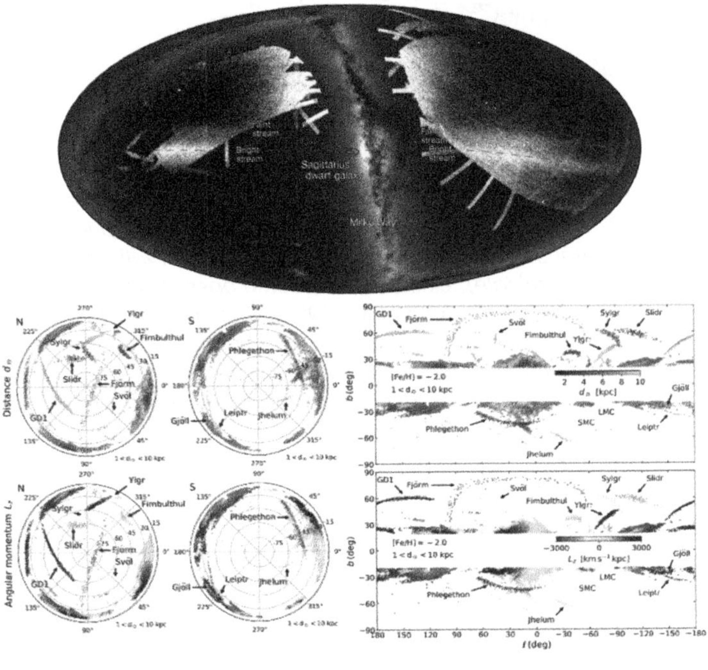

Figure 2.8. Top figure: *A map of the sky showing the numbers of stars counted in the Sagittarius stream. Colors indicate the distances to the stars: stars located in red areas are further away from the Sun, while stars in the blue areas are closer. The dotted red lines trace the Sagittarius stream, and the blue ellipses in the center show the current location of the Sagittarius dwarf galaxy. In this figure, the Galactic disc crosses the image nearly perpendicularly. The Magellanic Clouds are also visible in the bottom-left part of the map. Figure credit: S. Koposov and the SDSS-III collaboration.* Bottom figure: *Example of stellar streams detected by analyzing Gaia data. The first and second columns show zenithal equal-area projections centered on the north and south Galactic poles, respectively. The third column shows the distributions of these stellar streams in a longitude–latitude plane. In the first row, colors indicate the distance of stars in the stream from the Sun. In the second row, colors indicate their z-component of angular momentum (L_z). Adapted from Ibata* et al. *(2019)*

The discovery of spatially coherent structures in the halo of our Galaxy brings clues on the recent past of the Galaxy, since stars that we see nowadays in streams have possibly escaped their progenitor system typically in the last billion year, at most. The search for stars accreted by the Milky Way at earlier times of its evolution is more difficult, since stellar streams associated with early accretions are now expected to be fully mixed with the field stars and to have lost their spatial coherence. The search for these ancient building blocks of the stellar halo necessitates other diagnostics, such as stellar kinematics and abundances. We will see in section 2.6 how, by combining these two, it has been possible to find stars associated with the last significant merger experienced by our Galaxy, occurred about 10 billion years ago.

The accretion of satellite galaxies is not the only channel to form stellar halos. When satellites are accreted, they damage the disc of the Milky Way and kinematically "heat" it. Because of this process, Milky Way stars that before the interaction were confined in the disc can be ejected at large heights, contributing to form the halo (see Figure 2.9). As a result, not all stars in the halo have an extragalactic origin, but a fraction of them formed *in situ*, in the Galactic disc and were then ejected into the halo. Since we have evidence that the most massive accretions in the Galaxy occurred at early times (i.e. in its first billion years of evolution), this *in situ*, kinematically heated component of the stellar halo must be made of stars of the ancient Galactic disc. Until a couple of years ago, this was the scenario traced by numerical simulations. As we will see in section 2.6, because of *Gaia* and complementary spectroscopic surveys, we can now demonstrate that these two channels (accretions and *in situ*, heated disc) are the main two modes of formation of the inner Galactic halo (inside about 15 kpc from the Galactic center).

Historically, the accretion scenario dates back to Searle and Zinn (1978), who suggested that the stellar halo, and in particular its globular cluster system, was formed by a number of "proto-galaxies that subsequently merged to form the present Galactic halo". This view was seen as antithetic to the scenario proposed a decade before by Eggen *et al.* (1962), where "the oldest stars were formed out of gas falling toward the galactic center in the radial direction and collapsing from the halo onto the plane" on time scales as short as 10^8years. While for several decades, it has been thought that the stellar halo was the result of the combination of these two processes (see, Helmi 2008, for a review) the most recent works seem to exclude a dominant contribution of the radially "collapsed" component to the inner stellar halo (see, Haywood *et al.* 2018; Di Matteo *et al.* 2019).

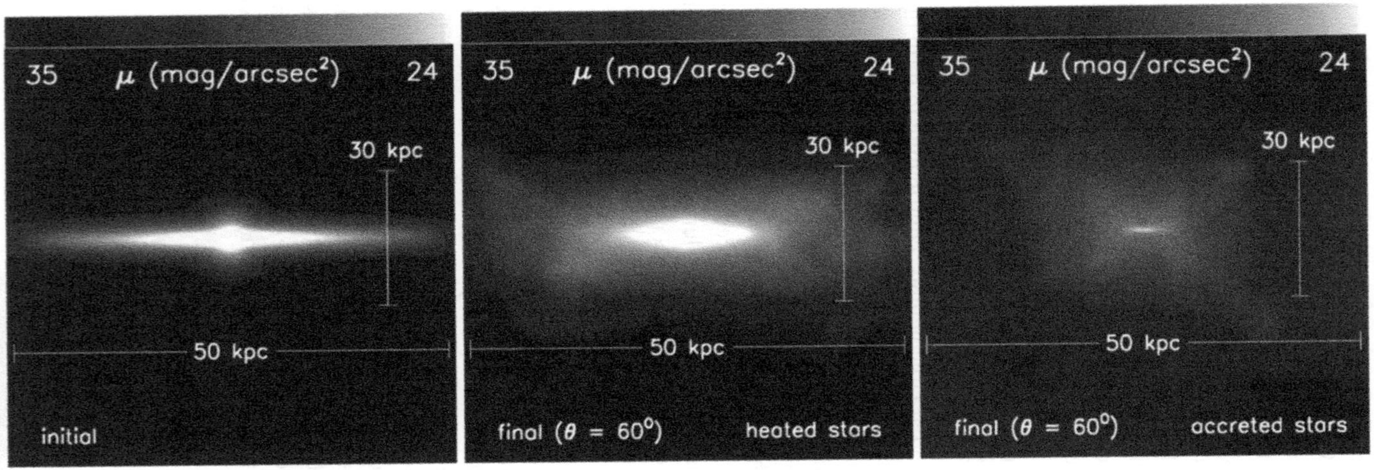

Figure 2.9. *Heated disc stars in the stellar halo, as found in a N-body simulation.* Left panel: *Surface brightness map of stars in a Milky Way–like galaxy, viewed edge-on, before the accretion of a satellite.* Middle panel: *Surface brightness map of stars in a Milky Way–like galaxy, after the accretion of the satellite. The disc of the galaxy has been heated by the interaction, and part of the stars confined close to the galaxy mid-plane in the left panel are now found at large heights, i.e. tens of kiloparsecs from it.* Right panel: *Surface brightness map of stars originally in the satellite, after its accretion onto the Milky Way–like galaxy. Adapted from Purcell* et al. *(2010)*

Interestingly, from star counts, the stellar halo shows a "break" at galactocentric distances of 20 – 30 kpc, that is the density profiles above and below this distance cannot be described as a single power law with the same slope. It appears indeed that the star counts fall off much more rapidly beyond the break radius than inside it. The form of this double profile turns out to be (Deason *et al.* 2011):

$$\rho(r_b) \propto \begin{cases} {r_q}^{-\alpha_{in}} & r_q \leq r_b \\ {r_q}^{-\alpha_{out}} & r_q > r_b \end{cases}$$

with $r_q = \sqrt{x^2 + y^2 + (z/q)^2}$, and x, y and z the Galactocentric rest-frame coordinates. The best-fitting model has an inner slope $\alpha_{in} = 2.3$ and an outer slope $\alpha_{out} = 4.6$, together with a break radius r_b occurring at about 27 kpc from the Galactic center and a constant halo flattening $q = 0.6$ (see, Deason *et al.* 2011). Because the halo flattening – that is the ratio of the minor to major axis – q is less than 1, the stellar halo appears oblate[4]. This "duality" of the halo density profile could be related to the dual origin of the Galactic halo described above, but additional work is needed to probe how these two channels of formation of the stellar halo impact the observed density profile.

2.5. On the dark matter content and shape, as inferred from rotation curves and stellar streams

The description given so far concerns the baryonic (gaseous and stellar) matter. However, in a Newtonian gravity framework, most of the mass in the Galaxy is made of "dark matter", matter than is not visible, nor it emits radiation, but which is detectable through the gravitational field it exerts on the visible mass. Gas and stellar tracers show indeed evidence that their motions cannot be explained by the visible matter alone: the mass of gas and stars together, indeed, would not be sufficient to explain the high rotational velocities observed in the disc of the Milky Way. An additional, massive component (about $10^{12} M_{\odot}$, see section 2.1) present at all radii, but which becomes dominant at large distances from the Galactic center, needs to be present to explain the measured kinematics of all tracers (see Figure 2.10). While the rotation curve tells us the presence of dark matter and its distribution within a distance R from the Galactic center, it cannot tell us about the shape of the dark matter halo outside the Galactic plane, such as its flattening. This is because to define the rotation curve, we generally use only visible matter (gas or stars) confined in the Galactic plane. To learn about the distribution of dark matter at large distances and outside the Galactic plane, we need other tracers. And, in this framework, the stellar

4 Oblate density distributions are characterized by a halo flattening $q < 1$, spherical distributions by $q = 1$ and prolate distributions by $q > 1$.

streams that we discussed in the previous section have an important role, since they are a probe of the overall (baryons and dark matter) gravitational potential of the Galaxy at different distances from its center.

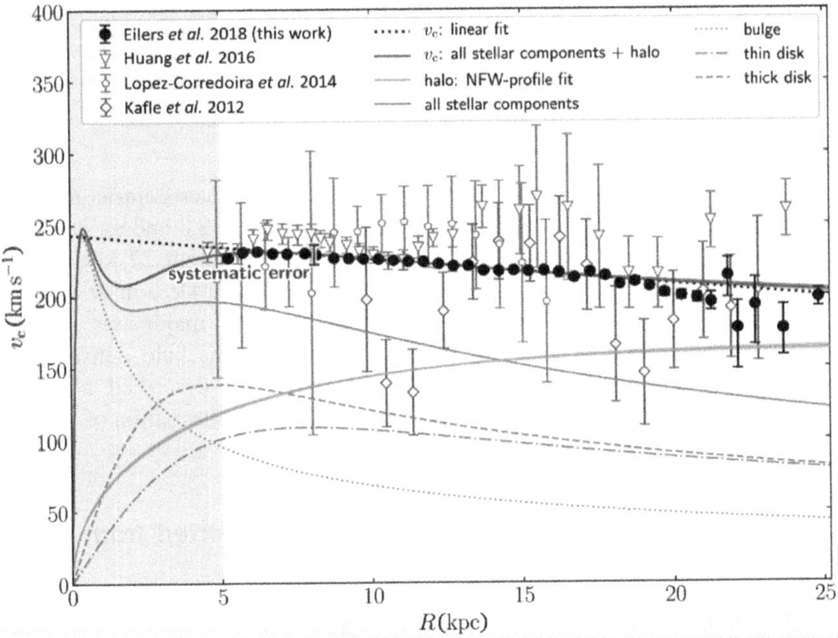

Figure 2.10. *The circular velocity curve of the Milky Way (black points), between 5 and 25 kpc, as derived by combining data from APOGEE with photometric information from WISE, 2MASS and Gaia. The velocity curve that would be obtained by taking into account only the visible stellar matter in the Galaxy (bulge, thin and thick discs) is shown by the solid gray line. The inclusion of a dark matter distribution, which gives rise to the velocity curve reported by the orange line, is thus necessary to obtain the total velocity curve (red line), which fits the data. Adapted from Eilers* et al. *(2019)*

As an example, the Sagittarius stream — the prominent stellar stream associated with the accretion of the Sagittarius dwarf galaxy, that we have discussed in the previous section – has been used to derive indications about the distribution and shape of the dark matter halo. By comparing the distribution on the sky of stars in the stream, and their line-of-sight velocities, it has been suggested that the Milky Way dark matter halo should have a triaxial shape, with the large and a small axis in the Galactic plane, and the intermediate axis in the direction perpendicular to the Galactic plane (Law *et al.* 2009). However, by making use of other Galactic stellar streams, much thinner than the Sagittarius one, and confined to the inner 20 kpc from the Galactic center, Bovy *et al.* (2016) deduced that the dark matter halo must be almost spherical in shape, and that a triaxial halo would not be able to reproduce stellar streams, as thin as those observed around globular clusters. These two results are clearly in tension. Still, it must be considered that streams at different distances from the Galactic center probe different regions of the Galactic potential and that, indeed, the shape of the dark matter halo could change with the distance. A number of numerical simulations show that, in fact, the shape of the dark matter halo may change with the distance from the Galactic center, with the shape becoming increasingly triaxial in the outer regions (Vera-Ciro *et al.* 2011; Vera-Ciro and Helmi 2013). These works may reconcile differences in the shape of the dark matter halo, obtained by using different streams.

Apart from being used to constrain the shape of the dark matter halo, stellar streams (especially the thinnest among them) can also inform us about the granularity of the halo, i.e. the number, mass and size of the thousands of dark matter subhaloes that are hypothesized to be gravitationally linked to our Galaxy. The passage of a dark matter subhalo near a stellar stream indeed disturbs the morphology of the latter, creating regions of overdensity and "holes" in it. Stellar streams thus appear to be very promising tools not only to study the current and past accretion events experienced by our Galaxy, and hence its mass growth over time, but also because they provide fundamental information about its "invisible" side, by constraining the shape and mass spectrum of the dark matter. They are thus some of the best tools that we have, at the Galaxy scale, to test cosmological theories about the nature and origin of the dark matter.

2.6. Dissecting the global structure: stellar kinematics, abundances and ages

A number of observational surveys of external disc galaxies are allowing us to map their stellar kinematics, metallicity and age distributions from the galaxy centers to the peripheral regions, bringing fundamental clues to the integrated (along the line-of-sight) properties of their stars. In the Milky Way, because we can resolve individual stars, we can go beyond this description, and investigate star-by-star chemical abundances, kinematics and ages (when possible). In this section, we will see how these studies can help us in characterizing further the main stellar structures described in the previous sections (thin and thick discs, bar/bulge and stellar halo). Chemo-kinematic-age relations can indeed highlight the mutual links between the different Galactic components, when they exist, or reveal the presence of stars with characteristics significantly different from those of most of the stars in the Galaxy. It is by studying these links that we are achieving many new results about the recent and old past of our Galaxy, and that we can provide a first, tentative sketch of its evolution.

2.6.1. *Setting the scene: the solar vicinity*

The solar vicinity is defined as a region centered on the Sun, and contained within a radius of a few hundred parsecs from it. Despite being a very small volume, studies of stars at the solar vicinity have provided fundamental information about the disc and halo evolution, which, in these last years, we have been able to verify and extend also to larger spatial scales. One of the way to start exploring the solar vicinity is to plot stars confined in this volume in the $[\alpha/\mathrm{Fe}] - [\mathrm{Fe/H}]$. What is the meaning of this plane? First, we recall that, for each star, the abundance of its generic chemical element X is defined as:

$$[\mathrm{X/H}] = \log_{10}\left(\frac{\mathrm{N_X}}{\mathrm{N_H}}\right)_{\mathrm{star}} - \log_{10}\left(\frac{\mathrm{N_X}}{\mathrm{N_H}}\right)_{\mathrm{Sun}} \qquad [2.2]$$

where $\mathrm{N_X}$ and $\mathrm{N_H}$ are the number of atoms per unit of volume, of the generic element X and hydrogen, respectively.

Similarly, we can define the abundance of the element X relative to iron, Fe, as:

$$[\mathrm{X/Fe}] = [\mathrm{X/H}] - [\mathrm{Fe/H}] \qquad [2.3]$$

α Elements are chemical elements as Mg, Si, Ca, Ti, mostly produced and released in the interstellar medium by massive stars ($M_{star} > 8M_{\odot}$), at the end of their evolution, when they explose as supernovae type II (SNeII). Iron, and iron-peak elements, are mostly (but not exclusively) produced by supernovae type Ia (SNeIa), which are the by-product of the explosion of a white dwarf star that accretes mass from a companion red giant star, the two being part of a close binary system. The evolutionary time scales of SNeII and SNeIa are significantly different: a few tens of million years for SNeII, and several billion years for SNeIa. The $[\alpha/\mathrm{Fe}]$ ratio thus informs us about the time scale of formation of stars in the Galaxy, it is in fact a "clock" of the star formation: stars which have high $[\alpha/\mathrm{Fe}]$ ratios (typically higher than solar) have formed out from an ISM not significantly enriched in iron, but significantly polluted by SNeII, and thus on relatively short time scales (less than the few billion years time scale necessary for SNeIa explosions); stars with nearly solar, or subsolar, $[\alpha/\mathrm{Fe}]$ ratios have been formed out from an ISM that has been significantly enriched in iron (thus allowing the $[\alpha/\mathrm{Fe}]$ ratio to decrease). The ISM they formed from must have time to be enriched by SNeIa, hence the longer formation time scales. With these elements in mind, we can now discuss and start to interpret the $[\alpha/\mathrm{Fe}] - [\mathrm{Fe/H}]$ plane described by stars at the solar vicinity, as reported in Figure 2.11.

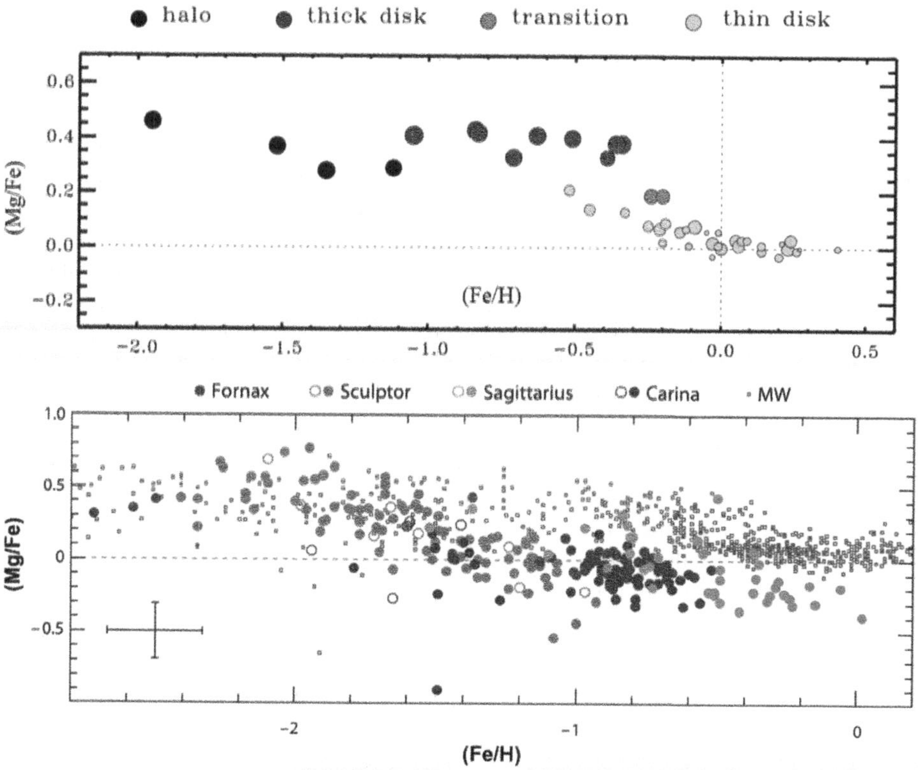

Figure 2.11. Top panel: *[Mg/Fe] abundances versus [Fe/H] for a sample of stars at the solar vicinity. Circle diameters denote age estimates, the smallest sizes indicating the youngest stars. Stars of the halo, thick and thin discs have different grayscale symbols, as indicated on the top of the figure. Adapted from Fuhrmann (1998).* Bottom panel: *[Mg/Fe] abundances versus [Fe/H] for a sample of stars at the solar vicinity (gray dots), compared to stars belonging to some of the Milky Way satellite galaxies: Fornax (blue), Sculptor (green), Sagittarius (yellow), Carina (purple). Adapted from Tolstoy* et al. *(2009)*

This plot shows that stars at the solar vicinity can cover all [Fe/H] values from -2 to super-solar (i.e. $[Fe/H] > 0$). Their [Mg/Fe] ratios, however, do not redistribute homogeneously in this plane, but there is a tendency for metal-poor stars ($[Fe/H] \leq -0.5$) to have also high-α abundances relative to iron. The [Mg/Fe] ratios indeed decrease only for $[Fe/H] > -0.5$ and rapidly attain solar values for $[Fe/H] \geq 0$. This is a fundamental plot for the understanding of how the star formation proceeded over time in the Galaxy, and it tells us that in the initial phases, the star formation must have been so intense to rapidly drive the metallicity of stars in the Galaxy to values as high as [Fe/H] $= -0.5$, without decreasing the [Mg/Fe]. In other words, the early chemical enrichment in the Milky Way must have proceeded very fast, and attained [Fe/H] $= -0.5$ in a few billion years ($2 - 3$ billion years at most, otherwise we should see a decline in the [Mg/Fe] ratios already at values of [Fe/H] < -0.5). This fast chemical enrichment is typical of galaxies as massive as the Milky Way, but it is not found for less massive systems. Figure 2.11 shows, for comparison, the [Mg/Fe] abundances of stars of several Milky Way satellites, that is of several dwarf galaxies orbiting around our Galaxy, at different distances from its center. All these galaxies show α-abundance ratios that start to decline at [Fe/H] values typically around -1.5, indicating that it took at least a couple of billion years in these systems to reach such levels of metallicities. In other words, in these dwarf galaxies the chemical enrichment proceeded much slower than in the Milky Way. Now that we have understood the fundamental information that is contained in the [α/Fe] – [Fe/H] plane, we can turn again to Figure 2.11 (top panel) to further discuss the $\alpha-$enhanced and the low-α sequences. By studying the kinematics and ages of stars in this diagram, we learn that high-α stars are mainly old and have "hot" kinematics, that is they have high kinetic energies, and also a rotation slower than that of the Local Standard of Rest (LSR)[5] (see, for example, Bensby *et al.* 2005). Because they rotate slower than the LSR, they are said to "lag" it, and in a reference frame centered on the LSR, they would appear to have negative rotation. As the [Fe/H] increases, and the [α/Fe] ratio decreases, the kinematics of stars becomes increasing cold: their velocity dispersions decrease, and their rotation increases, and reach values similar to those of the LSR, for stars with [Fe/H]$\sim$ 0. Thus, we see that chemical abundances, ages and kinematics are on average related in the Galaxy: the oldest, most metal-poor and most $\alpha-$enhanced stars tend also to be those with kinematics that differ the most from those of stars like the Sun. Because "hot kinematics" means high kinetic energy, also in the vertical direction, relative to the Galaxy midplane, stars having "hot kinematics" tend also to have the largest vertical excursions from the Galaxy midplane: they are thus stars of the thick disc and halo, since they can reach very far heights from the Galactic disc. Stars with "cold" kinematics, in turn, do not have such large excursions, and stay more confined in the disc midplane: they are thus stars of the thin disc. Figure 2.11 (top panel) thus also tells us that there is an apparent smooth transition, in this chemical space, from the metal-poor, $\alpha-$enhanced stellar halo, all the way down along the chemical sequence, up to the metal-rich, α-solar thin disc.

Not only the vertical, but also the in-plane excursions change with the abundances of these stars. Stars of the $\alpha-$enhanced thick disc tend to have radial excursions that make them oscillate between the innermost few kpc from the Galactic center, to about 10 kpc from it. Their vertical excursions make them reach maximal heights from the

5 The Local Standard of Rest is the imaginary circular orbit around the Galaxy center that pass through the Sun position.

Galactic plane as large as $2 - 3$ kpc. As the $[\alpha/\mathrm{Fe}]$ ratio decreases, the radial and vertical excursion of stars tends, on average, also to decrease. However, we observe a difference between thin disc stars that are metal poor ($[\mathrm{Fe/H}] \lesssim -0.3$) or metal rich ($[\mathrm{Fe/H}] \gtrsim 0.2$): the former tend to have pericenters at the solar radius (i.e. at a distance from the Galactic center similar to that of the Sun) and apocenters at larger distances, while the latter tend to have both pericenters and apocenters inside the solar radius. This trend reveals that some structuring must exist in the Galactic thin disc, and that possibly the chemical properties of stars in the inner regions of the Galaxy (inside the solar radius) are not the same as stars in the outer disc (that is, outside it). This structuring, anticipated by Haywood (2008) from the analysis of solar vicinity stars only, finds now definite confirmation from spectroscopic studies, which analyzed the chemical composition of disc stars at several kpc from the Sun, as we will see in section 2.6.2.

I would like to spend some more time discussing the stellar halo of our Galaxy, and what we can understand about it, already from solar vicinity samples. Observing halo stars at the solar vicinity is not an easy task, since they are rare: typically, only 1 in over 100 stars in the Galaxy has $[\mathrm{Fe/H}] < -1$. Thus, it took time to recognize that the stellar halo is more complex, and less homogeneous, than what is shown in Figure 2.11. Instead of being made of stars with high-α abundances only, a decade ago it was discovered that chemical abundances in the halo come in two flavors: a sequence that is $\alpha-$enhanced, and that appears in chemical continuity with the thick disc, and a sequence that departs from it, and that tends, at a given value of [Fe/H], to contain less $\alpha-$enhanced stars (see Figure 2.12). This duality in the chemical abundances of halo stars is reminiscent of the two channels from the halo formation (accretion versus *in situ*) that we have discussed in section 2.4. But let us not anticipate the conclusions, and let us proceed step-by-step. The existence of two chemical sequences in the $[\alpha/\mathrm{Fe}] - [\mathrm{Fe/H}]$ plane, with two different slopes, indicates that halo stars found nowadays at the solar vicinity have not all followed the same chemical enrichment pattern: the high-α halo stars must have experienced a more rapid star formation than that of halo stars, which, at the same [Fe/H], have however smaller $\alpha-$abundances (i.e. red and blue chemical patterns in Figure 2.12). Remember the discussion made about Figure 2.11. The $[\alpha/\mathrm{Fe}] - [\mathrm{Fe/H}]$ plane is a clock of the star formation: if two groups of stars with the same [Fe/H] are found to have different $\alpha-$abundances, those that have the highest α abundances are those which experienced a more rapid chemical enrichment, and most importantly, a chemical enrichment different from those of their low-α counterpart. It turns out, comparing Figure 2.12 with the bottom panel of Figure 2.11, that the low-α pattern found for halo stars at the solar vicinity is reminiscent of the chemical patterns observed in dwarf galaxies satellites of the Milky Way: as we have discussed previously, indeed, in these galaxies the $[\alpha/\mathrm{Fe}]$ abundances start to decrease at [Fe/H] values that are lower than those observed in the Milky Way. In these dwarf satellites, as in the low-α sequence of halo stars found in the Galaxy, it took longer to reach a given metallicity. Because of the similarity of the trends observed in the low-α halo sequence with those observed in dwarf satellites of the Milky Way, Nissen and Schuster (2010) concluded that the simplest interpretation for the former sequence was that it is made up of stars formed in a galaxy less massive than the Milky Way, and accreted some time into the past. What has become clear only recently, with surveys extending far away from the solar vicinity, is that this accretion event was the last significant merger experienced by our Galaxy. As we will show in section 2.6.2, we are now able to date the end of this accretion, and also to quantify the damages this satellite caused on the Galactic disc, at the time of its accretion. Interestingly, it was found that these accreted halo stars have also peculiar kinematics, significantly different to that of the majority of halo stars with high-α–abundances. While the former tend to have, on average, retrograde motions, that is they rotate in the direction opposite to disc stars, the latter tend to have prograde motions, even if with a significant lag with respect to the LSR. In fact, in a kinematic diagram like the so-called "Toomre diagram", where a quadratic combination of the vertical and radial velocity of stars is plotted against their rotational velocities (relative to the LSR), a continuity is found between accreted halo stars, high-α stars and thick disc stars (see Figure 2.12, right panel).

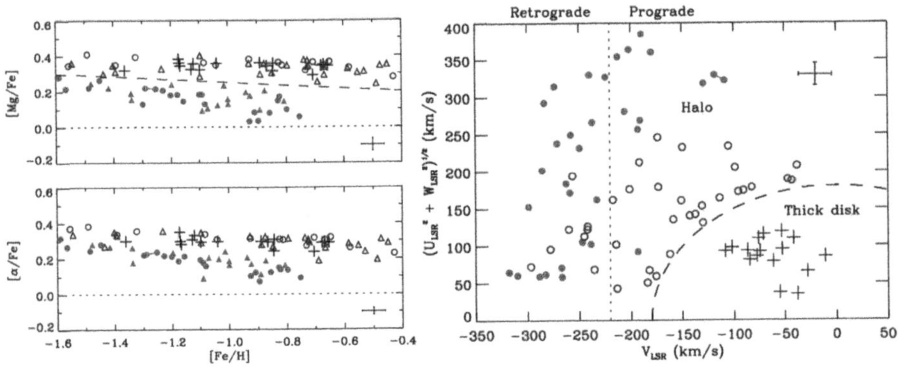

Figure 2.12. Left panels: *[Mg/Fe] and [α/Fe] abundances versus [Fe/H] for a sample of stars at the solar vicinity, kinematically defined as thick disc stars (black crosses), and halo stars (red and blue symbols). Two sequences are evident among the halo population: a "high-α" sequence (blue) and a "low-α" sequence (red).* Right panel: *Toomre diagram for the thick disc and halo stars shown on the left. Stars of the high and low-α sequence have, on average, kinematics that differ from that of the high-α stars. They are indeed characterized by a higher kinetic energy, and mostly retrograde motions (rotational velocities relative to the LSR,* V_{LSR}, *smaller than -220 km/s in this plot). Adapted from Nissen and Schuster (2010)*

The examples given above illustrate how powerful the use of chemical abundances and kinematics of individual stars can be to describe the Galaxy, to identify evolutionary sequences and to distinguish between stars with different origins. Before leaving the solar vicinity, we cannot forget to discuss ages, because they are an essential ingredient for any attempt to reconstruct the history of our Galaxy. Most of what we know of the ages of stars at the solar vicinity comes from dating stars through isophote-fitting techniques applied to main sequence and turn-off stars. Practically, the technique consists of (1) choosing a set of stellar isochrones from stellar evolution models, that is of curves in the Hertzsprung–Russell diagram[6] that represent a population of stars of the same age, and (2) finding the best isochrone (and thus age) that is the closest to the star position in the HR diagram. Practically this technique has its limitations: (1) the distances of stars must be known with good enough precision to be able to convert apparent magnitudes into absolute one; (2) isochrones tend to tight up in some regions of the HR diagram, as at the bottom of the main sequence (the lower the mass and luminosity, the closer isochrones are). This has limited for longtime the sample of stars for which ages could be estimated through this technique to upper main sequence and turn-off stars, and stars in the subgiant branch. Moreover, because the exact location of isochrones in the HR diagram depends on metallicity and $\alpha-$abundances, spectroscopic measurements are necessary to determine, through this technique, stellar ages with sufficient precision. With all these caveats, the dating of stellar ages through isochrone fitting has, however, allowed to better characterize stars at the solar vicinity and to show that they satisfy some very clean and neat relations, as the $[\alpha/\text{Fe}]-$age relation found by Haywood *et al.* (2013) and reported in Figure 2.13. This relation shows, first, $[\alpha/\text{Fe}]$ ratios are correlated to ages, and thus that, with the due caution, they can be used as a proxy of ages, when the latter are not available; second, the evolution with time of the $[\alpha/\text{Fe}]$ abundances is a two-phase process. In the first few billion years, the $[\alpha/\text{Fe}]$ of stars in the disc decreased very rapidly, of about 0.3 dex in 4 billion years, while in the subsequent 9 billion years, the decrease was much slower, overall less than 0.2 dex. Because, as we discussed at the beginning of this section, the $[\alpha/\text{Fe}]$ ratio is a clock of the star formation, the two slopes evidenced by this relation suggest that the star formation in the early phases (about first 4 billion years) of evolution of the Galaxy was significantly different from that experienced by the Milky Way later on (last 9 billion years). And also that the Galaxy must have been enriched very rapidly in its iron content to be able to form stars with $[\alpha/\text{Fe}] = 0.1$ only a few billion years after the Big Bang. We will comment on this further in section 2.7. Finally, because stars with metallicity typical of the thick disc are found in the steep part of the relation, while stars with metallicities typical of the thin disc are found in the flat part of the relation, this indicates that the two phases of formation of the Galactic disc also coincide with the formation, respectively, of the Galactic thick and thin discs.

6 We recall, for the reader, that the Hertzsprung–Russell diagram, abbreviated as the HR diagram, is a plot combining stars absolute magnitudes versus their effective temperature.

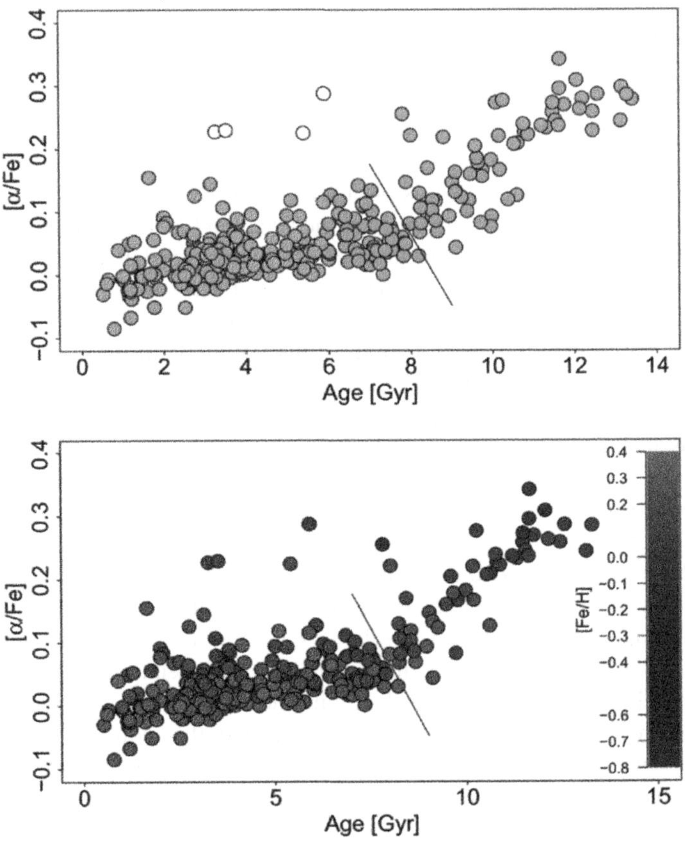

Figure 2.13. *[α/Fe] versus age relation for a sample of disc stars at the solar vicinity. In the bottom panel, the color codes the metallicity of the stars, as reported in the color bar on the right. Adapted from Haywood* et al. *(2013)*

2.6.2. *Zooming out on a several kpc scale*

2.6.2.1. *Chemical structure of the Galactic disc*

In the last decade, a number of large spectroscopic surveys – among others, SEGUE (Yanny *et al.* 2009), APOGEE (Majewski *et al.* 2017) and more recently Galah (De Silva *et al.* 2015)– have permitted to refine our understanding of the Galactic disc, by extending some of the studies described above to several kiloparsecs from the Sun. In this way, it has been possible to characterize the disc in chemical spaces, and to show how chemical abundances change as a function of the distance from the Galactic center, and height above the plane. Among the numerous results published these last years, it is worth discussing the work by Bovy *et al.* (2012), because, among the many merits of this work, it has introduced, for the first time, the concept of "mono-abundance" populations, which has been fundamental to describe and understand the Galactic disc on a large scale. What is a "mono-abundance" population in the Bovy *et al.* (2012) definition? It is an *ensemble* of stars that share similar [Fe/H] and $[\alpha/\mathrm{Fe}]$ abundances, and thus group together in the $[\alpha/\mathrm{Fe}] - [\mathrm{Fe/H}]$ plane (see Figure 2.14, top panels). To define "mono-abundance" populations it is thus necessary to have stellar samples made up of thousands of stars, in order to be able to divide them into subgroups on the basis of their chemical properties.

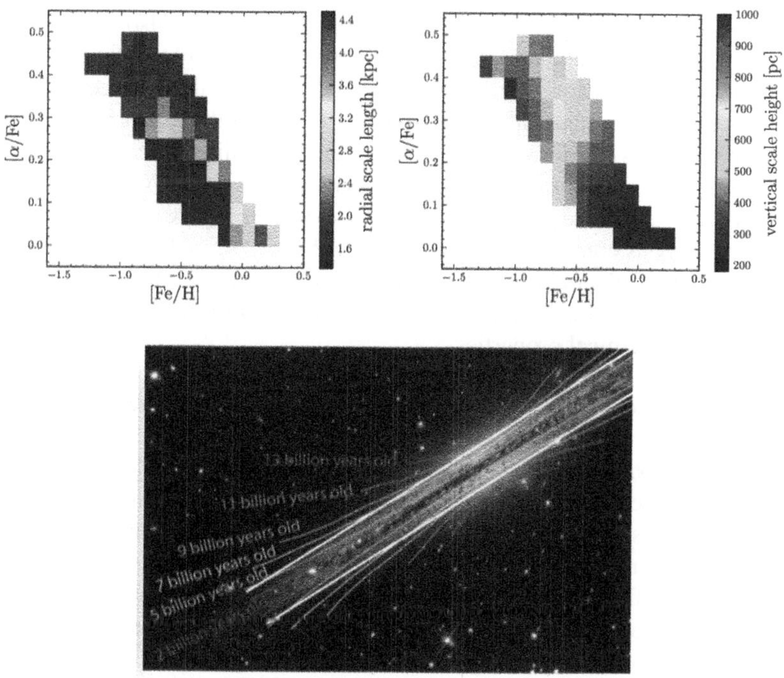

Figure 2.14. Top panels*: Mono-abundance populations as a function of their position in the [Fe/H] and [α/Fe] plane, color coded by their radial scale length (left) and vertical scale height (right). α-Abundant stars tend to have short scale lengths (blue colors on the left) and large-scale heights (yellow/red colors on the right). As the [α/Fe] ratio diminishes, the scale height of mono-abundant populations declines too. Metal-poor thin disc stars are those with the largest scale lengths (brown colors in the left panel). Adapted from Bovy* et al. *(2012).* Bottom panel: *The Milky Way analog galaxy, NGC 891, and, overlaid, color curves that show the flares for groups of stars with similar ages. Credit: Adam Block, Mt. Lemmon SkyCenter, University of Arizona/Ivan Minchev, AIP*

The giant step made possible by these spectroscopic surveys has been, and it is, to provide accurate abundances for a hundred thousand stars, thus allowing, for the first time, to statistically study the properties of stars with similar chemical properties. One of the major findings of these works (Bovy *et al.* 2012; Hayden *et al.* 2015; Mackereth *et al.* 2017) has been to show that the spatial distribution of stars in the disc depends on their chemical abundances (see Figure 2.14, top panels). Disc stars with high-$[\alpha/Fe]$ ratios (above 0.3) tend to be mostly redistributed in the inner regions of the Galactic disc, and hence their spatial distribution can be represented by an exponential density profile with a short-scale length, as short as 2 kpc. Stars with low $[\alpha/Fe]$ ratios, and low metallicities, tend to be much more widely spread in the disc, and because of this, they can be represented by an exponential density profile with a large-scale length, as large as 4 kpc. These "mono-abundance" populations show also different vertical extensions from the Galactic midplane (see Figure 2.14, top-right panel): in particular, the highest the $[\alpha/Fe]$ ratio, the larger the scale height of the exponential disc. All this implies that the $\alpha-$abundant, metal-poor stars have a thick and radially compact distribution, whereas the low-α, metal-poor stars are redistributed in a thinner and more extended disc configuration. The low-α, metal-rich stars – which constitute the high-metallicity tail of the thin disc sequence shown in Figure 2.11 – have a radial distribution nearly as compact as that of the $\alpha-$enhanced stars, but much thinner. This cartography of the Galactic disc, at a few kpc from the Sun, has thus allowed to show that there exists a radial and vertical structuring of stars in the Galaxy, and that this depends on their chemical properties.

The careful reader may now think that there is an incoherence between two results discussed in this chapter. In section 2.2.2, we have indeed said that, based on photometric studies, the thick disc appears to have a longer scale length than the thin disc, while here we state that the thick disc, or better, the $\alpha-$enhanced disc has a short-scale length, shorter than the average scale length of the thin disc. Where does this apparent contradiction come from? It comes from the fact that the photometrically defined thick disc contains also low-α stars, that is, it is not made only of $\alpha-$abundant populations. It contains indeed also many stars that are chemically associated with the thin disc (i.e. low-α), but that are on average thicker than stars at the solar vicinity, and thus contribute to the structural thick disc. This complexity of definitions (morphological, chemical or kinematical) may appear tedious or unnecessarily complicated. But in fact, it is fundamental to understand what the Galaxy is really made of, and how its different structures (or portion of them) assembled over time. To better understand this point, Figure 2.14 illustrates schematically what is going on, probably, in a Galaxy like ours. $\alpha-$Abundant populations, which are on average also the oldest (see Figure 2.13), tend to be distributed in the inner Galaxy, and – as the distance from the Galactic center increases – the morphologically thick populations are made up of stars increasingly younger, and thus (because of the relation shown in Figure 2.13) also with lower $\alpha-$abundances.

2.6.2.2. *Evidence of the last significant merger experienced by the Milky Way*

In section 2.6.1, we have provided evidence that a fraction of the halo stars at the solar vicinity may be the left over from a satellite galaxy, accreted by the Milky Way in its past (see Figure 2.12 and related discussion in the text). The advent of large spectroscopic surveys like APOGEE and the publication of the first, and second *Gaia* catalogues, have definitely confirmed these early findings, and showed that this accretion, whose debris are found also at the solar vicinity, has played a major role in shaping the Galactic disc. First, it has been shown that the low-α chemical sequence that was discovered by Nissen and Schuster (2010) among solar vicinity halo stars is still visible at several kpc from the Sun, thus indicating that the debris of the accreted satellite spread over a large portion of the Galaxy (see, Hayes *et al.* 2018; Haywood *et al.* 2018; Helmi *et al.* 2018, and Figure 2.15). As already anticipated by Nissen and Schuster (2010) (see also, Brook *et al.* 2003; Belokurov *et al.* 2018), these accreted stars have, on average, peculiar kinematic properties, since they tend to show no net rotation, or even retrograde motions. These kinematic properties make these accreted stars quite distinct from the large majority of stars in the Galaxy, as it shown in Figure 2.16: in a plane combining the rotational velocity of stars, v_Φ, to their radial velocities, v_R, they tend to redistribute in a kind of "Sausage-like" shape (Belokurov *et al.* 2018), with the major axis of the "Sausage" parallel to the v_R axis. In other words, these accreted stars have very radial orbits, and very low net rotation. Different arguments have led to conclude that the progenitor satellite must have been quite massive, relative to the mass of the Milky Way at the time of the accretion: the mass ratio of the satellite-to-Galaxy mass was indeed probably between 1:10 and 1:4 (Belokurov *et al.* 2018; Haywood *et al.* 2018; Helmi *et al.* 2018).

Numerical simulations predict that accretions with such large mass ratios should significantly damage the pre-existing Galactic disc, and kinematically heat it (see Figure 2.9), up to eject part of the disc stars into the halo. The kinematic imprint of this accretion event has been indeed found by coupling the kinematics of stars to their chemical abundances. More specifically, the disc of the Milky Way appears kinematically cold for stars with $[Fe/H] > -0.3$ and $[\alpha/Fe] < 0.2$, but its kinematics appears much hotter for stars with $[\alpha/Fe] > 0.2$. A fraction of stars (between 10% and 20%) with this level of $\alpha-$enhancement, indeed, despite having chemical properties consistent with the thick disc, have kinematic properties typical of the halo, that is they are too hot for their chemical abundances. This is the signature that these stars were formed in the Galactic disc (they have indeed α-abundances and kinematics typical of disc stars), but that they were subsequently heated to halo kinematics, at the time of this major accretion event. Because $[\alpha/Fe]$ ratios correlate with ages (see Figure 2.13), it is by identifying the lowest $\alpha-$abundances of these kinematically "heated" stars that it has been possible to date the end of the last major accretion event experienced by our Galaxy to 9–11 billion years ago.

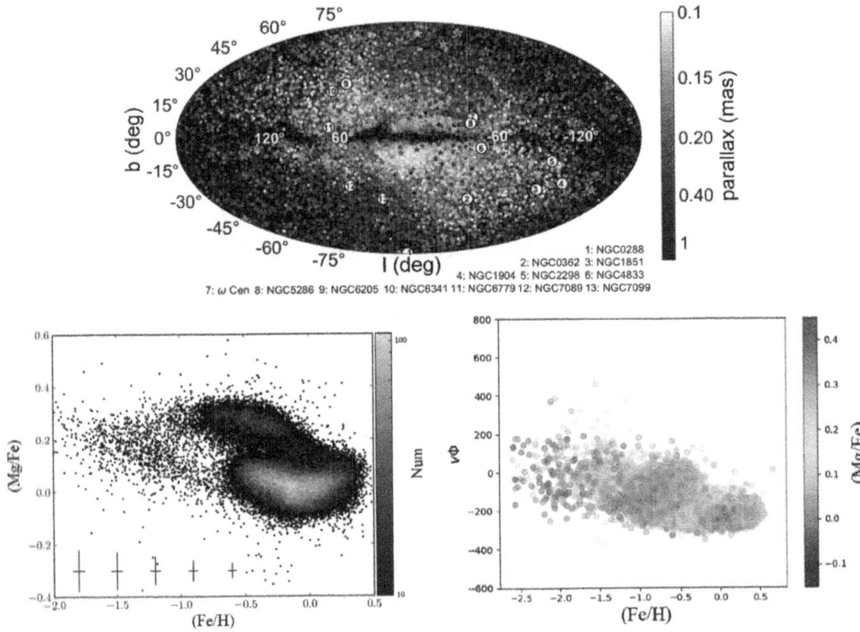

Figure 2.15. Top panel: *Sky distribution of stars possibly associated with Gaia Sausage/Enceladus, with colors coding their parallaxes. Adapted from Helmi* et al. *(2018).* Bottom-left panel: *[Mg/Fe] versus [Fe/H] distribution of stars at a few kpc from the Sun. Accreted stars are visible in this plane as the diagonal, low-[Mg/Fe] sequence that appears at* $[\mathrm{Fe/H}] < -0.5$. *Adapted from Hayes* et al. *(2018).* Bottom-right panel: *Rotational velocity versus [Fe/H] for stars at a few kpc from the Sun, with the colors coding their [Mg/Fe] abundances. In this plot, disc stars have the largest negative velocities, and stars on retrograde orbits have positive velocities. Moving from high [Fe/H] to lower values, a sudden change in the velocity distribution is observed for* $[\mathrm{Fe/H}] \sim -0.3$. *Below this metallicity, indeed, the number of stars with low, null or retrograde rotation increases significantly. This feature has been interpreted as due to the kinematic heating of the Galactic disc, due to the Gaia Sausage/Enceladus accretion and has allowed to date this accretion to about 10 Gyr ago. Adapted from Di Matteo* et al. *(2019)*

2.6.3. *Digging into the bulge*

In section 2.3, we have discussed the morphology of the Milky Way bulge and shown that it is typical of a class of bulges, called boxy/peanut-shaped bulges, which form out from disc material, during a phase of disc vertical instability.

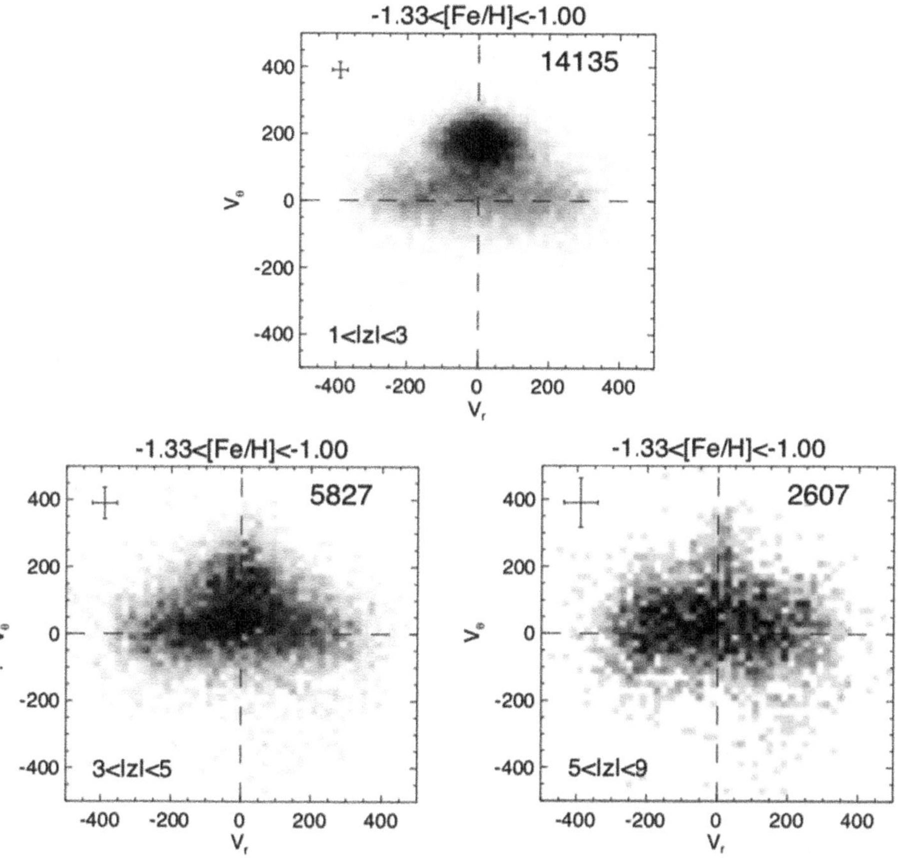

Figure 2.16. *Rotational versus radial velocity of metal-poor stars at a few kpc from the Sun. Accreted stars from Gaia Sausage/Enceladus, redistribute nearly horizontally in this plane, in an oval, "Sausage"-like, distribution, characterized by null mean rotational velocities. This structure does not dominate at low heights from the Galaxy midplane (top panel), where most of the stars have disc kinematics (rotational velocities as large as 200 km/s) but becomes dominant as the heights from the plane increase (bottom panel). Adapted from Belokurov* et al. *(2018)*

Many evidence have accumulated over the years to confirm this scenario. Not only the morphology, but also the kinematics of stars in the Galactic bulge indicate that all contribution of a classical spheroid, if present, must be small and should constitute not more than 10% of the disc mass. This is illustrated in Figure 2.17, where the bulge kinematics is shown, for different latitudes as a function of longitude. Data, from the spectroscopic BRAVA survey (Rich *et al.* 2007), are compared to numerical simulations of a Milky Way–type galaxy, where a boxy/peanut bulge co-exists with a classical bulge. It appears that, when the mass of the classical spheroid becomes too large, relative to the mass in the stellar disc, the simulations predict bulges whose kinematics is much hotter (that is, too high velocity dispersions) than what is observed (see, Shen *et al.* 2010). This is a first, very robust indication that the Galactic bulge cannot host a too massive spheroid and most of its mass must originate in the disc.

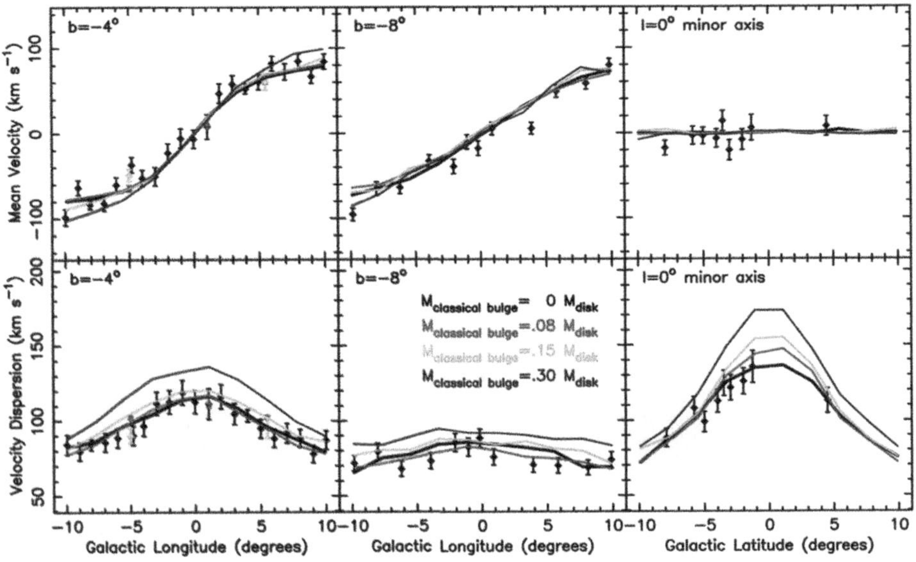

Figure 2.17. *Observed bulge kinematics (black points) compared to models of a Milky Way–type galaxy hosting a classical bulge whose mass varies from 0% to 30% of the disc mass (see legend). Rotational velocities of bulge stars are shown in the top panels, and velocity dispersions in the bottom panels. For classical bulges more massive than 10% of the disc mass, the agreement of the model with the data becomes poor. Adapted from Shen* et al. *(2010)*

A second confirmation to this scenario comes from the study of the chemical properties of stars in the bulge, in particular their metallicites, and their relative proportions at different distances from the Galactic center. Let us proceed step-by-step. We have mentioned several times in this chapter that the Galactic disc is structured, being made of a metal-poor, $\alpha-$enhanced population which is centrally concentrated, and which has hot kinematics (the thick disc), and a thin disc population, with lower $\alpha-$abundances, radially more extended and with colder kinematics (the thin disc). Because, in particular, thin and thick disc stars have different kinematics, their response to the disc vertical instabilities is not the same: the hotter the kinematics of the stars, the weaker the boxy/peanut structure they form, once the disc vertical instability phase takes place. In other words, thick disc stars trapped into the bar and boxy/peanut structure tend to have rounder, and less "peanuty" orbits than thin disc orbits. Because of this different response, depending on the region of the bulge we look at, how distant it is from the Galactic midplane or from the Galaxy center, the relative weight of metal-poor and metal-rich stars changes. Interestingly, the fraction of metal-poor and metal-rich stars predicted by numerical models, and the way it changes through the bulge region, is in very good agreement with what is derived from observational data, thus supporting, once more, the scenario where the Milky Way bulge is for its vast majority made of stars originating from the Galactic disc.

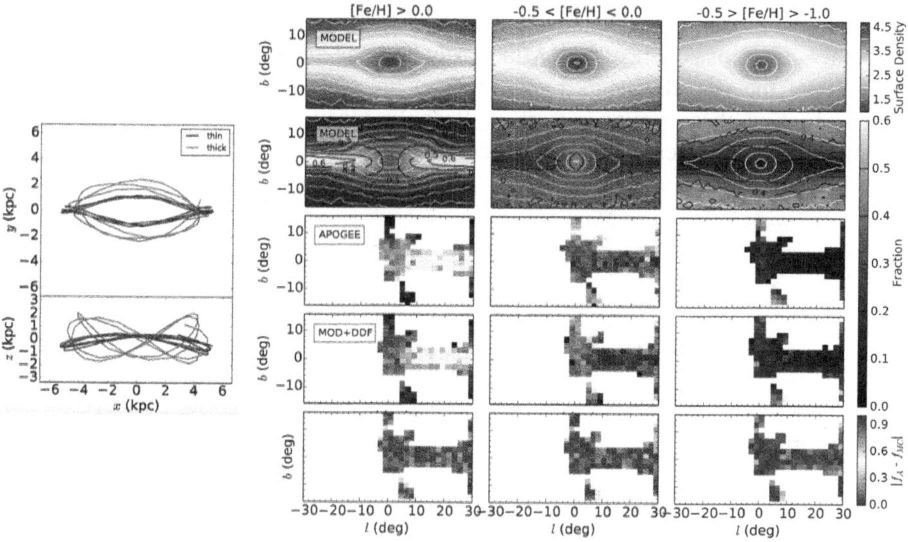

Figure 2.18. Left panels: *In-plane (top) and edge-on (bottom) projections of the orbits of a thick disc star (red curve) and of a thin disc star (blue curve) trapped into a boxy/peanut bulge. Adapted from Fragkoudi* et al. *(2017).* Right panels, first row: *Surface density maps in (longitude and latitude) of a N-body model of a Milky Way–like galaxy containing a boxy/peanut bulge. The metallicity range of each population is indicated at the top (from left to right we show the metal-rich, intermediate and metal-poor component, respectively).* Right panels, second row: *Fractional contribution maps of each population in the model.* Right panels, third row: *Fractional maps of the Milky Way bulge and inner disc from APOGEE data.* Right panels, fourth row: *Fractional maps of the model, corrected for observational distance biases, for each component.* Right panels, bottom row: *Difference in the fractional contribution between the model, corrected for observational biases, and the APOGEE data. The agreement between the model and the data demonstrates that the Milky Way bulge is, for the very large majority, made up of stars of the thin and thick discs trapped in a boxy/peanut structure. Adapted from Fragkoudi* et al. *(2018)*

Finally, it is worth noting that while there is mounting evidence from morphology, kinematics and chemical abundances that the Milky Way bulge is mostly a boxy/peanut-shaped bulge formed out of disc material, there is still no consensus in the community that this is supported by the ages of its stars. If the Milky Way bulge is made of disc stars, it should contain not only old disc stars, but also younger ones (a few billion years old). For long time, however, the bulge has been described as made exclusively of stars as old as 10 billion years (see, for example, Clarkson *et al.* 2011). Some recent works, however, have questioned this conclusion, by showing that a not negligible fraction (larger than 20%) of younger stars may be hidden in the Galactic bulge (Bensby *et al.* 2013; Haywood *et al.* 2016), and that – as a consequence – the Milky Way bulge is mostly, but not exclusively old. Because of the way disc stars redistribute in the Galactic bulge (Figure 2.18), the chance to find these younger stars is strongly sensitive to the bulge region (i.e. longitudes and latitudes) we look at (Ness *et al.* 2014).

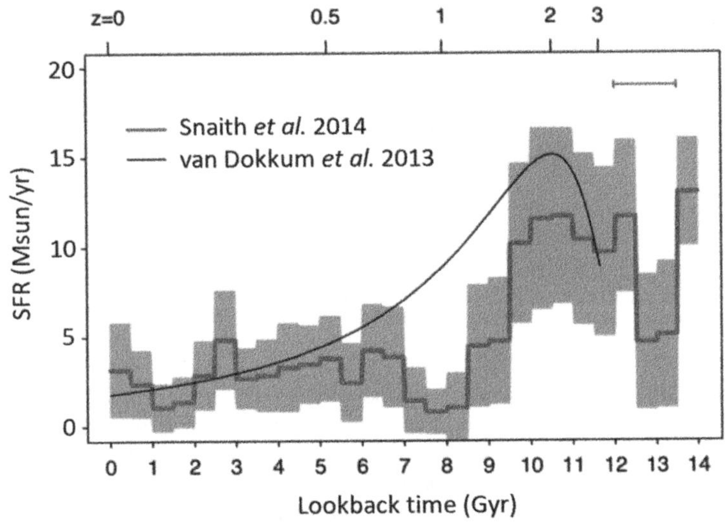

Figure 2.19. *The star formation history of the Galactic disc, as derived from the* α*-age relation shown in Figure 2.13. Adapted from Snaith* et al. *(2014)*

2.7. Reconstructing the Milky Way evolution

In the previous sections, we have made a journey through the main Galactic components, their chemo-kinematic properties, and ages (when available). We have seen that the Galaxy shows signatures of accretions, in its recent and far past. We have seen that the central regions of our Galaxy are dominated by a bulge, which is the thick part of an otherwise thin bar that extends for about 5 kpc from the Galactic center. A number of probably four, main, spiral arms extends through the disc, up to the solar radius and beyond. The outer regions of the gaseous and stellar discs are perturbed, showing signs of a flaring and warping. How and when all these components started to be in place? Can we sketch, even if tentatively, a timeline of the evolution of the Milky Way?

A way to start answering these questions is to come back to our Figure 2.13. Because, as already discussed, the $[\alpha/Fe]$ abundances depend on the number (and ratio) of SNeII and SNeIa exploded in the Galaxy over time, and this depends, among other things, on the intensity of the star formation, it is possible to make use the [α/Fe]-age relation presented in section 2.6.1 to look for the best star formation history that can reproduce this relation. The solution to this exercise (see, Snaith *et al.* 2014, 2015, which is based on a number of assumptions, for details) indicates that the history of the evolution of our Galaxy is marked by two distinct phases: a first phase, characterized by a sustained star formation rate, which lasted about 4 billion years. The Milky Way was a starbursting galaxy at that time (Lehnert *et al.* 2014), that is its star formation intensity, defined as the star formation rate per surface area, was probably intense enough to produce vigorous stellar winds, outflows and high turbulence in its interstellar medium. This is also the time when the most massive mergers took place: the "Gaia Sausage" satellite was indeed accreted at these epochs, and the last signatures of its impact on the Galactic disc date back to about 10 billion years ago. Interestingly, it is about 10 billion years ago that the star formation rate started to rapidly decrease (by about a factor 10, in about 1 billion years). This epoch of stars formation "quenching" (Haywood *et al.* 2016) marked the transition from the thick to thin disc formation. It is possibly also the time when the stellar bar formed, maybe driven during the last phases of accretion of the "Sausage", followed by the formation of the boxy/peanut-shaped bulge. While we still do not have direct estimates of the time of the bar formation in the Galaxy, it is plausible that it formed after the accretion of the "Gaia Sausage" since this latter had a significant impact onto the Galactic disc, and it is difficult to imagine that the central regions could otherwise survive, undamaged, such a perturbation. Moreover, the structuring of the Galactic bulge indicates that the metal-poor stars that make it must have originated from a disc, which was already kinematically hot, with properties similar to those of the current Milky Way thick disc. This suggests that the thick disc must have been mostly in place before the formation of the boxy/peanut bulge. Once the epoch of massive mergers is over, the Galaxy, with its stellar bar in its central regions, enters a new, more quiescent phase of evolution, characterized by a lower star formation (at a rate of $1 - 2M_{\odot}/year$), and a slower chemical enrichment. Spiral arms appear during this time, and satellites less massive than those accreted in the early phases keep orbiting around the Milky Way: the Sagittarius dwarf is an example of one of them. Others may have been fully absorbed by our Galaxy, but – because of their smaller masses – caused only limited damages to the disc. The warp and the disturbances found in the morphology of the outer Galactic regions may be the signatures of these gravitational perturbations experienced by the Galaxy over time.

2.8. Perspectives

Tracing back the history of our Galaxy and explaining its properties, as we observe them today, requires to combine a number of different tracers and techniques (photometry, astrometry, spectroscopy and numerical models). Because of the huge amount of data that have been made available to the community in the last few years, and to their extraordinary quality, we are discovering a new Galaxy in some ways: links between the different main Galactic components, unsuspected until a few years ago, are now becoming evident and reveal us that all regions of the Galaxy, from the innermost few kpc to the outer disc and the bulge to the halo, are indeed connected. The study of each of these components thus requires us to put them into the more general context of galaxy evolution and the dynamical processes that can reshape and mix stars over time. Many questions still remain, but certainly the years in front of us will be extremely exciting for Galactic science: new data releases from the *Gaia* astrometric mission are to come, and a number of spectroscopic surveys will soon see their light, such as WEAVE, 4MOST and MOONS, with the horizon, the Extremely Large Telescope[7] and its instruments, which will allow us to extend the kind of studies presented in this chapter to all galaxies of the Local Group, thus promising to unravel a global view of the evolution of the Milky Way, and its companion galaxies, through time.

2.9. References

Anderson, L.D., Wenger, T.V., Armentrout, W.P., *et al.* (2019). A Galactic plane defined by the Milky Way H II region distribution. *ApJ*, 871(2), 145.

Athanassoula, E., Misiriotis, A. (2002). Morphology, photometry and kinematics of N -body bars - I. Three models with different halo central concentrations. *MNRAS*, 330(1), 35–52.

Belokurov, V., Erkal, D., Evans, N.W., *et al.* (2018). Co-formation of the disc and the stellar halo. *MNRAS*, 478(1), 611–619.

Bensby, T., Feltzing, S., Lundström, I., *et al.* (2005). α-, r-, and s-process element trends in the Galactic thin and thick disks. *A&A*, 433(1), 185–203.

Bensby, T., Yee, J.C., Feltzing, S., *et al.* (2013). Chemical evolution of the Galactic bulge as traced by microlensed dwarf and subgiant stars. V. Evidence for a wide age distribution and a complex MDF. *A&A*, 549, A147.

7 https://www.eso.org/public/unitedkingdom/teles-instr/elt/?lang.

Binney, J., Gerhard, O.E., Stark, A.A., *et al.* (1991). Understanding the kinematics of Galactic Centre gas. *MNRAS*, 252, 210.

Binney, J., Tremaine, S. (1987). *Galactic dynamics*. Princeton Univ. Press, Princeton.

Bland-Hawthorn, J., Gerhard, O. (2016). The Galaxy in context: Structural, kinematic, and integrated properties. *ARA&A*, 54, 529–596.

Bovy, J., Bahmanyar, A., Fritz, T.K., *et al.* (2016). The shape of the inner Milky Way Halo from observations of the Pal 5 and GD–1 stellar streams. *ApJ*, 833(1), 31.

Bovy, J., Rix, H.-W., Liu, C., *et al.* (2012). The spatial structure of mono-abundance sub-populations of the Milky Way disk. *ApJ*, 753(2), 148.

Brook, C.B., Kawata, D., Gibson, B.K., *et al.* (2003). Galactic halo stars in phase space: A hint of satellite accretion? *ApJ*, 585(2), L125–L129.

Clarkson, W.I., Sahu, K.C., Anderson, J., *et al.* (2011). The first detection of blue straggler stars in the Milky Way bulge. *ApJ*, 735(1), 37.

Combes, F., Debbasch, F., Friedli, D., *et al.* (1990). Box and peanut shapes generated by stellar bars. *A&A*, 233, 82.

Combes, F., Sanders, R.H. (1981). Formation and properties of persisting stellar bars. *A&A*, 96, 164–173.

Dame, T.M., Hartmann, D., Thaddeus, P. (2001). The Milky Way in molecular clouds: A new complete CO survey. *ApJ*, 547(2), 792–813.

Dame, T.M., Thaddeus, P. (2011). A molecular spiral arm in the far outer Galaxy. *ApJ*, 734(1), L24.

De Silva, G.M., Freeman, K.C., Bland-Hawthorn, J., *et al.* (2015). The GALAH survey: Scientific motivation. *MNRAS*, 449(3), 2604–2617.

de Souza, R.E., Dos Anjos, S. (1987). Box-shaped galaxies: A complete list. *A&AS*, 70, 465–480.

de Vaucouleurs, G. (1963). Revised classification of 1500 bright galaxies. *ApJS*, 8, 31.

Deason, A.J., Belokurov, V., Evans, N.W. (2011). The Milky Way stellar halo out to 40 kpc: squashed, broken but smooth. *MNRAS*, 416(4), 2903–2915.

Dehnen, W. (1998). The distribution of nearby stars in velocity space inferred from HIPPARCOS data. *AJ*, 115(6), 2384–2396.

Di Matteo, P., Haywood, M., Lehnert, M.D., *et al.* (2019). The Milky Way has no in-situ halo other than the heated thick disc. Composition of the stellar halo and age-dating the last significant merger with Gaia DR2 and APOGEE. *A&A*, 632, A4.

Dobbs, C.L., Burkert, A. (2012). The myth of the molecular ring. *MNRAS*, 421(4), 2940–2946.

Dwek, E., Arendt, R.G., Hauser, M.G., *et al.* (1995). Morphology, near-infrared luminosity, and mass of the Galactic bulge from COBE DIRBE observations. *ApJ*, 445, 716.

Eggen, O.J., Lynden-Bell, D., Sandage, A.R. (1962). Evidence from the motions of old stars that the Galaxy collapsed. *ApJ*, 136, 748.

Eilers, A.-C., Hogg, D.W., Rix, H.-W., *et al.* (2019). The circular velocity curve of the Milky Way from 5 to 25 kpc. *ApJ*, 871(1), 120.

Englmaier, P., Pohl, M., Bissantz, N. (2011). The Milky Way spiral arm pattern. 3D distribution of molecular gas. *Memorie della Societa Astronomica Italiana Supplementi*, 18, 199.

Fragkoudi, F., Di Matteo, P., Haywood, M., *et al.* (2017). Bars and boxy/peanut bulges in thin and thick discs. I. Morphology and line-of-sight velocities of a fiducial model. *A&A*, 606, A47.

Fragkoudi, F., Di Matteo, P., Haywood, M., *et al.* (2018). The disc origin of the Milky Way bulge. Dissecting the chemo-morphological relations using N-body simulations and APOGEE. *A&A*, 616, A180.

Fuhrmann, K. (1998). Nearby stars of the Galactic disk and halo. *A&A*, 338, 161–183.

Gaia Collaboration, Brown, A.G.A., Vallenari, A., *et al.* (2018). Gaia data release 2. Summary of the contents and survey properties. *A&A*, 616, A1.

Gaia Collaboration, Brown, A.G.A., Vallenari, A., *et al.* (2016). Gaia data release 1. Summary of the astrometric, photometric, and survey properties. *A&A*, 595, A2.

Gilmore, G., Reid, N. (1983). New light on faint stars - III. Galactic structure towards the South Pole and the Galactic thick disc. *MNRAS*, 202, 1025–1047.

Hammer, F., Puech, M., Chemin, L., *et al.* (2007). The Milky Way, an exceptionally quiet Galaxy: Implications for the formation of spiral galaxies. *ApJ*, 662(1), 322–334.

Hayden, M.R., Bovy, J., Holtzman, J.A., *et al.* (2015). Chemical cartography with APOGEE: Metallicity distribution functions and the chemical structure of the Milky Way disk. *ApJ*, 808(2), 132.

Hayes, C.R., Majewski, S.R., Shetrone, M., *et al.* (2018). Disentangling the Galactic halo with APOGEE. I. Chemical and kinematical investigation of distinct metal-poor populations. *ApJ*, 852(1), 49.

Haywood, M. (2008). Radial mixing and the transition between the thick and thin Galactic discs. *MNRAS*, 388(3), 1175–1184.

Haywood, M., Di Matteo, P., Lehnert, M.D., *et al.* (2013). The age structure of stellar populations in the solar vicinity. Clues of a two-phase formation history of the Milky Way disk. *A&A*, 560, A109.

Haywood, M., Di Matteo, P., Lehnert, M.D., *et al.* (2018). In disguise or out of reach: First clues about in situ and accreted stars in the stellar halo of the Milky Way from Gaia DR2. *ApJ*, 863(2), 113.

Haywood, M., Di Matteo, P., Snaith, O., *et al.* (2016). Hiding its age: The case for a younger bulge. *A&A*, 593, A82.

Helmi, A. (2008). The stellar halo of the Galaxy. *A&A Rev.*, 15(3), 145–188.

Helmi, A., Babusiaux, C., Koppelman, H.H., *et al.* (2018). The merger that led to the formation of the Milky Way's inner stellar halo and thick disk. *Nature*, 563(7729), 85–88.

Hou, L.G., Han, J.L. (2014). The observed spiral structure of the Milky Way. *A&A*, 569, A125.

Hou, L.G., Han, J.L., Shi, W.B. (2009). The spiral structure of our Milky Way Galaxy. *A&A*, 499(2), 473–482.

Ibata, R.A., Gilmore, G., Irwin, M.J. (1994). A dwarf satellite galaxy in Sagittarius. *Nature*, 370(6486), 194–196.

Ibata, R.A., Malhan, K., Martin, N.F. (2019). The streams of the gaping abyss: A population of entangled stellar streams surrounding the inner galaxy. *ApJ*, 872(2), 152.

Jurić, M., Ivezić, Z., Brooks, A. *et al.* (2008). The Milky Way tomography with SDSS. I. Stellar number density distribution. *ApJ*, 673(2), 864–914.

Kalberla, P. M.W., Kerp, J. (2009). The Hi distribution of the Milky Way. *ARA&A*, 47(1), 27–61.

Khoperskov, S., Gerhard, O., Di Matteo, P., *et al.* (2019). Hic sunt dracones: Cartography of the Milky Way spiral arms and bar resonances with Gaia Data Release 2. *arXiv e-prints*, p.arXiv:1910.06335.

Kwee, K.K., Muller, C.A., Westerhout, G. (1954). The rotation of the inner parts of the Galactic system. *Bull. Astron. Inst.*, Netherlands, 12, 211.

Law, D.R., Majewski, S.R., Johnston, K.V. (2009). Evidence for a triaxial Milky Way dark matter halo from the Sagittarius stellar tidal stream. *ApJ*, 703(1), L67–L71.

Lehnert, M.D., Di Matteo, P., Haywood, M., *et al.* (2014). The Milky Way as a high-redshift galaxy: The importance of thick disk formation in galaxies. *ApJ*, 789(2), L30.

Levine, E.S., Blitz, L., Heiles, C. (2006). The spiral structure of the outer Milky Way in hydrogen. *Science*, 312(5781), 1773–1777.

Licquia, T.C., Newman, J.A., Bershady, M.A. (2016). Does the Milky Way obey spiral galaxy scaling relations? *ApJ*, 833(2), 220.

Licquia, T.C., Newman, J.A., Brinchmann, J. (2015). Unveiling the Milky Way: A new technique for determining the optical color and luminosity of our galaxy. *ApJ*, 809(1), 96.

Liszt, H.S., Burton, W.B. (1980). The gas distribution in the central region of the galaxy. III—A barlike model of the inner-galaxy gas based on improved H I data. *ApJ*, 236, 779–797.

López-Corredoira, M., Cabrera-Lavers, A., Garzón, F., *et al.* (2002). Old stellar Galactic disc in near-plane regions according to 2MASS: Scales, cut-off, flare and warp. *A&A*, 394, 883–899.

Mackereth, J.T., Bovy, J., Schiavon, R.P., *et al.* (2017). The age-metallicity structure of the Milky Way disc using APOGEE. *MNRAS*, 471(3), 3057–3078.

Majewski, S.R., Schiavon, R.P., Frinchaboy, P.M., *et al.* (2017). The Apache Point Observatory Galactic Evolution Experiment (APOGEE). *AJ*, 154(3), 94.

Majewski, S.R., Skrutskie, M.F., Weinberg, M.D., Ostheimer, J.C. (2003). A two micron all sky survey view of the Sagittarius dwarf Galaxy. I. Morphology of the Sagittarius core and tidal arms. *ApJ*, 599(2), 1082–1115.

Martin, N.F., Ibata, R.A., Bellazzini, M., *et al.* (2004). A dwarf galaxy remnant in Canis Major: The fossil of an in-plane accretion on to the Milky Way. *MNRAS*, 348(1), 12–23.

Momany, Y., Zaggia, S., Gilmore, G., *et al.* (2006). Outer structure of the Galactic warp and flare: Explaining the Canis Major over-density. *A&A*, 451(2), 515–538.

Ness, M., Debattista, V.P., Bensby, T., *et al.* (2014). Young stars in an old bulge: A natural outcome of internal evolution in the Milky Way. *ApJ*, 787(2), L19.

Ness, M., Lang, D. (2016). The X-shaped bulge of the Milky Way revealed by WISE. *AJ*, 152(1), 14.

Newberg, H.J., Yanny, B., Rockosi, C., *et al.* (2002). The ghost of Sagittarius and lumps in the halo of the Milky Way. *ApJ*, 569(1), 245–274.

Nissen, P.E., Schuster, W.J. (2010). Two distinct halo populations in the solar neighborhood. Evidence from stellar abundance ratios and kinematics. *A&A*, 511, L10.

Oort, J.H., Kerr, F.J., Westerhout, G. (1958). The galactic system as a spiral nebula (Council Note). MNRAS, 118, 379.

Poggio, E., Drimmel, R., Lattanzi, M.G., *et al.* (2018). The Galactic warp revealed by Gaia DR2 kinematics. *MNRAS*, 481(1), L21–L25.

Pohl, M., Englmaier, P., Bissantz, N. (2008). Three-dimensional distribution of molecular gas in the barred Milky Way. *ApJ*, 677(1), 283–291.

Portail, M., Gerhard, O., Wegg, C., *et al.* (2017). Dynamical modelling of the galactic bulge and bar: The Milky Way's pattern speed, stellar and dark matter mass distribution. *MNRAS*, 465(2), 1621–1644.

Purcell, C.W., Bullock, J.S., Kazantzidis, S. (2010). Heated disc stars in the stellar halo. *MNRAS*, 404(4), 1711–1718.

Raha, N., Sellwood, J.A., James, R.A., *et al.* (1991). A dynamical instability of bars in disk galaxies. *Nature*, 352(6334), 411–412.

Reid, M.J., Menten, K.M., Brunthaler, A., *et al.* (2019). Trigonometric parallaxes of high-mass star-forming regions: Our view of the Milky Way. *ApJ*, 885(2), 131.

Reid, M.J., Menten, K.M., Brunthaler, A., *et al.* (2014). Trigonometric parallaxes of high mass star forming regions: The structure and kinematics of the Milky Way. *ApJ*, 783(2), 130.

Renzini, A. (1999). Origin of bulges. *In* The Formation of Galactic Bulges, Carollo, C.M., Ferguson, H.C., Wyse, R.F.G. (eds). 9.

Reylé, C., Marshall, D.J., Robin, A.C., *et al.* (2009). The Milky Way's external disc constrained by 2MASS star counts. *A&A*, Cambridge University Press, Cambridge, 495(3), 819–826.

Rich, R.M., Reitzel, D.B., Howard, C.D., *et al.* (2007). The bulge radial velocity assay: Techniques and a rotation curve. *ApJ*, 658(1), L29–L32.

Rodriguez-Fernandez, N.J., Combes, F. (2008). Gas flow models in the Milky Way embedded bars. *A&A*, 489(1), 115–133.

Roman-Duval, J., Jackson, J.M., Heyer, M., *et al.* (2009). Kinematic distances to molecular clouds identified in the Galactic ring survey. *ApJ*, 699(2), 1153–1170.

Schönrich, R., Dehnen, W. (2018). Warp, waves, and wrinkles in the Milky Way. *MNRAS*, 478(3), 3809–3824.

Searle, L., Zinn, R. (1978). Composition of halo clusters and the formation of the galactic halo. *ApJ*, 225, 357–379.

Shen, J., Rich, R.M., Kormendy, J., *et al.* (2010). Our Milky Way as a pure-disk GalaxyA challenge for Galaxy formation. *ApJ*, 720(1), L72–L76.

Skowron, D.M., Skowron, J., Mróz, P., *et al.* (2019). A three-dimensional map of the Milky Way using classical Cepheid variable stars. *Science*, 365(6452), 478–482.

Snaith, O., Haywood, M., Di Matteo, P., *et al.* (2015). Reconstructing the star formation history of the Milky Way disc(s) from chemical abundances. *A&A*, 578, A87.

Snaith, O.N., Haywood, M., Di Matteo, P., *et al.* (2014). The dominant epoch of star formation in the Milky Way formed the thick disk. *ApJ*, 781(2), L31.

Stecker, F.W., Solomon, P.M., Scoville, N.Z., *et al.* (1975). Molecular hydrogen in the Galaxy and galactic gamma rays. *ApJ*, 201, 90–97.

Tolstoy, E., Hill, V., Tosi, M. (2009). Star-formation histories, abundances, and kinematics of dwarf galaxies in the local group. *ARA&A*, 47(1), 371–425.

Vallée, J.P. (2017). A guided map to the spiral arms in the galactic disk of the Milky Way. *The Astronomical Review*, 13(3-4), 113–146.

van de Hulst, H.C., Muller, C.A., Oort, J.H. (1954). The spiral structure of the outer part of the Galactic system derived from the hydrogen emission at 21 cm wavelength. *Bull. Astron. Inst.*, Netherlands, 12, 117.

Vera-Ciro, C.A., Sales, L.V., Helmi, A., *et al.* (2011). The shape of dark matter haloes in the Aquarius simulations: Evolution and memory. *MNRAS*, 416(2), 1377–1391.

Vera-Ciro, C.A., Helmi, A. (2013). Constraints on the shape of the Milky Way dark matter halo from the Sagittarius stream. *ApJ*, 773(1), L4.

Wegg, C., Gerhard, O., Portail, M. (2015). The structure of the Milky Way's bar outside the bulge. *MNRAS*, 450(4), 4050–4069.

Weiland, J.L., Arendt, R.G., Berriman, G.B., *et al.* (1994). COBE diffuse background experiment observations of the Galactic bulge. *ApJ*, 425, L81.

Yanny, B., Rockosi, C., Newberg, H.J., *et al.* (2009). SEGUE: A spectroscopic survey of 240,000 stars with g = 14-20. *AJ*, 137(5), 4377–4399.

3

Early-type Galaxies

Eric EMSELLEM
European Southern Observatory (ESO), Garching, Germany

3.1. Introduction

"Early-type galaxies" (ETGs) consist of the two historical classes of elliptical (E) and lenticular (S0) galaxies. As we shall see later, these classes are occasionally difficult to distinguish. ETGs only account for 33% of the light density emitted by the various populations of galaxies with redshift $z = 0$ but approximately 50% of their stellar mass density (Bernardi *et al.* 2010). Thus, they play a key role in the formation of stars in the Local Universe.

In general classifications of galaxies by morphology (Chapter 1), ETGs are often presented in opposition to spiral or late-type galaxies (Chapter 4) and are characterized by the lack of any visible spiral structure. This is the most frequently cited property to discriminate between ETGs (without a visible spiral) and late-type galaxies. Another common observation is that ETGs tend to have very low proportions of cold gas ($<$ 100 K) or warm gas ($< 10^4$ K); this criterion is often used to distinguish ETGs from other systems in cosmological simulations. Another common but incorrect approach to distinguish between ETG and late-type galaxy is to consider the relative proportion of the spheroid component, which supposedly dominates in ETGs and plays a minor role in spiral galaxies (however, there exist spiral galaxies with large spheroids and ETGs with small bulges, see Chapters 1 and 4).

For color versions of the figures in this chapter, see www.iste.co.uk/combes/galaxies.zip.

Galaxies,
coordinated by Françoise COMBES.

A prototypical example of an early-type galaxy is represented in Figure 3.1, showing the (visible) purely spheroid character (without a disc component) of the galaxy, with galaxies that would typically have a spiral disc structure in the foreground and background for contrast. The figure also shows the color difference between this ETG and a spiral galaxy (bottom right of Figure 3.1); the color of the ETG suggests that it is mainly composed of old stars (more than $\sim$ 8 billion years), contrasting once again with the blue color of the spiral galaxy, which indicates active star formation.

Figure 3.1. *Early-type (elliptical) galaxy ESO325-G0004 surrounded by other galaxies in the foreground and background [Blakeslee, NASA/ESA-HST]*

The difficulty of classifying this type of galaxy arises from two key aspects: first, the definition of ETGs is historical and has not always followed the same morphological or color criteria; second, most of the properties considered by the definition depend on the viewing angle of the galaxy (e.g. the flattening, whether a spiral structure or disc can be detected, or the overall color). Nevertheless, ETGs present a few broad trends:

– at equivalent stellar mass, spiral galaxies tend to have a characteristic radius (i.e. containing half of the light) that is larger than the characteristic radius of an ETG (see section 3.5.1). In other words, ETGs have more concentrated light profiles than spiral galaxies;

– the proportion of ETGs increases with the total stellar mass and local galaxy density;

– the proportion of gas and dust is significantly lower in ETGs, assuming similar total stellar mass;

– the stellar population of ETGs is considered to be dominated by old stars (older than 8 Gyr), with little residual star formation. When the stellar mass is high (above $5 \cdot 10^{11}$ $M_{\odot}$), most stars form with a redshift z of more than 2.

Note that these trends only describe populations and therefore cannot be applied to a given system individually. The trends relate to the classification itself and may vary as a function of other parameters (such as the stellar mass). Some trends are very strongly dependent on others (colors, star formation, gas and dust content) and can be jointly explained by the formation and assembly scenarios of these galaxies (see section 3.5). The separation of ETGs into elliptical (E) and lenticular (S0) galaxies is itself not very robust (see section 3.4.2) and is typically determined by whether or not a large-scale disc can be detected. Figure 3.3 illustrates the case of NGC 3115 (left), one of the closest S0 galaxies to us, with its very fine disc and bright central region. On the right, the so-called Sombrero Galaxy closely resembles an elliptical galaxy surrounded by a ring of dust (and gas), but it is sometimes classified as an Sa galaxy rather than an S0 galaxy. Thus, the stellar component of E galaxies is dominated by a spheroid, whereas S0 galaxies are disc galaxies whose spheroid is often called a "bulge".

The properties of these galaxies are signatures of the processes involved in their formation and evolution. In section 3.5.1, we will see how we can broadly retrace these mechanisms simply by considering the mass and the size. But first, let us review some of the global properties of ETGs.

3.2. General properties: components and morphology

ETGs span a very broad range of luminosity and mass profiles, ranging from galaxies with very low luminosity (10^7 $L_{\odot}$ in the visible band) to extreme systems

($> 10^{12} L_\odot$). For instance, the compact galaxy M32 has a luminosity of a few 10^8 $L_\odot$, whereas Messier 87, renowned for its supermassive black hole and radio jet, has a luminosity of just under 10^{12} $L_\odot$. Often, ETGs are restricted to galaxies more massive than 10^9 $M_\odot$, grouping lower mass galaxies into a category of "dwarf galaxies" that encompasses both "spheroid" galaxies and irregular galaxies; we will adopt this convention for the remaining chapters.

Figure 3.2. *Early-type galaxies NGC3115 (left) and NGC4594 (right; the Sombrero Galaxy)*

3.2.1. *Discs and bars*

ETGs are often wrongly thought to be dominated by a spheroid. In reality, the spectrum of possible spheroid proportions (and conversely disc proportions) is very broad, ranging from almost 0% (pure disc galaxy) to 100% (so-called elliptical galaxy), as was already observed by van den Bergh (1976). Thus, there are S0a and S0d galaxies with large and small bulges or spheroids, just as there are Sa and Sd spiral galaxies. This continuity of the Bulge/Disc ratio has been studied photometrically by various groups, often by decomposition of light profiles or images of the surface brightness into at least two principal components: a disc, typically satisfying an exponential distribution, and a spheroid, satisfying a Sersić law (see Chapter 1). ETGs with a large-scale disc satisfy the same trends as spiral galaxies; they also contain bars, rings and "lenses" (in at least 60% of cases).

3.2.2. *Gas and dust content*

Generally, ETGs contain little neutral gas (HI) or molecular gas (H_2), although HI gas can be detected in at least half of them, and molecular gas is found in around 25% of ETGs: a typical ETG with a stellar mass of 10^{10} $M_\odot$ contains less than 1% HI gas by mass and less than 0.1% H_2 gas, roughly 30 times less than late-type galaxies (see for example Calette *et al.* 2018; Lagos *et al.* 2014). Of course, this is just an average; there are many ETGs with a high mass of HI gas, especially in regions with low galaxy density (Serra *et al.* 2012). As a general rule, the dust content follows the molecular gas content, with a global mass ratio of 50–1,000 between gas and dust (Kaviraj *et al.* 2012; Davis *et al.* 2015). The overall gas content also varies as a function of the mass of the galaxy and, as we will see in section 3.4.2, the dynamical state (and therefore the history) of the galaxy.

The spatial distribution of gas and dust in ETGs is often restricted to a domain with a narrow radius, either very central, or in the outer regions, like for the Sombrero Galaxy. There are many examples of ETGs where the gas (and dust) are (spatially and kinematically) decoupled from the stellar components, despite being dynamically relaxed. This is more common when the galaxy density is lower, which seems to suggest an external origin. Recent cosmological simulations have focused on the accretion of gas by cooling the hot halo (see for example Lagos *et al.* 2014), but other ideas, such as accretion during an interaction or merger between galaxies, also remain to be tested.

3.2.3. *Dark matter and halo*

Within the paradigm that assumes the presence of dark matter to explain results such as flat rotation curves, the velocities of galaxies in clusters and the large structures observed today at cosmological scales, ETGs are no different from late-type galaxies. The shared lack of HI gas in the outer regions of ETGs means that measurements of the mass profile at large radii depend on other techniques (e.g. the stellar velocity dispersion, see section 3.4.3, or techniques based on gravitational lens effects). Like for spiral galaxies, the dark matter content of these galaxies seems in general to be (partially) correlated with their baryonic mass. The proportion of dark matter by mass at the center of an ETG (inside the effective radius R_e) is therefore smallest (0–20%) in galaxies with a mass of $10^{10.5}$ $M_\odot$ and increases in galaxies of lower and higher mass, reaching around 50% at masses below approximately $10^{9.5}$ and above $10^{11.7}$ $M_\odot$ (Cappellari *et al.* 2013). Like for late-type galaxies, the gravitational potential of ETGs seems to be dominated by dark matter at radii larger than around three or four times the effective radius R_e, although it should be borne in mind that some alternative hypotheses (e.g. MOND) can also account for the observations of most individual galaxies.

The three-dimensional morphology of the dark matter halos in ETGs is poorly constrained. Measurements are typically derived from cosmological simulations (e.g. Hayashi *et al.* 2007). In the ΛCDM paradigm, if only dark matter is taken into account, the halos evolve dynamically under the sole effect of gravity with a potential that has axis ratios of around 0.7. The influence of baryons seems primordial, especially in the central regions where the dark matter component would be noticeably more axisymmetric oblate. The shape of halos seems to clearly vary with radius, from the central regions where baryons dominate to the far outer regions where primordial interactions (torques between halos) and the time evolution of the global potential (group, cluster) become significant.

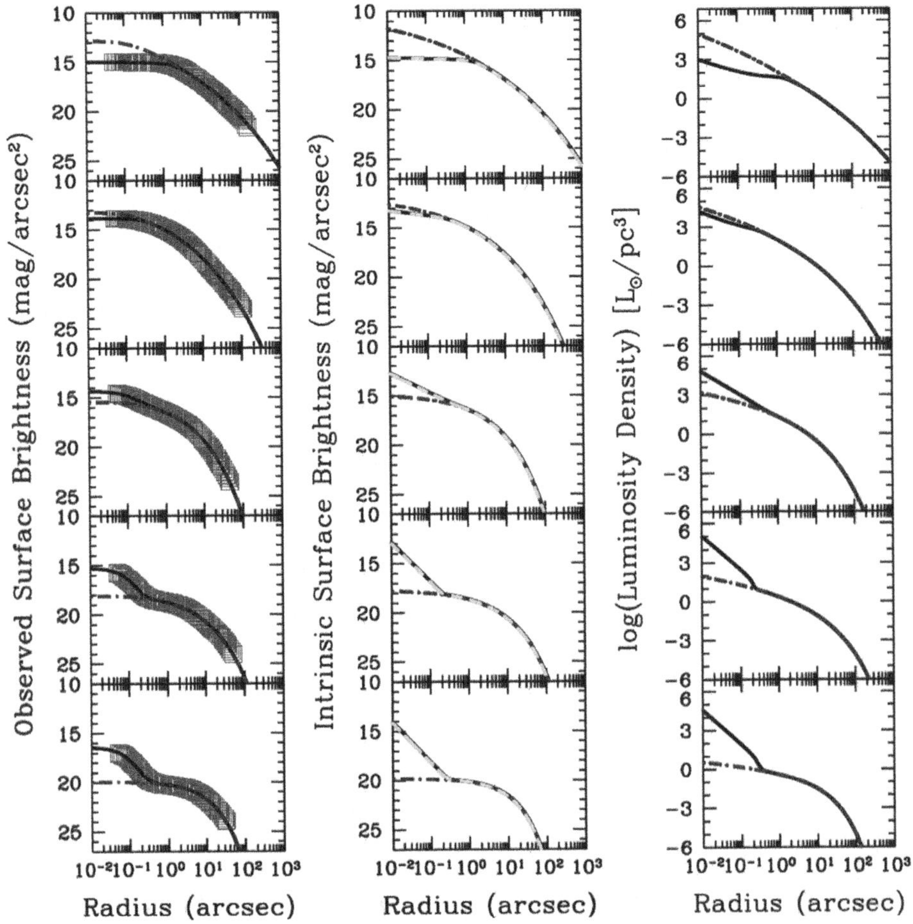

Figure 3.3. *Light profiles of 5 ETGs, illustrating the variety of central profiles of galaxies in the Fornax and Virgo clusters (ACSVCS project with the HST). Left: Surface brightness profiles. Center: Profiles after deconvolution of the seeing effect. Right: Deprojected profiles. The galaxies are ordered from brightest (top) to dimmest (bottom): VCC 1226 (= M49 = NGC 4472, with MB~21.9 mag), VCC 1231 (MB~19.9 mag), VCC 828 (MB~18.6 mag), VCC 1422 (MB~17.4 mag), VCC 1075 (MB~16.1 mag). Reproduced from Glass* et al. *(2011)*

3.2.4. *Globular clusters*

Another important component of ETGs is their population of globular clusters, which seems to have a high specific frequency at both extreme ends of the mass spectrum, for galaxies with low and high luminosity. The integrated mass of all globular clusters of a galaxy seems to correlate closely with the mass of its halo (Harris *et al.* 2013, 2017), representing around $2.9 \cdot 10^{-3}\%$ of the galaxy by mass for host galaxies of a large range of masses. This relation only seems to depend very weakly on the type of galaxy (early or late). Accordingly, globular clusters are excellent tracers of the external part of the potential for ETGs because they are present in very large numbers in the most massive galaxies (see the reviews by Brodie and Strader 2006; Harris 2010). We can often observe two (or sometimes even more) distinct populations of globular clusters: a blue population that is poor in metal with a typical metallicity of [Fe/H]~ -1.5 and a red population that is richer in metal ([Fe/H]~ -0.5). These two components have different spatial distributions and dynamics and are usually associated with two different formation or assembly processes. The blue population could result from the accretion of low-mass stellar systems, whereas the metal-rich clusters are thought to arise from violent and gas-rich mergers.

3.2.5. *Light and mass profiles*

The light profiles of disc-dominated galaxies are generally approximated by an exponential law (if we neglect the contribution of the bars), i.e. a Sersić index of approximately 1. The light profiles of ETGs are generally more complex, with multiple components, illustrating that both the spheroid and other structures (central discs, rings) contribute significantly. In Chapter 4, we will see that the profiles of ETGs have a steeper slope (larger Sersić index) than spiral galaxies on average, which is related to the fact that the luminosity distribution is generally more concentrated.

With modern high-resolution imagery and in particular the Hubble Space Telescope (HST), the light profiles at the galaxy centers (100 central parsecs) of the Local Universe have now been successfully resolved. By parameterizing these profiles with ad-hoc laws (Sersić profile or Nuker laws), it was observed that the most massive galaxies have profiles that flatten in the center with a slope that tends to 0 (Ferrarese *et al.* 1994; Lauer *et al.* 1995; Faber *et al.* 1997), whereas the less massive galaxies tend to have profiles with an "excess" of light relative to the Sersić law fitted to the outer regions. To describe these "flat core" profiles, new techniques have been introduced, such as parameterization by a so-called "Nuker law", as well as a core-Sersić law that adds an inflection to the Sersić profile, and occasionally even non-parametric analysis (Gebhardt *et al.* 1996). The discovery of these flat cores is an important evolutionary clue that is most often explained by the merger of two supermassive black holes. When two galaxies with massive central black holes merge, provided that certain conditions are satisfied, their black holes may form a binary system whose scale is of the order of $r_{\rm h} \sim GM_\bullet/\sigma^2$ (where $M_\bullet$ is the sum of the masses of the black holes and σ is the dispersion of the system). This can produce a slingshot effect on stars passing close to this couple ($\sim 3r_{\rm h}$), exchanging energy and reinforcing the binary system. While the merger is unfolding, a number of stars, primarily located near the center, are pushed back out to orbits with larger radii, flattening the stellar light profile and therefore the baryonic mass (Milosavljević and Merritt 2001). The quantity of displaced mass is predicted to be proportional to the mass of the binary system of black holes. The excess light, often associated with structures called nuclei, is likely related to the dissipative past of the galaxy (e.g role of gas in the central regions).

The global mass profile is obtained by adding together the various ingredients, most importantly the stars and dark matter (if it exists). Recent results appear to confirm that, at large scales, the global mass profiles of ETGs follow a power law $r^{-\gamma}$ with $\gamma \sim 2$ quite closely (for example Bellstedt *et al.* 2018). These profiles are constrained by a range of different techniques based on photometry, kinematics or gravitational lens effects (see references in Courteau *et al.* 2014). The resulting profile, close to the "isothermal profile", involves processes that balance the contributions of stars and dark matter so that they tend toward this value (Poci *et al.* 2017) and may be explained in terms of simple mechanisms of virialization and dynamical relaxation in ETGs. In general, several processes contribute to the evolution of the mass profile: the gas, namely the dissipative component, increases the concentration, while a feedback mechanism (for example associated with the active nuclei) decreases it (Wang *et al.* 2019).

3.2.6. *Extreme cases: brightest cluster galaxies and ultra-diffuse galaxies*

Some studies have focused on specific categories of extreme galaxies, such as central galaxies of very massive clusters or, at the other end of the mass spectrum, dwarf galaxies, with the goal of better understanding the evolutionary mechanisms of these galaxies. Two types of galaxies are worth mentioning to illustrate the links between structures, on the one hand, and formation and assembly processes, on the other hand.

Brightest cluster galaxies (BCGs) are extremely massive galaxies (of the order of 10^{12} $M_\odot$ or more in stars) in dense environments (clusters). They sometimes have diffuse halos of stars in excess of a simple extrapolation of the Sersić profile to the outer regions of the galaxy. BCGs are often brighter than the second-brightest galaxy in their cluster by around one order of magnitude, and the brightness difference seems to correlate with the dynamical age of the cluster. A large proportion of BCGs exhibit correlation between the direction of elongation of their light distribution and the principal axis of the cluster or larger superstructure. These properties suggest that mergers with other galaxies from the cluster contribute significantly to their growth; these mergers typically involve a very large number of low-mass galaxies but can also include a small number of more significant and later mergers (after $z \sim 1$; see section 3.5).

At the dwarf galaxy end of the spectrum, ultra-diffuse galaxies (UDGs) have recently been (re)discovered. UDGs are galaxies with a stellar mass of around 10^8 $M_\odot$ and an effective radius of 1 or 2 kpc, which is very high given their mass. These remarkably stretched out galaxies make up the tail of the size distribution of galaxies. It is interesting to study their dynamical properties and populations of globular clusters, which suggest a violent formation history and an evolution dominated by interactions with the other galaxies in their groups or clusters.

3.3. Zoom on the stellar component

In this section, we shed light on the global properties of the stellar component in ETGs, in particular by comparison with other galaxy populations (e.g. spiral galaxies). The low gas and dust content (by mass) and the young stars generally present in ETGs make studying the stellar component easier: low extinction along the line of sight, less intense emission lines, lower contribution from massive blue stars.

From the 1980s to the 2000s, small samples of a few ETGs were observed, mostly with slit spectrographs that typically cover the photometric major axis (and in some cases also the photometric minor axis) of the galaxy. With a sample of around 1 million galaxies, the Legacy Campaign of the Sloan Digital Sky Survey (Strauss *et al.* 2002) was the first study that allowed different types of galaxies to be characterized more systematically. The SDSS Legacy Survey made it possible to probe the fine properties of the stellar component more effectively, but it used a fixed aperture (one spectrum per galaxy) and therefore did not resolve any individual morphological or dynamical components. In the 2000s, new campaigns that used spectrographs equipped with so-called integral-field units (IFUs) sought to resolve and cover a bidimensional field: first SAURON (Bacon *et al.* 2001; de Zeeuw *et al.* 2002) and Atlas^{3D} (Cappellari *et al.* 2011), which focused entirely on ETGs with a few hundred systems, then SAMI (Croom *et al.* 2012) and MaNGA (Bundy *et al.* 2015), which collected samples from thousands of galaxies. These ambitious campaigns made it possible to establish various resolved properties of ETGs, especially relating to the kinematics of stars, their ages and metallicity, and their enrichment processes. After a brief overview of a few scaling relations that are important for ETGs, we will present some of the global characteristics of the stellar populations of ETGs below. The following section then discusses the dynamics and kinematics of ETGs based on observations and modeling.

3.3.1. *Scaling relations: Faber–Jackson, fundamental plane and virial plane*

Once spectra became available for larger samples of galaxies, it became possible to confirm and refine the global scaling relations that govern ETGs. In many cases, an a posteriori understanding of these relations can be established in terms of notions relating to the equilibrium and evolution of galaxy systems.

The first measurements of the velocity dispersion enabled a power law relation to be found between the global luminosity L of a galaxy and the velocity dispersion σ measured in the central region of the galaxy (Faber and Jackson 1976): $L \propto \sigma^4$. This relation, known as the Faber–Jackson relation, revealed a direct link between the quantity of light emitted by a galaxy and its dynamics for the first time. It also gives an elegant way to calculate a distance estimate based on spectroscopy and the apparent magnitude. The Faber–Jackson relation for stellar dynamics in central regions is often compared to the Tully–Fisher relation (Tully and Fisher 1977) for spiral galaxies, which relates the velocity of the gas in the outer regions of the disc to the luminosity. In both cases, the relations give a link between the luminosity and the velocity amplitude that can be used to calibrate distances. A so-called Tully–Fisher relation for ETGs has been studied with molecular gas as the dynamical tracer (Davis *et al.* 2011), although this tracer is typically restricted to the central regions of the galaxy.

Another relation, known as the $Dn - \sigma$ relation, relates the diameter of the galaxy to the reference magnitude of 20.75 mag/arcsec^2 (in the B band) and the velocity dispersion (Dressler 1987). This relation can also be used to calibrate distances. The $Dn - \sigma$ relation introduces the notion of characteristic radius to complement the measured luminosity (or surface brightness) and the dynamical amplitude (velocity or dispersion). Historically, the coupling between these three quantities predated the $Dn - \sigma$ relation itself (Dressler *et al.* 1987), using a concept that soon became known as the *fundamental plane* (Djorgovski and Davis 1987): ETGs occupy a two-dimensional region (plane) in the three-dimensional space defined by the effective radius R_e, the effective surface brightness I_e (at radius R_e) and the velocity dispersion σ. The relation observed at this epoch can be written as $I_e \propto \sigma^\alpha R_e^\beta$, where $\alpha \sim 1.6 \pm 0.17$ and $\beta \sim 0.9 \pm 0.1$ (Kormendy and Djorgovski 1989). The standard deviation observed around this plane is remarkably small, namely around 20%. Not only is the three-dimensional space defined by (R_e, I_e, σ) not fully occupied by galaxies, in fact, the distribution of galaxies is restricted to a limited region of this plane. This suggests the action of a structuring process in the formation and/or evolution of ETGs.

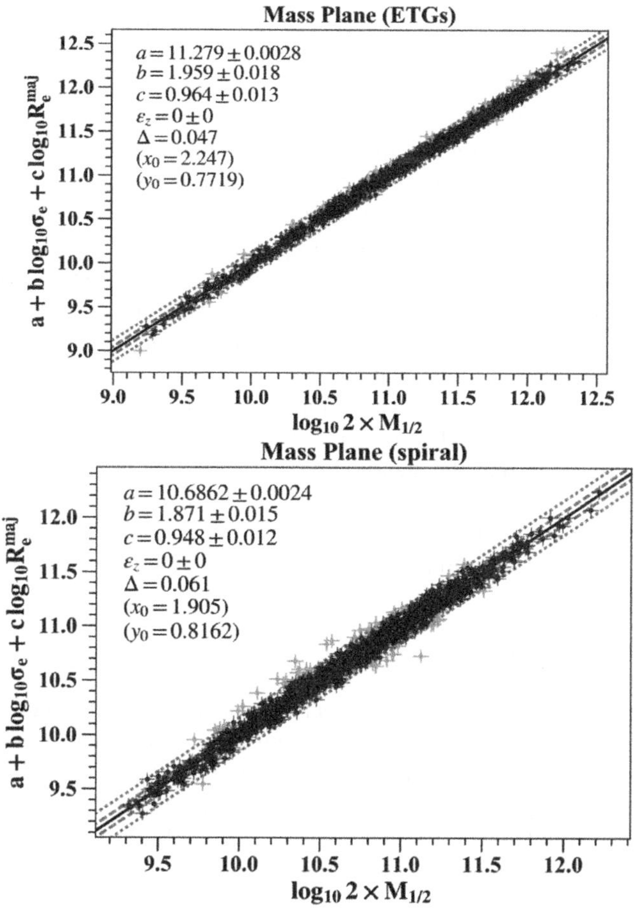

Figure 3.4. *Mass plane for ETGs (top) and spiral galaxies (bottom), showing the excellent correlation between σ_e, R_e and M with coefficients (here, b and c) very close to the values predicted by the virial theorem. Reproduced from Li* et al. *(2018)*

It was soon realised that an approximation of the fundamental plane can be predicted simply by applying the virial theorem's equilibrium between the kinetic and potential energies of a galaxy: any system in equilibrium may be expected to satisfy a relation of the form $M \propto \sigma^2 R$ (Faber *et al.* 1987). The deviation between the orientations of the observed and theoretical planes, or "tilt" of the fundamental plane, has been studied to understand whether it could be explained by alternative physics or some observational bias, and the conclusions that can be drawn from this tilt regarding the evolution of galaxies have also been investigated (Bernardi *et al.* 2003). Much work has focused on the fact that the measured tilt depends on the observed (filter) band when measuring the surface brightness, which likely indicates some variation effect of the mass-luminosity ratio M/L as a function of the velocity dispersion or the mass of the galaxy. This variation, caused by variations in the stellar population (age, metallicity and initial mass function [IMF]) or the relative distribution of dark matter, can be incorporated by writing $I_e \propto \sigma^2 R_e^{-1}(M/L)^{-1}$. Part of the tilt can also be related to differences in mass or light profiles, implying variation in the measured dispersion. A more modern version of the fundamental plane replaces I_e with the integrated luminosity $L = 2\pi R_e^2 I_e$ and the measured dispersion with an integrated measurement up to the effective radius based on IFUs: $L \propto \sigma_e^{1.25} R_e^{0.96}$. When ETGs are modeled in detail, allowing the dynamical mass of the galaxy to be measured robustly and directly, we almost exactly recover the equation of the virial theorem within the assumed systematics (see Cappellari 2016, for a more detailed discussion), namely $M \propto \sigma_e^{\alpha} R_e^{\beta}$ with $\alpha \sim 2$ and $\beta \sim 1$ (Figure 3.4). One of the secondary findings of these studies is that the velocity dispersion σ_e is the best predictor ("proxy") to (statistically) determine the M/L ratio of an ETG, given that the standard deviation around the relation $M/L - \sigma$ is less than 30%. Thus, remarkably, just two measurements (e.g. σ_e and R_e) can tell us about the dynamical mass, the color, or the M/L ratio as a direct consequence of the virial theorem and the formation processes of galaxies (and stars).

3.3.2. *Age and metallicity*

The stellar populations of ETGs reflect their lack of spiral structures and more generally the low contributions of gas or young stars. Thus, most ETGs have colors and spectral signatures that correspond to old stars (> 5 Gyr) and whose richness in metals depends directly on the total mass of the galaxy. Nevertheless, this perspective has proved to be simplistic, since many ETGs, especially for low-mass galaxies and in environments less dense in galaxies, contain residues of star formation, as well as molecular gas (see section 3.2.2).

Spectroscopy (typically in the visible region) has been systematically applied to constrain the properties of stars in external galaxies. Hydrogen absorption lines (Balmer: Hβ) are often used to determine the age; the absorption lines of magnesium (Mg), iron (Fe), and calcium (Ca) are used to measure the metallicity, and finally the relative abundance of so-called α elements (e.g. Mg, Ca and Si) associated with formation processes in type-II Supernovae is also considered. Line indices corresponding to the relative depth of these lines are calculated from the galaxy spectra and compared to reference models constructed from empirical or theoretical spectra (Worthey 1994). This method is generally robust for relative values but is fairly uncertain for absolute measurements of the age or metallicity (Conroy 2013).

3.3.2.1. *Integrated values*

Most line indices are correlated with the mass or velocity dispersion of ETGs. One of the strongest relations is found between the Mg_2 or Mg_b index and the dispersion σ, suggesting that more massive galaxies have stars that are richer in metals (Burstein *et al.* 1988; Gallazzi *et al.* 2006). The relative abundance of magnesium to iron [Mg/Fe] is also directly correlated with σ, indicating a more rapid enrichment for ETGs with high mass (Thomas *et al.* 2010; McDermid *et al.* 2015). Lower mass galaxies also seemingly tend to have a lower average age, sometimes indicating the presence of young stars ($<$ 2 Gyr), especially in central regions (Figure 3.5). Even if their mass contribution is low (a few percent at most), young stars are found in around 40% of ETGs.

3.3.2.2. *Gradients*

Some of the variational trends in the stellar population from galaxy to galaxy are also observed within the galaxies themselves. The age and metallicity gradients are a signature of the evolutionary processes of ETGs; they should be able to tell us about both assembly mechanisms and internal secular processes such as migration. If we assume that the average metallicity of a galaxy depends strongly on the mass of the system at the time of formation of its stars (and therefore on the galaxy's capacity to enrich the gas clouds that give birth to these stars), the metallicity or abundance gradients should allow us to partially deduce the merging sequence of the ETGs.

In general, ETGs have relatively low age and metallicity gradients as a function of the distance to the center. Additionally, the average age generally increases and the average metallicity decreases slightly with the radius, since the metallicity gradient is correlated with the total stellar mass, and the gradient is larger for galaxies with greater stellar mass (see Figure 3.6 and Goddard *et al.* 2017). The discs found at the heart of ETGs often have stellar components with different properties; they have greater enrichment and a larger contribution from young stars. This makes sense if we consider that the vast majority of stars form within discs over relatively long time scales in the course of the history of the Universe, making these regions privileged hotspots for metal enrichment. This interpretation is also consistent with the lower relative abundance values of alpha elements to iron in discs (Kuntschner *et al.* 2010). In general, the velocity dispersion σ_e is once again the best proxy for variations in the stellar population of ETGs (see for example Ferreras *et al.* 2019).

3.3.3. *Initial mass function*

Some studies have revealed a discrepancy between the dynamical mass predicted from the observed kinematics of ETGs and the total mass (stars, gas and dark matter) that cannot seemingly be explained by a variation in the proportion of dark matter. The two explanations suggested to explain this difference that still remain plausible today are either a poor calibration of the stellar population models used to account for the observed light or a variation in the stellar IMF.

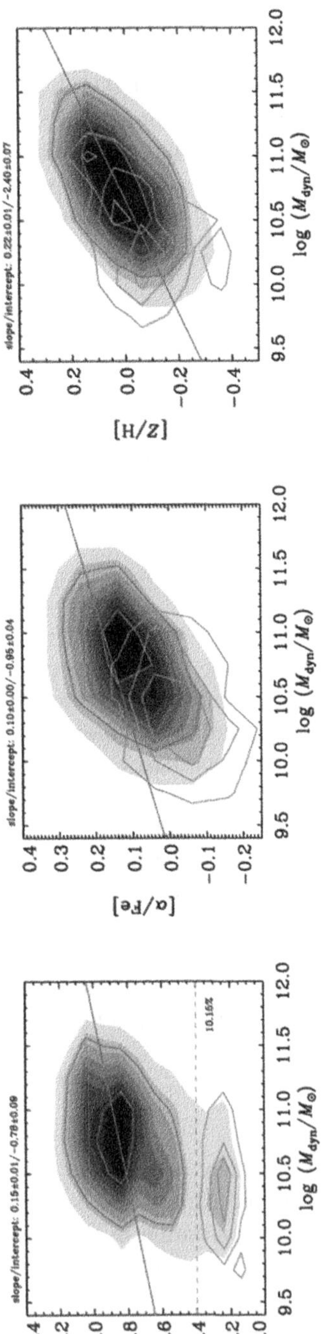

Figure 3.5. *Age, abundance α/Fe and metallicity (weighted by the light) as a function of the dynamical mass (calculated from the dispersion and the effective radius) for a sample of 3360 galaxies (SDSS). The red contours show the passive galaxies with old stellar populations, and the blue contours show galaxies whose age is below 2.5 Gyr. Reproduced from Thomas et al. (2010)*

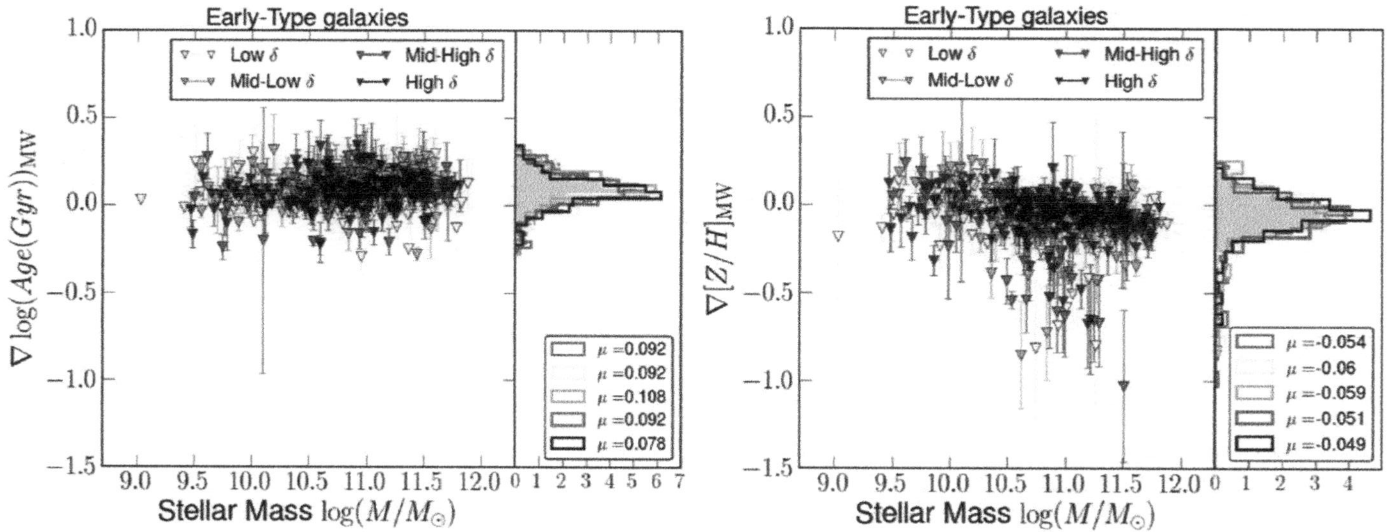

Figure 3.6. *Age and metallicity gradients (weighted by the stellar mass) as a function of the stellar mass of the galaxies observed by the MaNGA campaign (Bundy* et al. *2015). The colors correspond to different environments in terms of local galaxy density. Reproduced from Goddard* et al. *(2017)*

The IMF is a function that specifies the number of stars with any given mass. It is generally considered a universal function (Bastian *et al.* 2010) and is usually parameterized by so-called Chabrier (Chabrier 2003) or Kroupa functions (Kroupa 2001): a power law with the Salpeter index ($\gamma = 1.35$) for massive stars, and a more log-normal law for low-mass stars ($\gamma \sim 0{-}0.25$). These IMFs can only explain part of the stellar mass observed in ETGs, especially in the central regions (see for example Sarzi *et al.* 2018). Spectral signatures (Conroy and van Dokkum 2012; McConnell *et al.* 2016) and dynamical modeling suggest that the slope of the IMF in the low mass regime approaches and occasionally even exceeds that of a pure Salpeter law, implying a significantly higher proportion of dwarf stars (see Figure 3.7 and Posacki *et al.* 2015). This may be related to the rapid and violent mode of stellar formation at the heart of massive galaxies, for example involving a bias in the assembly of dense gaseous substructures (see for example Chabrier *et al.* 2014).

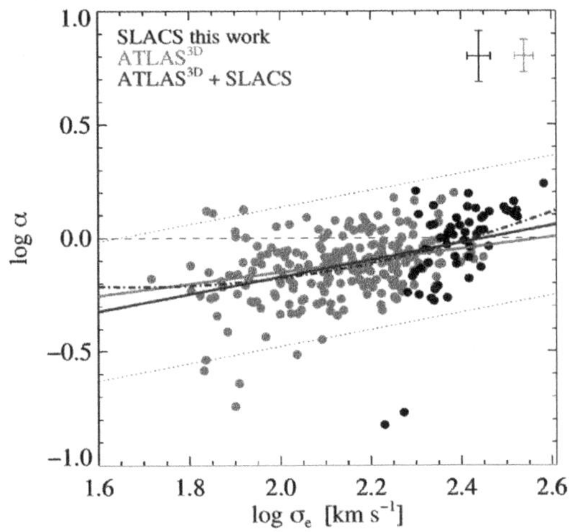

Figure 3.7. *Difference (log) between the dynamical mass and the stellar mass, assuming a so-called Salpeter initial mass function ($\alpha \equiv (M_\star/L)_{\rm syn}/(M_\star/L)_{\rm Salp}$), as a function of the velocity dispersion σ_e for two samples (Atlas^{3D}, in pink, and SLACK, in black). Reproduced from Posacki* et al. *(2015)*

3.4. Dynamics of ETGs

If we only consider the stellar component of a given stellar system arbitrary stellar system, the Holy Grail for some astronomers is to obtain a precise count of stars not only for any given interval of the position (x, y, z) and velocity (V_x, V_y, V_z), but also for properties relating to the age and metallicity, as well as more general abundance properties (e.g. $[Fe/H]$, $[Mg/Fe]$ and $[Ca/Fe]$). This function f is commonly called the "distribution function" and naturally also depends on time t.

For external galaxies whose stars have not been resolved, or even galaxies outside of the Local Group, we typically drop the time variable t, and the stellar kinematics are limited to the radial velocity component, which simultaneously eliminates the two other velocity components (proper motion in the sky plane). Furthermore, we observe these galaxies by integrating along the line of sight, which only leaves us with two spatial components, (x', y'). This results in a function $f(x', y', V_{\rm los}, [Fe/H], [Mg/H], ...)$ (where $V_{\rm los}$ is the velocity along the line of sight) that is usually separated into a "dynamical" component and a "stellar population" component. This section presents a few observations and modeling methods associated with the kinematics of ETGs and cites some potential paths for future development.

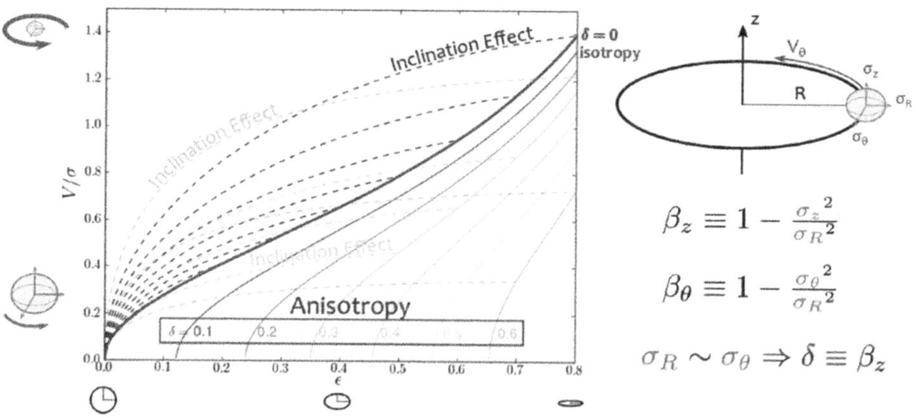

Figure 3.8. *Illustration of the V/σ-ϵ diagram, explaining the concept of anisotropy. Right: The velocity dispersion tensor includes the three spatial components of the dispersion (assumed to be diagonal in this cylindrical frame of reference). The anisotropy (for example β_z) is defined in terms of the ratio of the squares of the two dispersion amplitudes. If we assume that $\sigma_R = \sigma_\theta$, the anisotropy is defined by $\delta = \beta_z$. Left: An isotropic stellar system viewed edge-on follows the thick blue line and is dependent on intrinsic flattening. As the anisotropy δ increases, the curves (side view) move toward the right of the diagram. The inclination will affect the apparent flattening (further left corresponds to lower apparent ellipticities) and the value of V/σ (further down)*

3.4.1. *Observations*

In the late 1970s, long-slit spectroscopic observations of ETGs along one or multiple axes of the same galaxy revealed some fundamental differences in the stellar kinematics of ETGs by comparison with spiral galaxies (Bertola and Capaccioli

1975). At equal mass, ETGs have a lower average rotational velocity than spiral galaxies. The $V/\sigma - \epsilon$ diagram has been proposed to quantify this result (see Figure 3.8) by showing the contribution of ordered motion within the stellar dynamics of a flattened system (with ellipticity ϵ) compared to the predictions of the tensor virial theorem assuming known mass and light profiles and simple symmetry hypotheses (Binney 1978). In practice, observers more frequently use the maximum velocity V_m along the major axis of the galaxy and the central velocity dispersion σ_0 to determine the V/σ ratio.

One of the first samples (of 11 galaxies with low luminosity, see Davies *et al.* 1983) suggested an intrinsic difference between the dynamics of low-mass galaxies, more isotropic, and those of high-mass galaxies, likely triaxial and anisotropic. This fundamental difference was later identified by measuring the quantity $(V/\sigma)_\star$ as the ratio between the observed value of V/σ and the predicted value for an isotropic system. Correlation between the shape of the isophotes and $(V/\sigma)_\star$ suggests (see for example Kormendy and Bender 1996) that, first, the apparent morphology of galaxies (e.g. presence of isophotes with a disc component) is an indicator of the dynamics of a system (its anisotropy), and, second, there are indeed two families of galaxies, each spanning a different range of masses: massive galaxies, which tend to be anisotropic, and low-mass galaxies, which tend to be nearly isotropic.

Although the existence of two major dynamical families has now been confirmed, the anisotropy seems to have been interpreted incorrectly. Indeed, $(V/\sigma)_\star$ does not depend linearly on the anisotropy; this dependence is a function of the intrinsic flattening of the galaxy (neglecting the effect of the inclination, often not fully taken into account; see for example the discussion in Cappellari *et al.* 2007). A galaxy with $(V/\sigma)_\star << 1$ can therefore be close to isotropy if it is intrinsically spherical, or very anisotropic if it is intrinsically flattened. The distribution of galaxies in the V/σ-ϵ diagram therefore seems to account for massive galaxies that tend to be less flattened and are likely less isotropic than the population of flattened ETGs.

3.4.2. *Toward a kinematic classification of ETGs*

In an article that gives a clear summary of the application hypotheses of the virial theorem, Binney (2005) showed that a measurement of the anisotropy of a stellar system cannot be derived simply from the observed flattening or V/σ parameters when the latter are calculated using long-slit spectroscopic data. A rigorous measurement requires a bidimensional covering of the galaxy, which was finally made possible by the first observation campaigns of ETGs using IFUs (see the start of section 3.3). The formula $V/\sigma = \sqrt{\langle V^2 \rangle / \langle \sigma^2 \rangle}$ involves integration ($<>$) over the entire system (weighted by the flux). Observational measurements therefore still require an aperture correction to account for the limited field of the instruments. Thus, we typically resort to measuring up to some fixed fraction or multiple of the effective radius R_e.

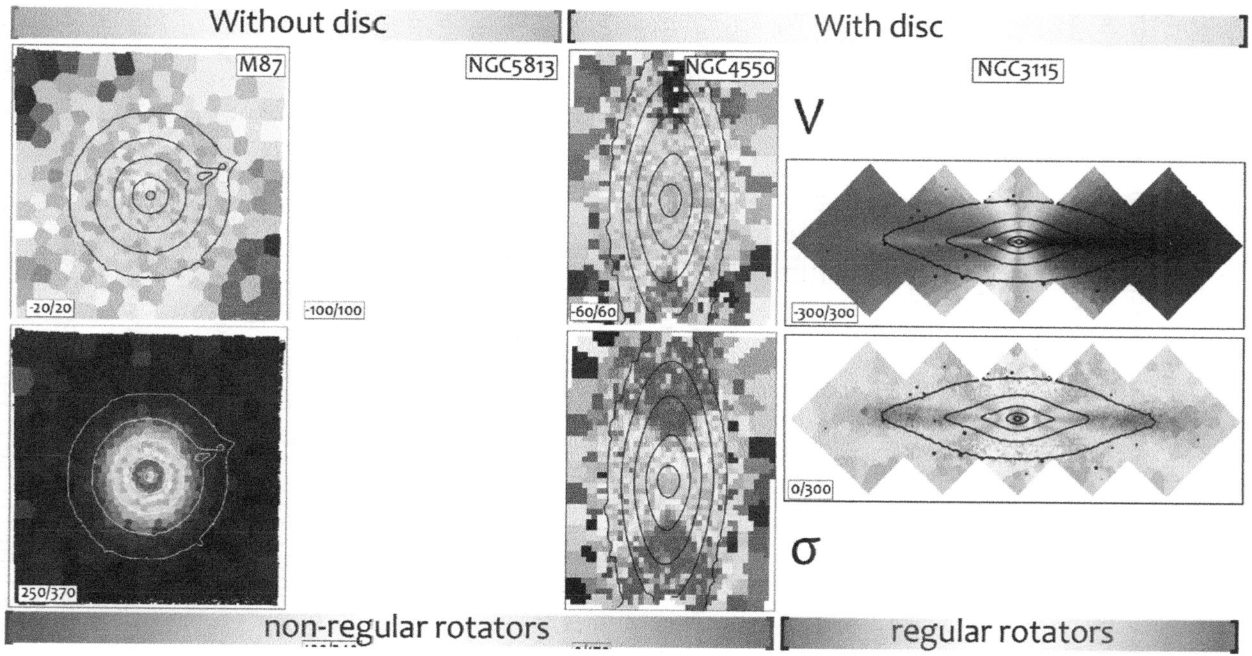

Figure 3.9. *Velocity (top) and velocity dispersion fields (bottom) for regular and non-regular rotators, ETGs with and without discs. NGC4550 is an example of a non-regular rotator with two discs counter-rotating relative to one another, implying a low average velocity and two dispersion peaks along the major axis (this category of galaxy is sometimes described as "2-sigma"). Data taken from (left to right, Emsellem* et al. *2014; Krajnović* et al. *2015; Emsellem* et al. *2007; Guérou* et al. *2016) and inspired by Cappellari (2016)*

By taking advantage of the first sample of 48 ETGs, the SAURON project confirmed the previously observed trends, this time with spectroscopic data covering a bidimensional field. Establishing stellar velocity fields for these ETGs revealed an apparent dichotomy between galaxies with a so-called "regular" velocity field and galaxies with an irregular velocity field (with angular twists in the kinematic axis, or decoupled components). An empirical method of discrimination was formulated to separate these two families into "slow rotators" (SRs) and "fast rotators" (FRs) based on measurements of the $\lambda_R(R_e)$ parameter and the ellipticity $\epsilon = 1 - b/a$, where a and b are the characteristic lengths of the photometric major and minor axes. The parameter λ_R is simply a refinement of the integrated measurement of $(V/\sigma)_{R_e}$ as recommended by Binney (2005), satisfying $\lambda_R = \langle R|V|\rangle / \langle R\sqrt{V^2+\sigma^2}\rangle$. Here, $R \cdot |V|$ serves as a proxy for the angular momentum of the stars, and the denominator is a normalization term proportional to the dynamical mass. Weighting by the distance to the center of the galaxy has the advantage of reducing the role of central structures, which poorly represent the global dynamics of the system. Although it is more robust, λ_R is not fundamentally different from V/σ, and these quantities satisfy the approximate relation $\lambda_R \approx k \times (V/\sigma)/\sqrt{1+k^2 \times (V/\sigma)^2}$ (where $k \sim 1.1$).

A noteworthy result is that it does not seem possible to reconcile these two "kinematic" families with a single morphological and dynamical family. Indeed, fast-rotating ETGs (and spiral galaxies) can be represented by an axisymmetric oblate stellar system (possibly with added bars) with maximum anisotropy $\delta = k_\epsilon \cdot \epsilon$, where $k_\epsilon \sim 0.6$ (magenta lines in Figure 3.10; note that $k_\epsilon \sim 0.8$ is the theoretical maximum for these stellar systems). But this family (shown as a magenta surface in Figure 3.10) certainly does not account for the SRs that do not match this description. Another way to illustrate this point is to perform a simple dynamical model. If we consider a model with a fixed ratio $\kappa \equiv V_{\rm obs}/V_{\rm theo}$ between the observed velocity $V_{\rm obs}$ and the theoretical velocity $V_{\rm theo}$, constrained by $\sigma_R = \sigma_\theta$, it can be shown that the regular rotators (E or S0) are noticeably grouped around $\kappa = 1$ and the non-regular rotators have values of κ between 0 and 0.8 (Figure 3.10). This result has two major consequences: the separation between E and S0 does not seem relevant (for their kinematics), and the families of regular and non-regular rotators, associated with FRs and SRs, respectively, cannot be reconciled dynamically. FRs are galaxies that tend to be more "oblate" (with or without bars), flattened and anisotropic, whereas SRs are marginally flattened, tending to be more triaxial and not very anisotropic.

SR and FR, therefore, seem to correspond to two large and distinct families of ETG, as had already been suggested in the 1970s by the first spectroscopic studies, now studied in more detail through various observation campaigns, in particular based on IFUs. The study of kinematics has allowed this empirical classification to be made more robust and linked to the physics of stellar systems. The majority of fast-rotating ETGs have significant disc contributions, often have bars and can be associated more directly with spiral galaxies (which are themselves FRs, see

Figure 3.11). FRs span a larger range of values of the apparent flattening (or ellipticity), and their anisotropy correlates relatively well with their intrinsic flattening (which constrains the vertical dispersion σ_z). Most SRs are massive galaxies with limited flattening ($\epsilon < 0.4$), irregular apparent dynamics and negligible disc contributions. The differences between these two families imply that their predominant formation and assembly processes are distinct (see section 3.5). These results have made it possible to propose a new general classification (see Figure 3.12) that reiterates and refines the terms developed by van den Bergh (1976). FRs now include spiral galaxies and ETGs, as well as galaxies with moderate gas content and stellar formation rates ("anemic galaxies"). The branch of SRs therefore only accounts for a minority of ETGs ($\sim$ 5–15%) based on the environment.

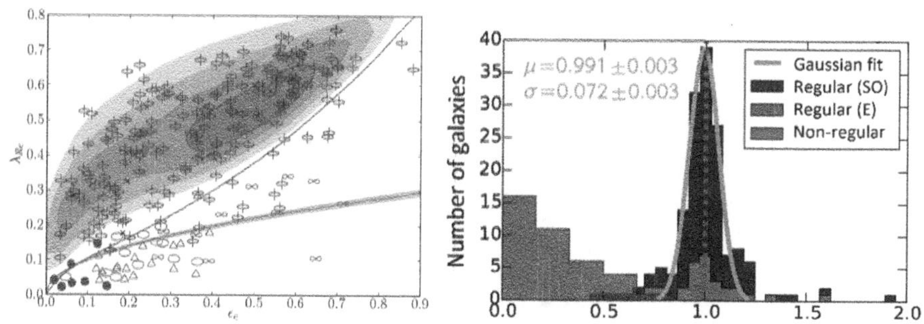

Figure 3.10. *$\lambda_R - \epsilon$ diagrams. Left: Version with 260 ETGs from the Atlas^{3D} sample. The magenta line plots the relation $\delta = k_\epsilon \cdot \epsilon$, where $k_\epsilon = 0.65$. The separation between SRs and FRs is given by $\lambda_{R_e} = 0.31 \cdot \sqrt{\epsilon}$. The vast majority of fast rotators are consistent with the location of axisymmetric oblate galaxies (for a random inclination: magenta-colored surface). Reproduced from Emsellem* et al. *(2011). Right: Histogram of values of $\kappa = V_{\rm obs}/V_{\rm theo}$ for $\sigma_\theta = \sigma_R$, for regular rotators (S0 and E, in blue) and non-regular rotators (red). Adapted from Cappellari (2016)*

3.4.3. *Modeling*

Now that a classification of galaxies has been established on the basis of kinematics; it is important to use suitable tools for the dynamical modeling of these stellar systems. In this section, we give a brief overview of the various techniques used to account for the observed kinematics of ETGs.

Information about dynamics is incorporated into the distribution function f. For a gravitational system like an ETG, we can consider as an initial approximation that the system consists of stars and possibly dark matter. Dynamically speaking, these are non-dissipative components, unlike gas (or dust), which can dissipate energy by radiation. If we restrict our attention to a distribution function f that only depends on the positions and velocities (namely, which does not depend on the time t), we obtain a six-dimensional function that can be written as $f(x, y, z, V_x, V_y, V_z)$ for a Cartesian space. Integrating f over the velocities gives the volumic mass density $\rho(x, y, z)$. The gravitational force enables us to relate ρ to the gravitational potential $\Phi(x, y, z)$ using Poisson's equation $\nabla^2\Phi = 4\pi G\rho$. The potential Φ itself appears in the conservation relation of the distribution function f, known as the collisionless Boltzmann equation, which simply states that the distribution function is conserved along trajectories determined by gravity, or in other words that the total derivative of f is zero (see Figure 3.13).

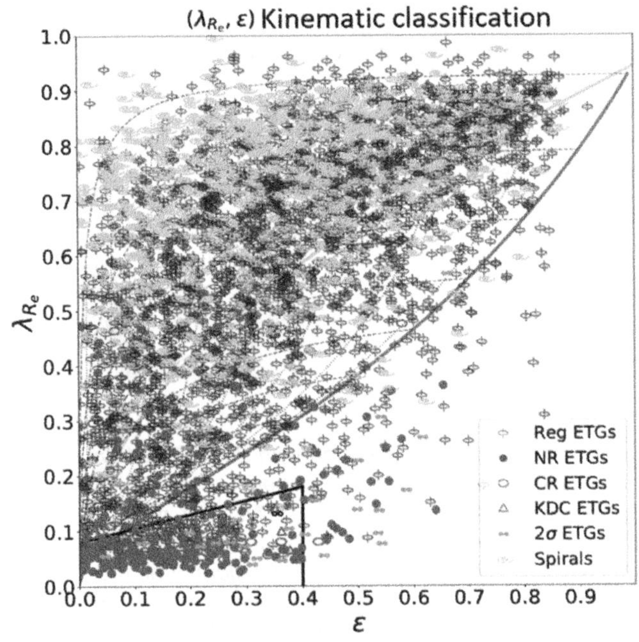

Figure 3.11. $\lambda_R - \epsilon$ *diagram for a sample of more than 4,000 galaxies (Graham* et al. *2018), including both ETGs and spiral galaxies, from the MaNGA survey (the magenta line corresponds to* $k_\epsilon = 0.6$*). Here, the separation between SRs and FRs follows the relation* $\lambda_{R_e} < 0.08 + \epsilon/4$*, and* $\epsilon < 4$*. The various symbols represent different categories of ETG: regular rotators (Reg), non-regular rotators (NR), with a counter-rotating component (CR), or a decoupled component (KDC), 2-*σ *galaxies and spiral galaxies. Reproduced from Graham* et al. *(2018)*

This gives us a system of integral and differential equations relating f, Φ and ρ that we need to solve. A critical aspect of dynamical modeling is to ensure consistency between these quantities. Thus, the distribution function of a given tracer (a population of stars, dark matter) satisfies Boltzmann's equation, but this equation must also include the overall potential, the sum of all massive contributions, in order to be valid. A system whose tracer generates the total potential is said to be self-consistent, which means that its mass distribution fully determines the potential that constrains the trajectories (and therefore the conservation of the distribution function). In this way, the gravitational potential determines which trajectories (orbits) are possible. This distribution function specifies the number of stars (or dark matter particles) for each orbit: the combination of these orbits is called the orbital structure. Galaxy modeling is based on the reconstruction of these quantities, either directly or indirectly, or possibly by working in terms of averages. This reconstruction is important because it allows us to constrain the orbital structure and the dynamics of the galaxy and in principle to draw conclusions about its formation history.

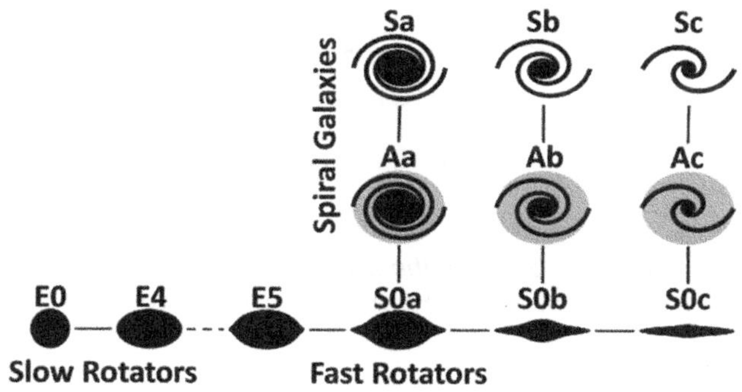

Figure 3.12. *Kinematic classification of galaxies refining the terms developed by van den Bergh (1976) with the introduction of slow and fast rotators. Reproduced from Cappellari et al. (2011)*

Distribution functions, integrals of motion, action angles: One possible method is to directly parameterize the distribution function in terms of known quantities (e.g. integrals of motion, often the energy or the angular momentum) and perform the necessary integrals to obtain quantities such as the projected density or the velocity field that can be compared to observables. The difficulty lies in accounting for the complex structures of ETGs with simple functions. A few techniques have nevertheless delivered results (Dejonghe and Laurent 1991; Emsellem *et al.* 1999; van de Ven *et al.* 2008), but this method remains marginal because it is laborious to apply.

For constrained hypotheses (for example with the assumption that the distribution function only depends on the energy E and the angular momentum L), there are also direct mathematical tools to deduce the distribution function from the density ρ (see for example Hunter and Qian 1993). Unfortunately, these strong hypotheses on f do not appear to reflect the dynamical structure of real galaxies (for instance, $f(E, L_z)$ implies that $\sigma_z = \sigma_R$, among other things). Other representations of the distribution function based on so-called "action" and "angle" coordinates can also be used to account for the dynamics of galaxy systems (Sanders and Binney 2016; Trick *et al.* 2019), but these techniques are generally only used for systems whose stars have been resolved (e.g. the Milky Way).

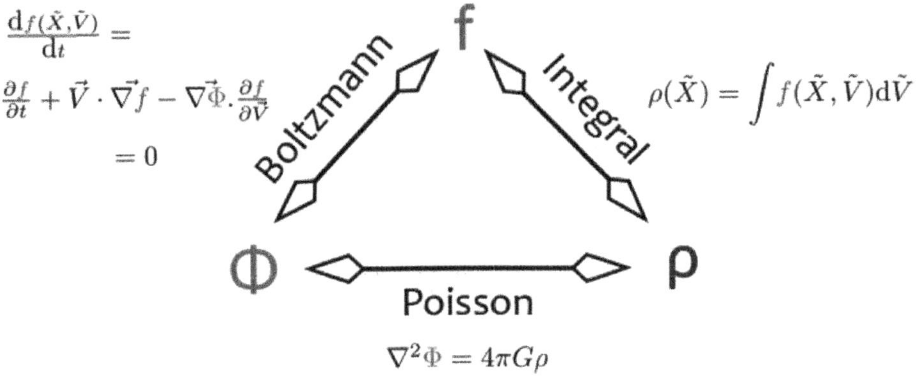

Figure 3.13. *Relations between the distribution function f, the gravitational potential Φ and the density ρ. Here, $\tilde{X} \equiv (x, y, z)$ and $\tilde{V} \equiv (V_x, V_y, V_z)$*

Hydrodynamical equations: One widespread technique uses a shortcut to predict the kinematics of gravitational systems: the so-called Jeans or hydrodynamical equations. The goal is to eliminate the dependency of the distribution function on the velocities by integrating it over the velocity space (V_x, V_y, V_z). By taking the integral of order 0 of the collisionless Boltzmann equation ($\int f d\tilde{V} = 0$), we obtain a simple continuity equation. By taking the integral of order 1 of the Boltzmann equation over the velocities ($\int f V_i d\tilde{V} = 0$, where V_i is one of the three velocity components), we obtain three equations involving the first and second moments of the velocity (the average velocity, and the dispersion along the three axes). In principle, serial integration at every order n is necessary to solve the system. Adding symmetry constraints to the problem allows us to close the system of equations and obtain three so-called Jeans equations (Jeans 1915). Choosing simple symmetries (spherical, cylindrical) leads to viable practical applications. In this way, from a model of the mass distribution ρ, we can obtain a set of predictions of the desired observables. Nevertheless, the theoretical values depend on a range of parameters if we start with an observable quantity such as the distribution of projected light: the inclination of the galaxy, the M/L ratio being considered (fixed or variable) and the anisotropy. Practical applications with ETGs have been systematically performed on sufficiently large samples (for example Li *et al.* 2019).

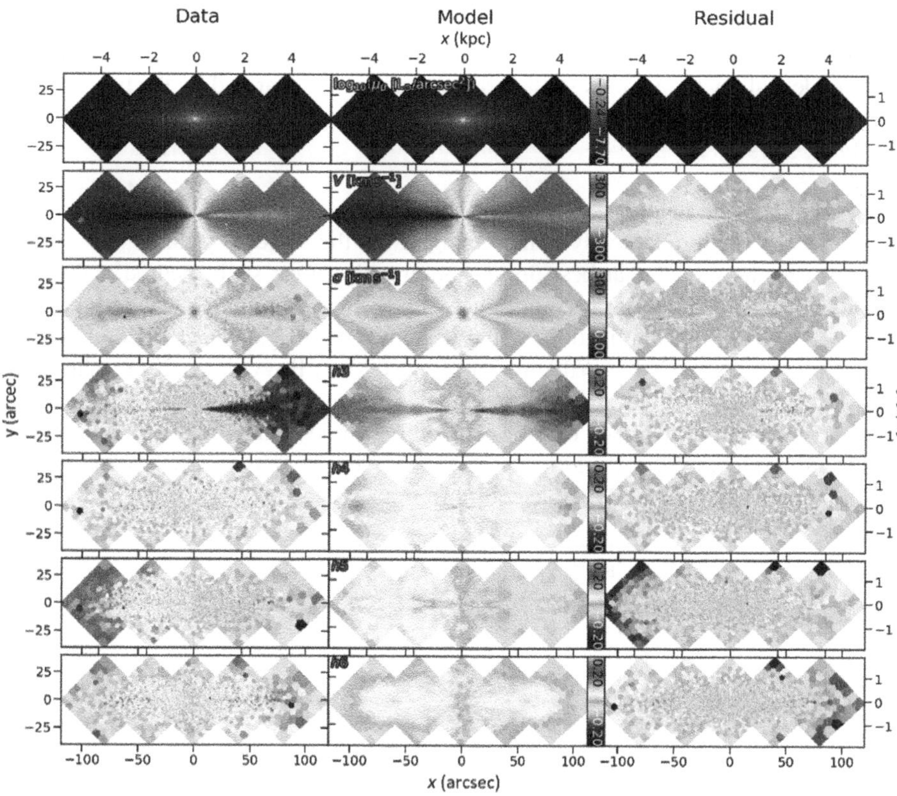

Figure 3.14. *Schwarzschild model of NGC 3115. From top to bottom: Velocity, dispersion, higher order velocity moments (Gauss–Hermite h_3, h_4, h_5 and h_6. Left: Symmetrized data. Center: Model. Right: Residuals. Reproduced from Poci et al. (2019)*

Schwarzschild, "made-to-measure", N-body methods: The method introduced by Schwarzschild (1979, 1982) uses the fact that dynamical time scales are sufficiently large to approach the construction of a stellar system by superimposing a set of trajectories or orbits. For a given potential, the admissible (but possibly non-realized) trajectories are fixed, so we must find the weight of each orbit for the system in question. To fit any density and velocity observables, we simply need to know the (density and velocity) contribution of each orbit on the sky from a library constructed from the potential whose combination minimizes the difference with the observed values. This is a linear problem that can be solved with simple minimization algorithms. There are various difficulties: choosing the library, handling uncertainties, establishing an initial estimate of the potential. Nevertheless, this method is very powerful, as it can be adapted to both simple systems (e.g. spherical) and complex systems (triaxial, with bars, etc.). The Schwarzschild method has been successfully applied to samples of several hundreds of galaxies (e.g. Zhu *et al.* 2018): these models allow the various stellar components to be separated by the type of orbit that supports them, or alternatively by their angular momentum (see Figure 3.14 and Poci *et al.* 2019).

Another method involves calculating the sum of the contributions of each orbit by fitting their weight adaptively when integrating the trajectory until the system converges to the model that best fits the observables. One such method, called the "made-to-measure" method, was introduced by Syer and Tremaine (1996) and more recently developed for applications with different types of galaxies (de Lorenzi *et al.* 2008; Long and Mao 2012; Williams and Evans 2015; Bovy *et al.* 2018; Zhu *et al.* 2018; Blaña Díaz *et al.* 2018). The "made-to-measure" technique lies halfway between the Schwarzschild method and the N-body method, which involves consistently tracking the evolution of a system of N gravitational particles and therefore goes beyond just dynamic equilibrium.

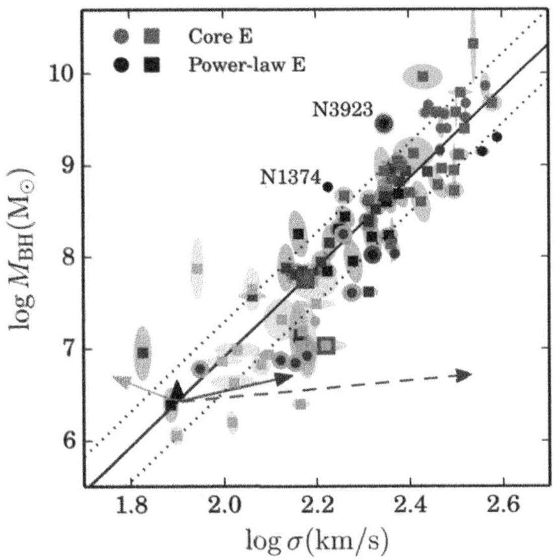

Figure 3.15. *Compilation of mass measurements of supermassive black holes in nearby galaxies, illustrated with the black hole mass-dispersion relation* $M_{\rm BH}$-σ *. Reproduced from Saglia* et al. *(2016)*

3.4.4. *Supermassive black holes*

Studies of the centers of galaxies have almost systematically revealed the presence of supermassive black holes ranging from 10^5 to 10^{10} $M_\odot$ and have shown the role of these black holes in the evolution of galaxies (Heckman and Best 2014). Most of these studies used Jeans or Schwarzschild models based on stellar kinematics (long-slit or field integral data) and occasionally gas kinematics (Ford *et al.* 1994) or even maser emissions (Miyoshi *et al.* 1995) and, very recently, direct imaging (Event Horizon Collaboration *et al.* 2019). The very elevated central rotation or velocity dispersion values often signify high mass: modeling these kinematics makes it possible to constrain whether a point mass may indeed be contributing to the central potential well. Adaptive optical data are often needed to obtain the resolution required by this kind of study, because, beyond a few parsecs, the gravitational contribution of the black hole is typically small enough to be undetectable, unlike the nuclear stellar component.

Dynamical modeling of kinematic data has allowed more than a hundred of these black holes to be measured in ETGs (see the review by Kormendy and Ho 2013). As soon as small samples became available, a relation was observed between the mass of the black hole and the host galaxy, first as a relation with the mass of the spheroid (Magorrian *et al.* 1998) (the black hole accounts for a few thousandths of this mass), then with less scatter using the velocity dispersion (Ferrarese and Merritt 2000). This relation has now been confirmed and does not seem to depend on the type of galaxy (LTG or ETG). The correlation with the spheroid is controversial, even though galaxy/black hole coevolution processes often appeal to the coupled growth of the spheroid and the central mass. Recent compilations have illustrated this correlation across four orders of magnitude of black hole mass from around 10^6 to 10^{10} $M_\odot$ (Figure 3.15, Saglia *et al.* 2016). Since ETGs are mostly massive galaxies and have a spheroid contribution that frequently dominates these masses, they populate the upper mass region of the M_{BH}-σ relation. The most massive galaxies also have stellar "cores" at their centers (see section 3.2.5), which seems to indicate that, at large masses, a significant fraction of the growth of ETGs (and their black holes) arises from mergers of galaxies.

3.5. Formation and evolution processes

This section examines the major formation processes of ETGs in more detail within a general context. We begin with a few simple characteristics of galaxies: the effective radius and the stellar mass.

3.5.1. *Perspective at $z = 0$: the mass–radius plane*

Figure 3.16 illustrates the distribution of different types of galaxies with redshift $z \sim 0$: late-type (spiral), early-type (ETGs), compact or low-luminosity (dwarf).

ETGs are therefore the most compact stellar systems at a given mass (excluding compact galaxies, which are themselves often classified as ETGs), or alternatively the most massive galaxies of a given characteristic scale. In this diagram, the contribution of the disc increases orthogonally to the curves of constant dispersion in the direction of larger radii and smaller masses.

There is a principal sequence that goes from dwarf galaxies up to the most luminous galaxies at around $3 \cdot 10^{10}$ $M_{\odot}$ along which the variation in the effective radius is positive (galaxies are generally more extended at larger masses) but relatively minor (factor of 3 for a mass variation of more than 3 orders of magnitude). Above $3 \cdot 10^{10}$ $M_{\odot}$, we observe an abrupt change of slope in the distribution of galaxies, which then converges to a power law relation $R_e \propto M_{\star}^{0.75}$. This slope is slightly lower than predicted by the scalar virial theorem for curves of constant dispersion ($R_e \propto M$) if the mass is dominated by the stellar component. In this intermediate mass region, we mostly observe FRs. Above around $2 \cdot 10^{11}$ $M_{\odot}$, the galaxy population is almost exclusively made of ETGs, and largely dominated by SRs. Recent theoretical and observational studies have suggested that the most massive galaxies ($M_{\star} \sim 10^{12}$ $M_{\odot}$ and above) tend to have more "prolate" kinematics (average rotation around the major axis), suggesting that, to reach such an extreme mass, a major merger of two already massive galaxies (SRs) is necessary.

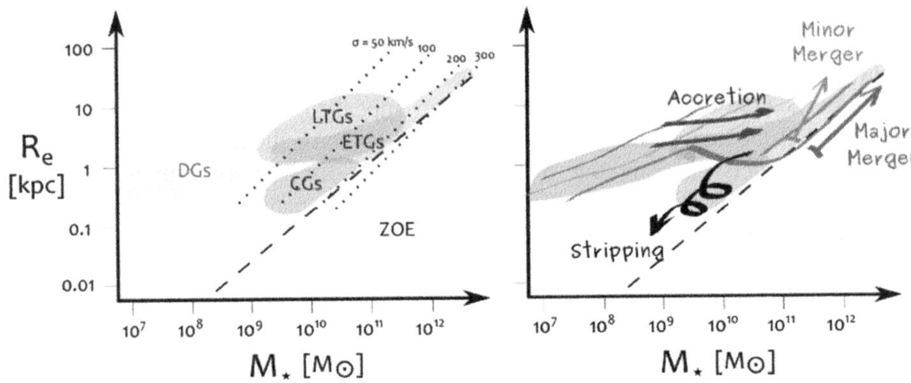

Figure 3.16. *Illustration of the distribution of different types of galaxies as a radius–mass diagram (left: LTG = late-type galaxies; CG = compact galaxies; DG = dwarf galaxies; ZOE = zone of exclusion), as well as a simplified diagram (since the distribution of galaxies in this plane evolves over time, see Figure 3.17) of the impact of certain evolutionary processes (right: accretion, minor or major merger, stripping by interaction). The dotted lines (left) correspond to the curves of constant dispersion σ*

It is interesting to note that compact galaxies are found in the prolongation of ETGs along the lines of constant dispersion σ. A simple prediction based on the metallicity–dispersion relation is that compact galaxies should have metallicities similar to ETGs of equal dispersion, despite being significantly less massive; this result has been confirmed by spectroscopic studies (e.g. Guérou *et al.* 2015).

3.5.2. *Growth, mergers and transformations*

Over time, there is a generally increasing trend in the stellar mass of galaxies. In the left region (low masses) of the $R_e - M_\star$ plane, growth generally occurs through stellar formation within the system, the vast occurring in the galaxy disc. This growth may be slow or fast depending on the secular processes of the galaxy (e.g. formation of bars) and the rate of accretion of gas (e.g. reservoir of neutral gas and interactions). The abrupt change in slope above $3 \cdot 10^{10}$ $M_\odot$ indicates a significant contribution from galaxy mergers (minor or major). Indeed, a merger between purely stellar systems will tend to rapidly increase the mass and characteristic radius of the system according to a law $R \propto M_\star^\gamma$, where γ varies depending on the type of merger: for a major merger, γ is close to 1 (but less than 1 due to the non-total relaxation of the system), and for several minor mergers, γ will be closer to 2 (Naab *et al.* 2009; Bezanson *et al.* 2009). In other words, the radii and masses increase almost equivalently when two galaxies of similar mass merge, and the radius increases approximately like the square of the mass in the case of accretion of small satellites, with the average stellar density of the final system diminishing accordingly. For interactions or mergers between galaxies with significant gas contributions, the dissipative properties of this gas may lead to the formation of stars in the more central regions and will therefore tend to increase the average stellar density. In general, this concentration is associated with a "nuclear" component, or a spheroid component if it is more spread out.

Galaxies can also lose mass in several ways. Interactions with other galaxies may lead to the stripping of the least bound (outermost) stellar or gaseous components. Interactions with the hot gas of a galaxy cluster may lead to sweeping induced by the ram pressure of the environment. Modeling the stripping process shows that the radius and mass can be significantly reduced, even down to the scales of a galaxy nucleus with a characteristic radius of a few dozen parsecs (e.g. Pfeffer and Baumgardt 2013).

This view of the transformation processes of galaxies allows us to draw several conclusions, as illustrated in Figure 3.16. ETGs of low or intermediate mass must have grown primarily by internal stellar formation; thus, their distribution may be superimposed with spiral galaxies. The low proportion of gas in ETGs can therefore be explained either as a reservoir that has been exhausted by stellar formation (given that the typical characteristic time of exhaustion for a galaxy is ~ 2 Gyr) without any significant renewal from external accretion or by a more active transformation through interactions with other galaxies or the inter-galactic environment (for example within clusters). At masses above $\sim 3 \cdot 10^{10}$ $M_\odot$.

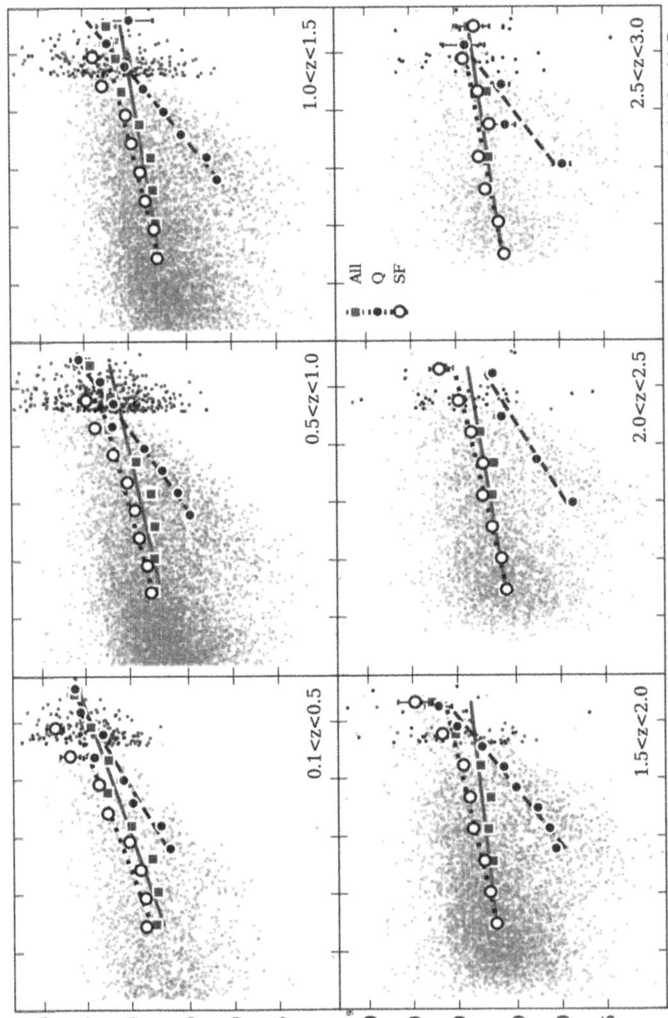

Figure 3.17. *Passive galaxies (red points) and galaxies with star formation (blue points) in a radius–mass plane as a function of the redshift (shown in the bottom right of each panel) and the average relations. Reproduced from Mowla et al. (2019)*

ETGs must have assembled a significant part of their stellar mass through mergers, either major or minor. Finally, above 10^{11} $M_\odot$, the radius interval at a given mass becomes narrower, which indicates that major mergers of massive galaxies (and therefore primarily ETGs) must have played an important role.

Figure 3.16 shows the evolution of galaxies in at least three ways:

– The distribution of galaxies in the $R_e - M_\star$ plane has evolved significantly over time (Figure 3.17). At $z = 2$, for example, the population of ETGs was generally represented by systems that were more compact by a factor of 2–10 but which already spanned masses greater than $10^{10.5}$ $M_\odot$.

– The galaxies populating the $R_e - M_\star$ plane depend on the redshift themselves (e.g. gas content, morphology and internal dynamics). As a result, we cannot identify a galaxy of the plane at $z = 0$ as the progenitor of another galaxy at the same redshift.

– Although the stellar mass $M_\star$ of ETGs generally tends to increase with time, $M_\star$ is not a monotonic function with respect to time. Furthermore, the evolutionary paths of galaxies are complex (see Figure 3.18), and the impact of a merger of galaxies depends on more than just the mass ratio of the two progenitors; it also depends on the gas content and the orbital angular momentum (which is often greater than the sum of the internal angular momenta of the galaxies themselves).

3.5.3. *Ex situ versus in situ*

Today, the formation and evolutionary history of ETGs is often characterized by two phases: a formation phase with a large redshift, therefore likely to be rapid and involve a large fraction of gas, and an assembly phase that is much more stretched out over time. The first phase corresponds to a growth of the stellar mass by accretion of external gas and the internal, or *in situ*, formation of new stars. The second, longer phase is dominated by the accumulation of stars formed in other systems, or *ex situ* (Oser *et al.* 2010). The fraction of *ex situ* stellar mass correlates strongly with the total mass at $z = 0$; thus, more than half of the stars of the ETGs that dominate above $2 \cdot 10^{11}$ $M_\odot$ formed *ex situ* at $z = 0$. This is perfectly consistent with the fact that the majority of stars in the most massive galaxies formed above $z = 2$; these stars form at an early stage in the history of the Universe, in lower mass systems, assembled after $z = 2.5$ on average. This illustrates the effect of archeological “downsizing”: even though the lower mass galaxies begin to form earlier (their halos contract more quickly), stellar formation is more stretched out and the age of the average stellar populations is lower. Nevertheless, there is a relation between the fraction of *ex situ* stellar mass and the angular momentum in the central regions (within $1R_e$): up to masses of $M_\star \sim 2 \cdot 10^{11}$ $M_\odot$, the population of ETGs is dominated by FRs. Above this point, it is dominated by SRs.

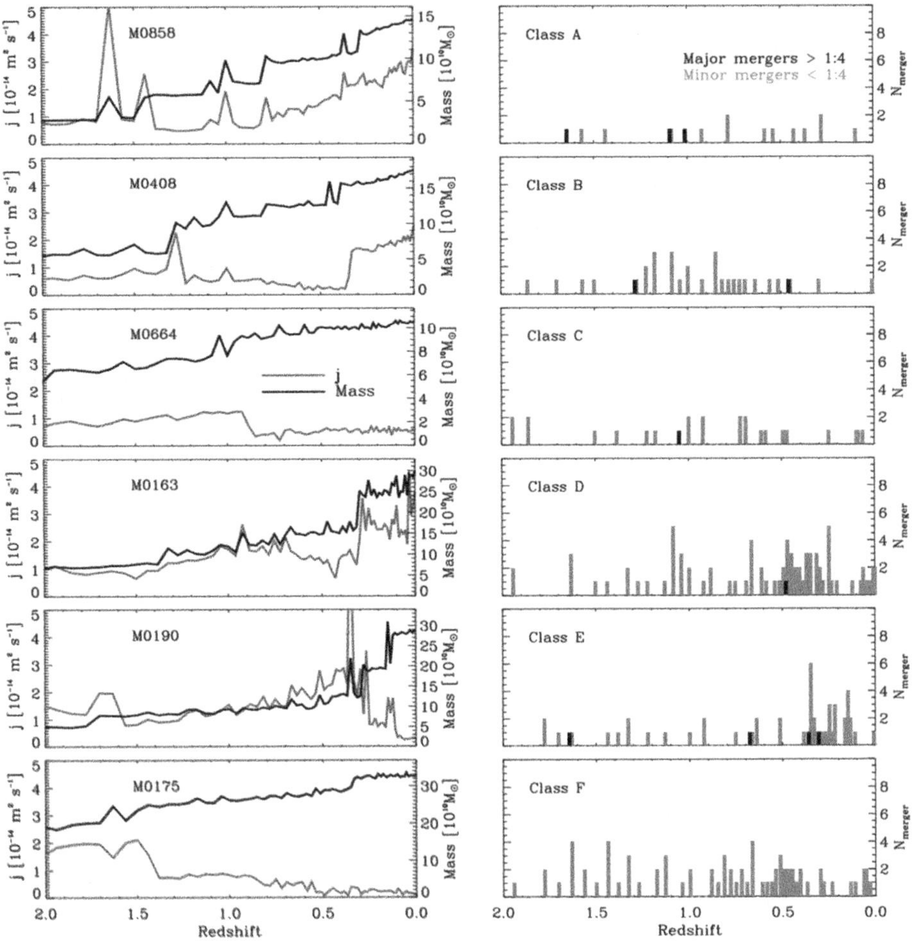

Figure 3.18. *Evolution of galaxies in a simulation with a cosmological context. Left: Evolution of the angular momentum (green line) and the stellar mass (black line) as a function of the redshift for six prototypical galaxies of a simulation with a cosmological context. Right: Minor mergers (orange) and major mergers (black) as a function of the redshift. This figure illustrates the complexity of the evolutionary paths and the many-layered relationship between mergers and variations of the angular momentum. Reproduced from Naab* et al. *(2014)*

3.5.4. *Environment*

The fraction of ETGs depends strongly on their environment, which is usually measured in terms of the local density in galaxies. Thus, a decrease in the fraction of spiral galaxies and an increase in the fraction of ETGs is observed as the density increases (Dressler 1980). This "morphology-density" relation was originally interpreted to suggest a direct transformation between spiral galaxies and lenticular (S0) galaxies. Today, it is not clear what fraction of ETGs with discs is simply "passive" transformations of spiral galaxies where the majority of the gas reservoir have been removed. Given that the ratio of ETGs seems to have increased by a factor of around 3 since $z = 1$, this transformation must have unfolded over time scales of the order of Gyrs. ETGs with discs have similar $\lambda_R(R_e)$ values to Sa spiral galaxies, but they are generally more concentrated. The transformation of spiral galaxies into ETGs within clusters or groups of galaxies therefore implies both a restriction of the gas reservoir and an increase in the global concentration of the galaxy. According to the morphological classification of ETGs (E and S0), a significant fraction of elliptical galaxies are observed at every galaxy density. Together with the kinematic classification (SR/FR), we obtain a more robust relation called the "kinematic morphological-density" relation: the fraction of spiral galaxies in a cluster decreases remarkably regularly as the galaxy density increases, a decrease that is almost exactly compensated by the increasing fraction of fast-rotating ETGs, and SRs are found almost exclusively only at the center of local overdensities of galaxies (potential wells in clusters, center of groups, see Figure 3.20).

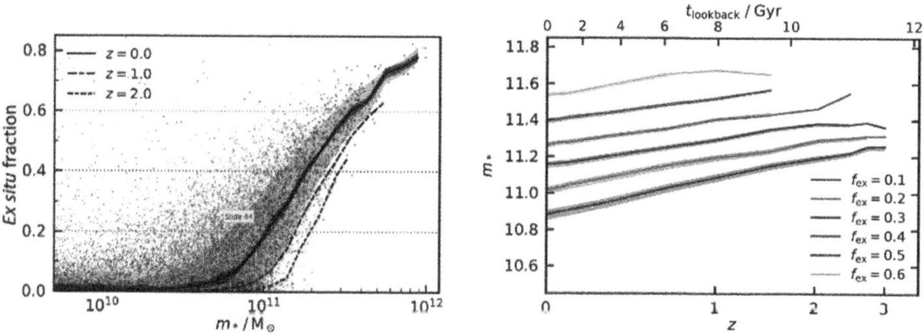

Figure 3.19. *Fraction of ex situ mass in ETGs as a function of the stellar mass (left) and the redshift (right). Reproduced from Moster* et al. *(2019)*

Recent studies based on large samples of galaxies available because of IFU campaigns (e.g. MaNGA) seem to suggest that the segregation of SRs is strictly determined by the relative density of galaxy overdensities; within a given group or cluster, SRs are found at the minimum of the potential well (Graham *et al.* 2019). This is a key insight to understand the formation of these systems.

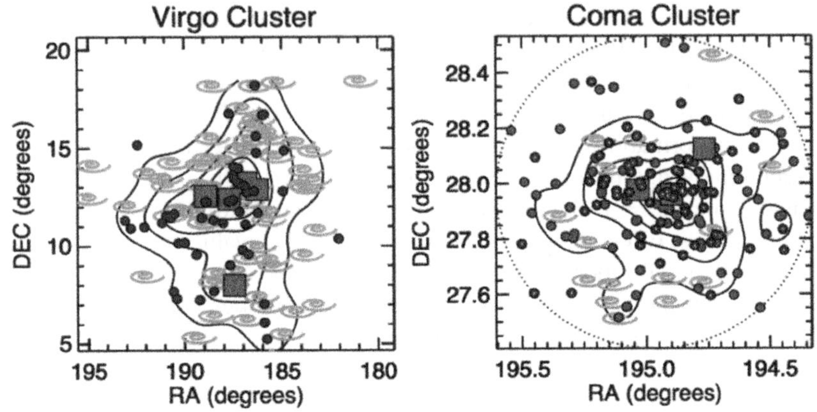

Figure 3.20. *Two clusters with different richnesses (Virgo on the left, Coma on the right), showing the distribution of spiral galaxies (green symbols), fast rotators (blue points) and slow rotators (red squares). These two diagrams clearly show the discrimination into types as a function of the local density of galaxies. Reproduced from Cappellari* et al. *(2013)*

3.6. Conclusion

In this chapter, we briefly discussed the properties of ETGs. Figure 3.21 summarizes the broad trends of these systems at different scales and masses, once again with the radius–mass plane (R_e-$M_\star$). Measuring the dispersion of the characteristic velocities (with a simple application of the scalar virial theorem) seems to represent the best "proxy" for the variation of the average age, metallicity, mass–luminosity ratio (M/L), color and even the IMF or the relative abundance of α elements. This diagram also excellently illustrates the above-mentioned transition from spiral galaxies to fast-rotating ETGs, then SRs, as the systems become more and more massive. As we pass from galaxies with a disc morphology to disc galaxies with an increasing spheroid contribution, we transition from axisymmetric systems (with or without bars) to slightly triaxial systems (and likely prolate systems at very large masses).

To complement this chapter, the reader can also refer to various excellent reviews of the formation and evolution of ETGs, including the following articles:

– "An updated view of the structure and kinematics of early-type galaxies from integral-field spectroscopy" Cappellari (2016).

– "Galaxy masses" Courteau *et al.* (2014).

– "General review on the structure and formation of ellipticals" Kormendy *et al.* (2009).

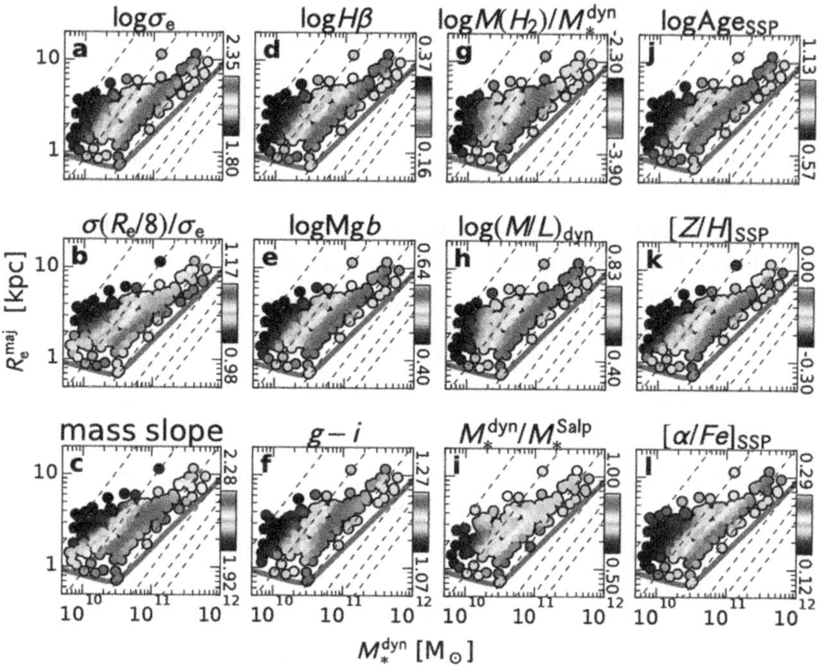

Figure 3.21. *Evolution of measured quantities for a sample of ETGs in the plane defined by the effective radius R_e and the dynamical mass $M_{\rm dyn}$: for each galaxy (circle), the values are locally averaged to extract general trends. Top to bottom, left to right: Dispersion σ_e, H_β index, fraction of H_2, average stellar age, dispersion ratio between $R_e/8$ and R_e, Mg_b index, dynamical M/L ratio, average metallicity, slope of the mass profile, color $g-i$, fraction of dynamical mass relative to stellar mass (Salpeter IMF), abundance of alpha elements relative to iron α/Fe. Reproduced from Cappellari (2016)*

3.7. References

Bacon, R., Copin, Y., Monnet, G., *et al.* (2001). The SAURON project-I. The panoramic integral-field spectrograph. *MNRAS*, 326, 23–35.

Bastian, N., Covey, K.R., Meyer, M.R. (2010). A universal stellar initial mass function? A critical look at variations. *Annual Review of Astronomy and Astrophysics*, 48, 339.

Bellstedt, S., Forbes, D.A., Romanowsky, A.J., *et al.* (2018). The SLUGGS survey: A comparison of total-mass profiles of early-type galaxies from observations and cosmological simulations, to 4 effective radii. *Monthly Notices of the Royal Astronomical Society*, 476(4), 4543.

Bernardi, M., Shankar, F., Hyde, J.B., *et al.* (2010). Galaxy luminosities, stellar masses, sizes, velocity dispersions as a function of morphological type. *Monthly Notices of the Royal Astronomical Society*, 404(4), 2087.

Bernardi, M., Sheth, R.K., Annis, J., *et al.* (2003). Early-type galaxies in the Sloan Digital Sky Survey. III. The fundamental plane. *The Astronomical Journal*, 125(4), 1866.

Bertola, F., Capaccioli, M. (1975). Dynamics of early type galaxies. I. The rotation curve of the elliptical galaxy NGC 4697. *The Astrophysical Journal*, 200, 439.

Bezanson, R., van Dokkum, P.G., Tal, T., *et al.* (2009). The relation between compact, quiescent high-redshift galaxies and massive nearby elliptical galaxies: Evidence for hierarchical, inside-out growth. *The Astrophysical Journal*, 697(2), 1290.

Binney, J. (1978). On the rotation of elliptical galaxies. *Monthly Notices of the Royal Astronomical Society*, 183, 501.

Binney, J. (2005). Rotation and anisotropy of galaxies revisited. *Monthly Notices of the Royal Astronomical Society*, 363(3), 937–942.

Blaña Díaz, M., Gerhard, O., Wegg, C., *et al.* (2018). Sculpting Andromeda - made-to-measure models for M31's bar and composite bulge: Dynamics, stellar and dark matter mass. *Monthly Notices of the Royal Astronomical Society*, 481, 3210.

Bovy, J., Kawata, D., Hunt, J.A.S. (2018). Made-to-measure modelling of observed galaxy dynamics. *Monthly Notices of the Royal Astronomical Society*, 473, 2288.

Brodie, J.P., Strader, J. (2006). Extragalactic globular clusters and galaxy formation. *Annual Review of Astronomy and Astrophysics*, 44(1), 193.

Bundy, K., Bershady, M.A., Law, D.R., *et al.* (2015). Overview of the SDSS-IV MaNGA survey: Mapping nearby galaxies at Apache point observatory. *The Astrophysical Journal*, 798, 7.

Burstein, D., Davies, R.L., Dressler, A., *et al.* (1988). Global stellar populations of elliptical galaxies: A. Optical properties. *Towards Understanding Galaxies at Large Redshift*, 141, 17.

Calette, A.R., Avila-Reese, V., Rodríguez-Puebla, A., *et al.* (2018). The HI- and H^2-to-stellar mass correlations of late- and early-type galaxies and their consistency with the observational mass functions. *Revista Mexicana de Astronomia y Astrofisica*, 54, 443.

Cappellari, M. (2016). Structure and kinematics of early-type galaxies from integral-field spectroscopy. *Annual Review of Astronomy and Astrophysics*, 54(1), 597–665.

Cappellari, M., Emsellem, E., Bacon, R., *et al.* (2007). The SAURON project–X. The orbital anisotropy of elliptical and lenticular galaxies: Revisiting the (V/ ,) diagram with integral-field stellar kinematics. *Monthly Notices of the Royal Astronomical Society*, 379(2), 418–444.

Cappellari, M., Emsellem, E., Krajnović, D., *et al.* (2011). The ATLAS3D project–I. A volume-limited sample of 260 nearby early-type galaxies: Science goals and selection criteria. *Monthly Notices of the Royal Astronomical Society*, 413(2), 813.

Cappellari, M., Emsellem, E., Krajnović, D., *et al.* (2011). The ATLAS3D projectVII. A new look at the morphology of nearby galaxies: The kinematic morphologydensity relation. *Monthly Notices of the Royal Astronomical Society*, 416(3), 1680–1696.

Cappellari, M., McDermid, R.M., Alatalo, K., *et al.* (2013). The ATLAS3D projectXX. Masssize and massσ distributions of early-type galaxies: Bulge fraction drives kinematics, mass-to-light ratio, molecular gas fraction and stellar initial mass function. *Monthly Notices of the Royal Astronomical Society*, 432(3), 1862–1893.

Chabrier, G. (2003). Galactic stellar and substellar initial mass function. *Publications of the Astronomical Society of the Pacific*, 115(809), 763.

Chabrier, G., Hennebelle, P., Charlot, S. (2014). Variations of the stellar initial mass function in the progenitors of massive early-type galaxies and in extreme starburst environments. *The Astrophysical Journal*, 796(2), 75.

Conroy, C. (2013). Modeling the panchromatic spectral energy distributions of galaxies. *Annual Review of Astronomy and Astrophysics*, 51(1), 393–455.

Conroy, C., van Dokkum, P.G. (2012). The stellar initial mass function in early-type galaxies from absorption line spectroscopy. II. Results. *The Astrophysical Journal*, 760(1), 71.

Courteau, S., Cappellari, M., de Jong, R.S., *et al.* (2014). Galaxy masses. *Reviews of Modern Physics*, 86(1), 47–119.

Croom, S.M., Lawrence, J.S., Bland-Hawthorn, J., *et al.* (2012). The Sydney-AAO multi-object integral field spectrograph. *Monthly Notices of the Royal Astronomical Society*, 421, 872.

Davies, R.L., Efstathiou, G., Fall, S.M., *et al.* (1983). The kinematic properties of faint elliptical galaxies. *The Astrophysical Journal*, 266, 41.

Davis, T.A., Bureau, M., Young, L.M., *et al.* (2011). The ATLAS^{3D} project–V. The CO Tully-Fisher relation of early-type galaxies. *Monthly Notices of the Royal Astronomical Society*, 414, 968.

Davis, T.A., Rowlands, K., Allison, J.R., *et al.* (2015). Molecular and atomic gas in dust lane early-type galaxies–I. Low star formation efficiencies in minor merger remnants. *Monthly Notices of the Royal Astronomical Society*, 449(4), 3503.

de Lorenzi, F., Gerhard, O., Saglia, R.P., *et al.* (2008). Dark matter content and internal dynamics of NGC 4697: NMAGIC particle models from slit data and planetary nebula velocities. *Monthly Notices of the Royal Astronomical Society*, 385(4), 1729.

de Zeeuw, P.T., Bureau, M., Emsellem, E., *et al.* (2002). The SAURON project–II. Sample and early results. *Monthly Notices of the Royal Astronomical Society*, 329, 513–530.

Dejonghe, H., Laurent, D. (1991). Abel models for triaxial systems. *Monthly Notices of the Royal Astronomical Society*, 252, 606.

Djorgovski, S., Davis, M. (1987). Fundamental properties of elliptical galaxies. *The Astrophysical Journal*, 313, 59.

Dressler, A. (1980). Galaxy morphology in rich clusters: Implications for the formation and evolution of galaxies. *The Astrophysical Journal*, 236, 351.

Dressler, A. (1987). The D n- sigma relation for bulges of disk galaxies: A new, independent measure of the Hubble constant. *The Astrophysical Journal*, 317, 1.

Dressler, A., Lynden-Bell, D., Burstein, D., *et al.* (1987). Spectroscopy and photometry of elliptical galaxies. I. New distance estimator. *The Astrophysical Journal*, 313, 42.

Emsellem, E., Cappellari, M., Krajnović, D., *et al.* (2011). The ATLAS3D project–III. A census of the stellar angular momentum within the effective radius of early-type galaxies: Unveiling the distribution of fast and slow rotators: Census of the stellar angular momentum in ETGs. *Monthly Notices of the Royal Astronomical Society*, 414(2), 888–912.

Emsellem, E., Cappellari, M., Krajnović, D., *et al.* (2007). The SAURON projec–IX. A kinematic classification for early-type galaxies. *MNRAS*, 379, 401–417.

Emsellem, E., Dejonghe, H., Bacon, R. (1999). Dynamical models of NGC 3115. *Monthly Notices of the Royal Astronomical Society*, 303, 495–514.

Emsellem, E., Krajnović, D., Sarzi, M. (2014). A kinematically distinct core and minor-axis rotation: The MUSE perspective on M87. *Monthly Notices of the Royal Astronomical Society*, 445, L79–L83.

Event Horizon Collaboration, T.E.H.T., Akiyama, K., Alberdi, A., *et al.* (2019). First M87 event horizon telescope results. I. The shadow of the supermassive black hole. *The Astrophysical Journal Letters*, 875(1), L1.

Faber, S.M., Dressler, A., Davies, R.L., Burstein, D., Lynden Bell, D., Terlevich, R., Wegner, G. (1987). Global Scaling Relations for Elliptical Galaxies and Implications for Formation. *Nearly Normal Galaxies. From the Planck Time to the Present.* The Eighth Santa Cruz Summer Workshop in Astronomy and Astrophysics, July 21-August 1, 1986, Lick Observatory. Faber, S.M. (ed.), Publisher, Springer-Verlag, New York.

Faber, S.M., Jackson, R.E. (1976). Velocity dispersions and mass-to-light ratios for elliptical galaxies. *The Astrophysical Journal*, 204, 668–683.

Faber, S.M., Tremaine, S., Ajhar, E.A., *et al.* (1997). The centers of early-type galaxies with HST. IV. Central parameter relations. *The Astronomical Journal*, 114, 1771.

Ferrarese, L., Merritt, D. (2000). A fundamental relation between supermassive black holes and their host galaxies. *The Astrophysical Journal*, 539(1), L9.

Ferrarese, L., van den Bosch, F.C., Ford, H.C., *et al.* (1994). Hubble space telescope photometry of the central regions of Virgo Cluster elliptical galaxies. III. Brightness profiles. *The Astronomical Journal*, 108, 1598.

Ferreras, I., Scott, N., La Barbera, F., *et al.* (2019). The SAMI galaxy survey: Stellar population radial gradients in early-type galaxies. *Monthly Notices of the Royal Astronomical Society*, 489, 608.

Ford, H.C., Harms, R.J., Tsvetanov, Z.I., *et al.* (1994). Narrowband HST images of M87: Evidence for a disk of ionized gas around a massive black hole. *The Astrophysical Journal*, 435, L27.

Gallazzi, A., Charlot, S., Brinchmann, J., *et al.* (2006). Ages and metallicities of early-type galaxies in the Sloan Digital Sky Survey: New insight into the physical origin of the colour-magnitude and the Mg^2-Σ^V relations. *Monthly Notices of the Royal Astronomical Society*, 370(3), 1106.

Gebhardt, K., Richstone, D., Ajhar, E.A., *et al.* (1996). The centers of early-type galaxies with HST. III. Non-parametric recovery of stellar luminosity distribution. *The Astronomical Journal*, 112, 105.

Glass, L., Ferrarese, L., Côté, P., *et al.* (2011). The ACS Fornax cluster survey. IV. Deprojection of the surface brightness profiles of early-type galaxies in the Virgo and Fornax clusters: Investigating the core/power-law dichotomy. *The Astrophysical Journal*, 726(1), 31.

Goddard, D., Thomas, D., Maraston, C., *et al.* (2017). SDSS-IV MaNGA: Stellar population gradients as a function of galaxy environment. *Monthly Notices of the Royal Astronomical Society*, 465, 688.

Graham, M.T., Cappellari, M., Bershady, M.A., *et al.* (2019). SDSS-IV MaNGA: New benchmark for the connection between stellar angular momentum and environment: A study of about 900 groups/clusters. *arXiv e-prints*, arXiv:1910.05139.

Graham, M.T., Cappellari, M., Li, H., *et al.* (2018). SDSS-IV MaNGA: Stellar angular momentum of about 2300 galaxies: Unveiling the bimodality of massive galaxy properties. *Monthly Notices of the Royal Astronomical Society*, 477, 4711.

Guérou, A., Emsellem, E., Krajnović, D., *et al.* (2016). Exploring the mass assembly of the early-type disc galaxy NGC 3115 with MUSE. *Astronomy and Astrophysics*, 591, A143.

Guérou, A., Emsellem, E., McDermid, R.M., *et al.* (2015). The next generation Virgo Cluster Survey. XII. Stellar populations and kinematics of compact, low-mass early-type galaxies from Gemini GMOS-IFU spectroscopy. *The Astrophysical Journal*, 804(1), 70.

Harris, W.E. (2010). Massive star clusters in galaxies. *Philosophical Transactions of the Royal Society of London Series A*, 368, 889.

Harris, W.E., Blakeslee, J.P., Harris, G.L.H. (2017). Galactic dark matter halos and globular cluster populations. III. Extension to extreme environments. *The Astrophysical Journal*, 836(1), 67.

Harris, W.E., Harris, G.L.H., Alessi, M. (2013). A catalog of globular cluster systems: What determines the size of a Galaxy's globular cluster population? *The Astrophysical Journal*, 772, 82.

Hayashi, E., Navarro, J.F., Springel, V. (2007). The shape of the gravitational potential in cold dark matter haloes. *Monthly Notices of the Royal Astronomical Society*, 377(1), 50.

Heckman, T.M., Best, P.N. (2014). The coevolution of galaxies and supermassive black holes: Insights from surveys of the contemporary universe. *Annual Review of Astronomy and Astrophysics*, 52, 589.

Hunter, C., Qian, E. (1993). Two-integral distribution functions for axisymmetric galaxies. *Monthly Notices of the Royal Astronomical Society*, 262(2), 401–428.

Jeans, J.H. (1915). On the theory of star-streaming and the structure of the universe. *Monthly Notices of the Royal Astronomical Society*, 76, 70.

Kaviraj, S., Ting, Y.-S., Bureau, M., *et al.* (2012). Galaxy zoo: Dust and molecular gas in early-type galaxies with prominent dust lanes. *Monthly Notices of the Royal Astronomical Society*, 423(1), 49.

Kormendy, J., Bender, R. (1996). A proposed revision of the Hubble sequence for elliptical galaxies. *The Astrophysical Journal*, 464, L119.

Kormendy, J., Djorgovski, S. (1989). Surface photometry and the structure of elliptical galaxies. *Annual Review of Astronomy and Astrophysics*, 27, 235.

Kormendy, J., Fisher, D.B., Cornell, M.E., *et al.* (2009). Structure and formation of elliptical and spheroidal galaxies. *The Astrophysical Journal Supplement Series*, 182(1), 216.

Kormendy, J., Ho, L.C. (2013). Coevolution (or not) of supermassive black holes and host galaxies. *Annual Review of Astronomy and Astrophysics*, 51(1), 511.

Krajnović, D., Weilbacher, P.M., Urrutia, T., *et al.* (2015). Unveiling the counter-rotating nature of the kinematically distinct core in NGC 5813 with MUSE. *Monthly Notices of the Royal Astronomical Society*, 452(1), 2.

Kroupa, P. (2001). On the variation of the initial mass function. *Monthly Notices of the Royal Astronomical Society*, 322(2), 231.

Kuntschner, H., Emsellem, E., Bacon, R., *et al.* (2010). The SAURON project–XVII. Stellar population analysis of the absorption line strength maps of 48 early-type galaxies. *Monthly Notices of the Royal Astronomical Society*, 408(1), 97.

Lagos, C. d.P., Davis, T.A., Lacey, C.G., *et al.* (2014). The origin of the atomic and molecular gas contents of early-type galaxies–I. A new test of galaxy formation physics. *Monthly Notices of the Royal Astronomical Society*, 443(2), 1002.

Lauer, T.R., Ajhar, E.A., Byun, Y.-I., *et al.* (1995). The centers of early-type galaxies with HST.I.An observational survey. *The Astronomical Journal*, 110, 2622.

Li, H., Mao, S., Cappellari, M., *et al.* (2018). SDSS-IV MaNGA: The intrinsic shape of slow rotator early-type galaxies. *The Astrophysical Journal*, 863, L19.

Li, R., Li, H., Shao, S., *et al.* (2019). SDSS–IV MaNGA: The inner density slopes of nearby galaxies. *Monthly Notices of the Royal Astronomical Society*, 490(2), 2124–2138.

Long, R.J., Mao, S. (2012). Made-to-measure galaxy models–II. Elliptical and lenticular galaxies. *Monthly Notices of the Royal Astronomical Society*, 421, 2580.

Magorrian, J., Tremaine, S., Richstone, D., *et al.* (1998). The demography of massive dark objects in galaxy centers. *The Astronomical Journal*, 115, 2285.

McConnell, N.J., Lu, J.R., Mann, A.W. (2016). Radial trends in IMF-sensitive absorption features in two early-type galaxies: Evidence for abundance-driven gradients. *The Astrophysical Journal*, 821(1), 39.

McDermid, R.M., Alatalo, K., Blitz, L., *et al.* (2015). The ATLAS3D projectXXX. Star formation histories and stellar population scaling relations of early-type galaxies. *Monthly Notices of the Royal Astronomical Society*, 448(4), 3484–3513.

Milosavljević, M., Merritt, D. (2001). Formation of Galactic nuclei. *The Astrophysical Journal*, 563, 34.

Miyoshi, M., Moran, J., Herrnstein, J., *et al.* (1995). Evidence for a black hole from high rotation velocities in a sub-parsec region of NGC4258. *Nature*, 373(6510), 127.

Moster, B.P., Naab, T., White, S.D.M. (2019). Empirical constraints on the formation of early-type galaxies. *arXiv:1910.09552 [astro-ph]*.

Mowla, L.A., van Dokkum, P., Brammer, G.B., *et al.* (2019). COSMOS-DASH: The evolution of the galaxy size-mass relation since z $\sim$ 3 from new wide-field WFC3 imaging combined with CANDELS/3D-HST. *The Astrophysical Journal*, 880(1), 57.

Naab, T., Johansson, P.H., Ostriker, J.P. (2009). Minor mergers and the size evolution of elliptical galaxies. *The Astrophysical Journal*, 699, L178.

Naab, T., Oser, L., Emsellem, E., *et al.* (2014). The ATLAS3D project–XXV. Two-dimensional kinematic analysis of simulated galaxies and the cosmological origin of fast and slow rotators. *Monthly Notices of the Royal Astronomical Society*, 444, 3357–3387.

Oser, L., Ostriker, J.P., Naab, T., *et al.* (2010). The two phases of galaxy formation. *The Astrophysical Journal*, 725(2), 2312.

Pfeffer, J., Baumgardt, H. (2013). Ultra-compact dwarf galaxy formation by tidal stripping of nucleated dwarf galaxies. *Monthly Notices of the Royal Astronomical Society*, 433, 1997.

Poci, A., Cappellari, M., McDermid, R.M. (2017). Systematic trends in total-mass profiles from dynamical models of early-type galaxies. *Monthly Notices of the Royal Astronomical Society*, 467, 1397.

Poci, A., McDermid, R.M., Zhu, L., *et al.* (2019). Combining stellar populations with orbit-superposition dynamical modelling: The formation history of the lenticular galaxy NGC 3115. *Monthly Notices of the Royal Astronomical Society*, 487, 3776.

Posacki, S., Cappellari, M., Treu, T., *et al.* (2015). The stellar initial mass function of early-type galaxies from low to high stellar velocity dispersion: Homogeneous analysis of ATLAS^{3D} and Sloan Lens ACS galaxies. *Monthly Notices of the Royal Astronomical Society*, 446(1), 493.

Saglia, R.P., Opitsch, M., Erwin, P., *et al.* (2016). The SINFONI Black Hole Survey: The black hole fundamental plane revisited and the paths of (co)evolution of supermassive black holes and bulges. *The Astrophysical Journal*, 818, 47.

Sanders, J.L., Binney, J. (2016). A review of action estimation methods for galactic dynamics. *Monthly Notices of the Royal Astronomical Society*, 457(2), 2107.

Sarzi, M., Spiniello, C., La Barbera, F., *et al.* (2018). MUSE observations of M87: Radial gradients for the stellar initial-mass function and the abundance of sodium. *Monthly Notices of the Royal Astronomical Society*, 478(3), 4084.

Schwarzschild, M. (1979). A numerical model for a triaxial stellar system in dynamical equilibrium. *The Astrophysical Journal*, 232, 236.

Schwarzschild, M. (1982). Triaxial equilibrium models for elliptical galaxies with slow figure rotation. *The Astrophysical Journal*, 263, 599.

Serra, P., Oosterloo, T., Morganti, R., *et al.* (2012). The ATLAS^{3D} project–XIII. Mass and morphology of H I in early-type galaxies as a function of environment. *Monthly Notices of the Royal Astronomical Society*, 422(3), 1835.

Strauss, M.A., Weinberg, D.H., Lupton, R.H., *et al.* (2002). Spectroscopic target selection in the Sloan Digital Sky Survey: the main galaxy sample. *The Astronomical Journal*, 124, 1810.

Syer, D., Tremaine, S. (1996). Made-to-measure N-body systems. *Monthly Notices of the Royal Astronomical Society*, 282(1), 223.

Thomas, D., Maraston, C., Schawinski, K., *et al.* (2010). Environment and self-regulation in galaxy formation. *Monthly Notices of the Royal Astronomical Society*, 404, 1775.

Trick, W.H., Coronado, J., Rix, H.-W. (2019). The Galactic disc in action space as seen by Gaia DR2. *Monthly Notices of the Royal Astronomical Society*, 484, 3291.

Tully, R.B., Fisher, J.R. (1977). Reprint of 1977A&A....54..661T. A new method of determining distance to galaxies. *Astronomy and Astrophysics*, 500, 105.

van de Ven, G., de Zeeuw, P.T., van den Bosch, R.C.E. (2008). Recovery of the internal orbital structure of galaxies. *Monthly Notices of the Royal Astronomical Society*, 385(2), 614.

van den Bergh, S. (1976). A new classification system for galaxies. *The Astrophysical Journal*, 206, 883.

Wang, Y., Vogelsberger, M., Xu, D., *et al.* (2019). Early-type galaxy density profiles from IllustrisTNG: II. Evolutionary trend of the total density profile. *Monthly Notices of the Royal Astronomical Society*, stz2907.

Williams, A.A., Evans, N.W. (2015). Made-to-measure dark matter haloes, elliptical galaxies and dwarf galaxies in action coordinates. *Monthly Notices of the Royal Astronomical Society*, 448, 1360.

Worthey, G. (1994). Comprehensive stellar population models and the disentanglement of age and metallicity effects. *The Astrophysical Journal Supplement Series*, 95, 107.

Zhu, L., van de Ven, G., Méndez-Abreu, J., *et al.* (2018). Morphology and kinematics of orbital components in CALIFA galaxies across the Hubble sequence. *Monthly Notices of the Royal Astronomical Society*, 479, 945.

4

Spiral Galaxies

Françoise COMBES
Observatoire de Paris (LERMA) and Collège de France, Paris, France

Our vision of galaxies has changed significantly since the era of large galaxy surveys like the Sloan, which gave us extensive statistics with millions of galaxies. The Hubble sequence classification described in Chapter 1 still remains very widely used but has been enriched with broad categories based on color that indicate the recent formation of stars: the red sequence of passive galaxies, consisting solely of old stars, and the blue cloud of galaxies with active star formation. Chapter 3 focused on galaxies with a dominant spheroid, which are generally found on the red sequence. One of the key questions about the evolution of galaxies that remain to be answered is to understand how a galaxy can suddenly pass from one category to another.

4.1. Introduction

This chapter is dedicated to spiral galaxies, also known as "late-type galaxies", which are dominated by their disc component. Some of them also have a spheroid component, or bulge, and the bulge/disc ratio increases along the Hubble sequence, from Sc to Sa. At this point, they then transition toward early-type galaxies (ETGs) dominated by a spheroid component, the majority of which are found on the red sequence. How do these two components form?

For color versions of the figures in this chapter, see www.iste.co.uk/combes/galaxies.zip.

Galaxies,
coordinated by Françoise COMBES. © ISTE Ltd. 2020.

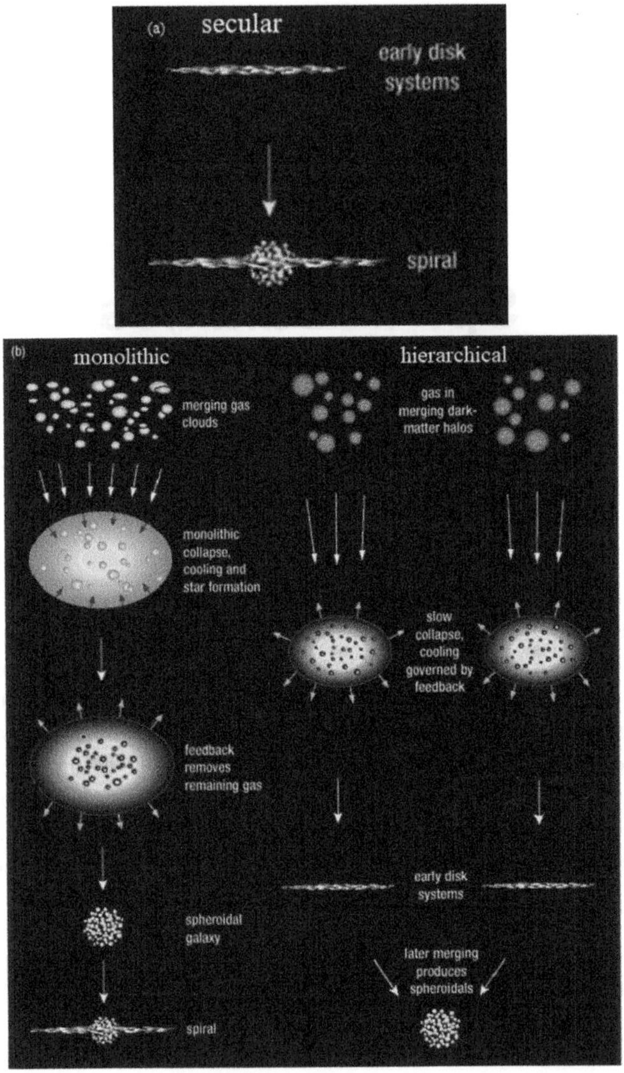

Figure 4.1. *(a) Secular evolution of galaxies: when gas falls into a dark matter halo, the first systems to form are discs if stars begin to form only slowly, following gas dissipation. Internal evolution and the formation of a bar then form a pseudo-bulge by vertical resonance. (b) Two other scenarios that may create spheroids. On the right, the hierarchical scenario: once both discs have formed, the interaction and merger of the two spiral galaxies may create an elliptical galaxy, or at least a galaxy dominated by its spheroid component. On the left, if the star formation occurs more quickly than the dissipation of gas, the stars form with a spheroidal geometry straight away. This is the monolithic collapse scenario*

Galaxies begin by forming their dark matter component: the halos collapse under their own gravity, well before the recombination of the Universe 380,000 years after the Big Bang. Before this date, ordinary matter (or baryons) is ionized and interacts closely with photons, which stabilize it. Only the dark matter itself does not interact directly with the photons; it may collapse once matter dominates the radiation in terms of density. During recombination, the baryons decouple from the photons, possibly falling into the potential wells of the halos. These baryons primarily take the form of hydrogen gas, with 25% helium by mass. As it dissipates, the gas quickly forms rotating discs that conserve a large proportion of their initial angular momentum, acquired by the halos due to tidal torques. Local gravitational instability causes the gas to form stars, which also spread throughout the disc. Thus, the first baryonic components naturally form as discs, as shown in Figure 4.1.

Figure 4.1(b) shows the hierarchical scenario in which spheroid components may form. When two disc galaxies with randomly oriented angular momenta merge, the final angular momentum is logically expected to be smaller, as a vector sum of randomly oriented momenta. Multiple minor mergers can create an increasingly massive bulge, whereas major mergers of spiral galaxies (i.e. mergers with a mass ratio close to 1) will form elliptical galaxies that are dominated by their spheroid. However, another scenario is possible, depending on the speed and efficiency of star formation during the gas collapse in the halos. If the star formation occurs more quickly than the collapse time, there is a so-called "monolithic" scenario, where the spheroid component forms first, and the spheroidal galaxy then potentially forms a disc later by the external accretion of gas originating from cosmic filaments (see Figure 4.1(b), left).

Monolithic collapse was in fact one of the first scenarios considered to explain the formation of our own galaxy, the Milky Way, by Eggen *et al.* (1962). The idea came from observing the galaxy's stellar halo, which consists of very old stars that are poor in heavy elements. These metals (like any elements that did not form in the Big Bang, such as carbon, oxygen and iron) are ejected by the earliest stars, enriching the interstellar medium where the stars of second and later generations then form. Metallicity is therefore an indicator of age, or number of astration cycles. Furthermore, Eggen *et al.* (1962) observed that, for stars in the halo, there is a correlation between the metallicity, indicated by [Fe/H], and the eccentricity. They interpreted this correlation as a change in the dynamics as the gas slowly collapses into a disc. The stars that form the earliest have a three-dimensional geometry and are poor in metals. Gradually, their orbits become less radial, tending toward circular orbits in the equatorial plane of the galaxy. In this scenario, the gravitational potential only evolves slowly, which allows us to consider the eccentricity and the angular momentum L_z perpendicular to the disc as invariants. However, Searle and

Zinn (1978) were not convinced by these hypotheses, objecting that the globular clusters of the stellar halo show large metallicity discrepancies that do not correlate with the distance to the center of the galaxy. Instead, they favored the explanation of a merger of small protogalaxies with a rapidly varying gravitational potential. Today, this scenario has prevailed, and it is widely accepted that many of the stars in the halo of the Milky Way originate from minor mergers. Numerical simulations have shown that stellar streams can remain visible even 10 billion years after the destruction of a small satellite galaxy (e.g. Helmi and White 1999).

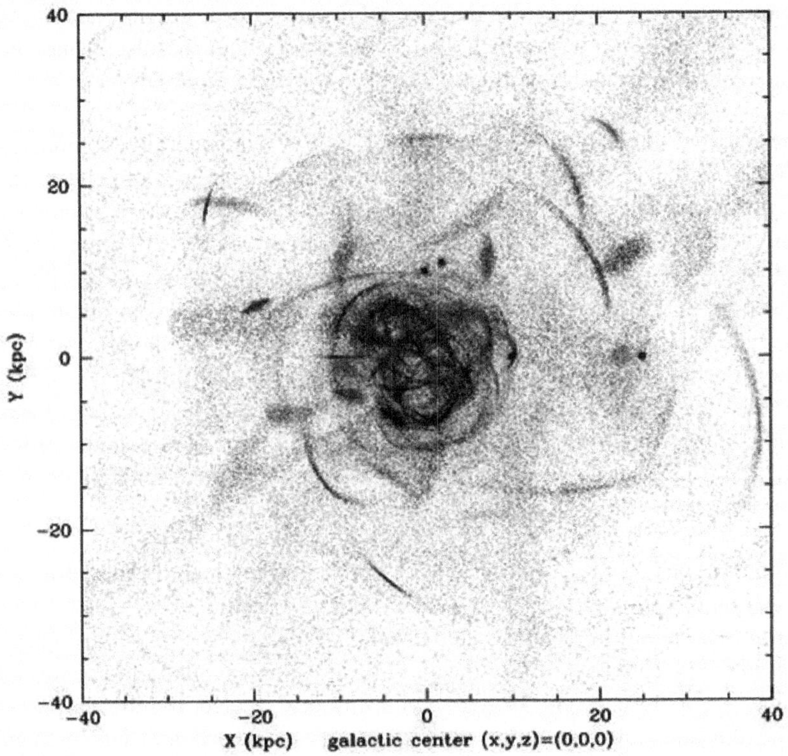

Figure 4.2. *The stellar halos of the Milky Way and Andromeda reveal evidence of their formation by accretion and destruction of satellite dwarf galaxies and globular clusters in the form of tidal loops and streams. They are entirely comparable to this simulation, which shows the accretion of 50 dwarf galaxies over a period of 10 billion years (credit: P. Harding 2001, PhDT)*

The original "monolithic" scenario has been abandoned, but another more elaborate version is still invoked to explain the extreme formation speed of some massive elliptical galaxies observed at high redshifts. It is not yet clear what proportion of galaxies form in this way (monolithic scenario) as opposed to by mergers of spiral galaxies (hierarchical scenario), which could unfold very quickly in overdense environments, ancestors of the dense galaxy clusters of today. However, it has been clearly established that spiral galaxies completely dominated by their disc component, either with a very small bulge or no bulge at all, are the field galaxies most frequently observed today, and that they could only have formed according to the secular scenario shown in Figure 4.1(a).

Today, observations from large surveys of local galaxies, together with deep cartographies sampling distant galaxies, have revealed to us the history of the cosmic star formation (e.g. Madau and Disckinson 2014). Starting from the Big Bang, the star formation density increases regularly for around 4 billion years, up to a peak at z = 2, then decreases regularly by a factor of 20 since this peak. Today, most galaxies are passive, while they had very active star formation for the first half of the history of the Universe. This also reflects the bimodality between the red sequence and the blue cloud highlighted below. Clearly, active star formation suddenly stopped in many galaxies, and it is essential for us to understand the mechanism of this change.

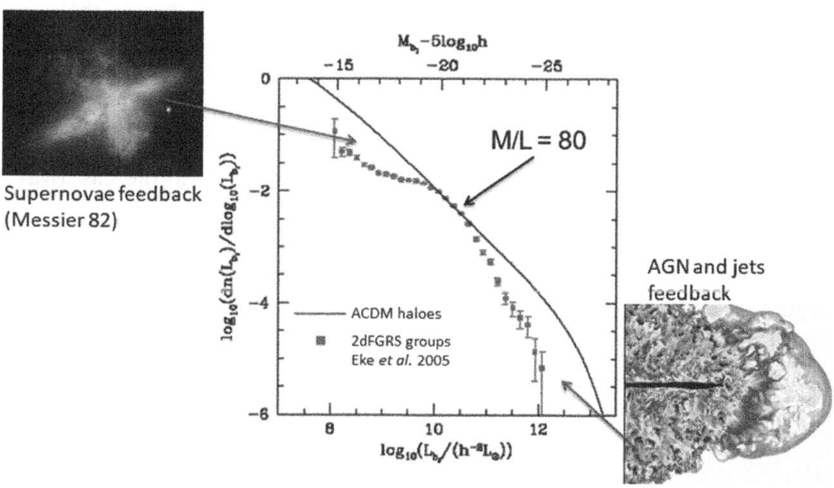

Figure 4.3. *Distribution of galaxies by luminosity (log-log diagram): the red points with error bars are the observations (Eke* et al. *2006), and the blue curve shows the predictions of the ΛCDM standard cosmological model. The two distributions are calibrated at their point of intersection, with a mass-to-luminosity ratio of M/L = 80. The model predicts too many small galaxies, which can be corrected with supernovae feedback (left), and too many massive galaxies, which could be prevented by feedback from active galaxy nuclei (right)*

This discussion is followed by another observation: the luminosity function of galaxies, which is shown in Figure 4.3 (see Jenkins *et al.* 2001; Baugh 2006). The number of galaxies with low stellar mass (up to $M_* = 3 \times 10^{10} M_\odot$) is smaller than the number predicted by numerical simulations performed within the framework of the standard cosmological model (ΛCDM). The number of galaxies that are more massive than this limit is also lower than expected. The central point is used for calibration and corresponds to a mass-to-luminosity ratio of 80, much higher than expected from the universal ratio of total matter (dark + ordinary) to baryons, which is equal to 6. This calibration reveals that most baryons have exited the galaxies and are now located in intergalactic space, more precisely in cosmic filaments. These observations all tend to show that galaxy formation is accompanied by violent events that are capable of ejecting the majority of the visible matter out of the halos. For low-mass galaxies, feedback from star formation, i.e. supernova explosions and stellar wind, can eject the surrounding gas (Dekel and Silk 1986). But this mechanism is not sufficient for massive galaxies, as the gas does not reach escape velocity. The energy deployed by the active nuclei around supermassive black holes at the center of each galaxy might give us the answer (Croton *et al.* 2006).

These two processes must certainly play an important role in the formation of galaxies. However, empirically, the two observational parameters that seem to control the star formation activity of galaxies are the total mass and the environment, namely whether the galaxy belongs to a group or cluster or has an average galaxy density in its immediate neighborhood (Peng *et al.* 2010). Is the total mass of baryons that accumulates at the center of a galaxy capable of stabilizing the disc and preventing the formation of stars? Or does the environment rip the gas away from the galaxy, or perhaps prevent cosmic filaments from resupplying it?

We will examine these various mechanisms that control the star formation, and then turn our attention to the spiral arm structure of the discs, as well as the formation of bars, which are key drivers of the evolution of these discs and the assembly of their mass. We will also see how the environment can play a crucial role.

4.2. Blue and red galaxies: quenching star formation

4.2.1. *Definition of bimodality*

If we plot the hundreds of thousands of local galaxies of the Sloan survey on a color-magnitude diagram, we find that they are not distributed homogeneously. Instead, they occupy certain clearly distinct regions. This is called bimodality

(Baldry *et al.* 2004). In the stellar mass-color diagram of Figure 4.4, we can clearly see a red sequence, which encompasses the most massive galaxies, and a blue cloud, in the region with lower masses. This diagram can be decomposed by galaxy morphology according to whether the spheroid component dominates the disc component and whether the disc has spiral arms. The red sequence clearly corresponds to morphological types that are dominated by their spheroid, namely the so-called ETGs (see Chapter 3). The blue cloud corresponds to spiral galaxies and late-type galaxies, although it is worth noting that some of the most massive spirals edge back toward the red sequence. These galaxies have the most massive bulges, and their star formation activity has stopped. For these galaxies, the divide between the two sequences, known as the green valley (Schawinski *et al.* 2014), is less pronounced than for all the galaxies combined.

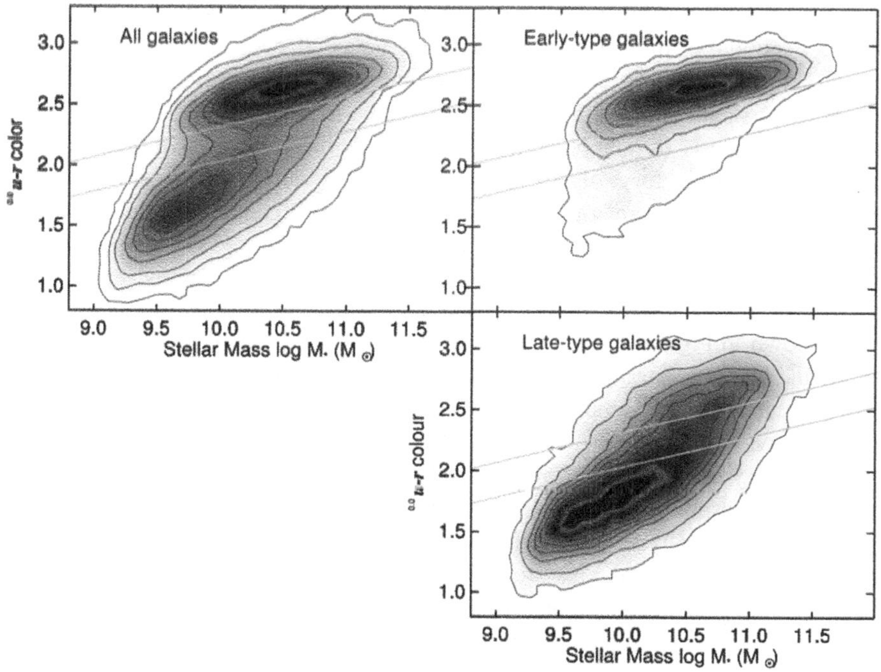

Figure 4.4. *Color (u-r)-stellar mass diagram for all galaxies combined (top left) and for early-type and late-type galaxies separately. The green lines show the position of the green valley, which separates the red sequence above it from the blue cloud below it (reproduced from Schawinski* et al. *2014)*

The bimodality of the color-magnitude diagrams shows the key importance of the star formation rate and its history in the evolution of a galaxy, as well as the importance of the quantity of gas and dust, which is linked to the metallicity. It sheds light on two mechanisms of galaxy formation, with a separation corresponding to the critical stellar mass of $3 \times 10^{10} M_{\odot}$.

Large-scale statistics about the properties of galaxies have been collected over the past few years, not just from large surveys such as Sloan and DECaLS (Dark Energy Camera Legacy Survey), but also due to voluntary classification by the keen amateur astronomers who participated in the great "Galaxy Zoo" adventure (Lintott *et al.* 2008). Each person classifies a few hundred galaxies. Ultimately, each galaxy receives around a dozen responses, which allows any mistakes to be eliminated. On a small test sample, the classification produced by these amateurs coincided with that of experts. A million galaxies have been added to the classification since July 2017 in version 4 of the Zoo. For more distant galaxies, deep surveys by the Hubble Space Telescope (HST) are used, revealing a major evolution as the redshift increases. Distant galaxies become increasingly irregular, with clumps, and the Hubble sequence is only applicable to a small minority of massive galaxies in the first half of the history of the Universe. Today, automatic classifications are becoming more and more sophisticated. Artificial intelligence will represent a very powerful tool as soon as deep learning programs are ready (e.g. Dominguez Sanchez *et al.* 2018). Learning is often performed on cosmological simulations.

As shown in Figure 4.4, the colors do not fully match the Hubble sequence. We can define various parameters, such as the surface density μ (in magnitude per second squared), or the Sersic index n, i.e. the slope of the radial distribution of light in the galaxy. The index n is defined by $I(R) = I0 \exp(-kR^{1/n})$, where I(R) is the light surface density averaged along the azimuth, and I0 is its value at R = 0. Figure 4.5 shows the correspondence between a simplified Hubble sequence (elliptical, early-type spiral, and late-type spiral), the color (u-r), the Sersic index n and the surface brightness μ. Clearly, galaxies are distributed bimodally by color, but also by the index n and the surface brightness μ. The red and passive galaxies are for the most part elliptical galaxies, which are traditionally characterized by the De Vaucouleurs light profile with n = 4, and which are more compact (low μ). Late-type spiral galaxies are blue, and their discs satisfy the relation described by Freeman in 1970 (μ_0 = 21 mag/arcsec^2). The separation is less pronounced for early-type spiral galaxies, which are found at the transition.

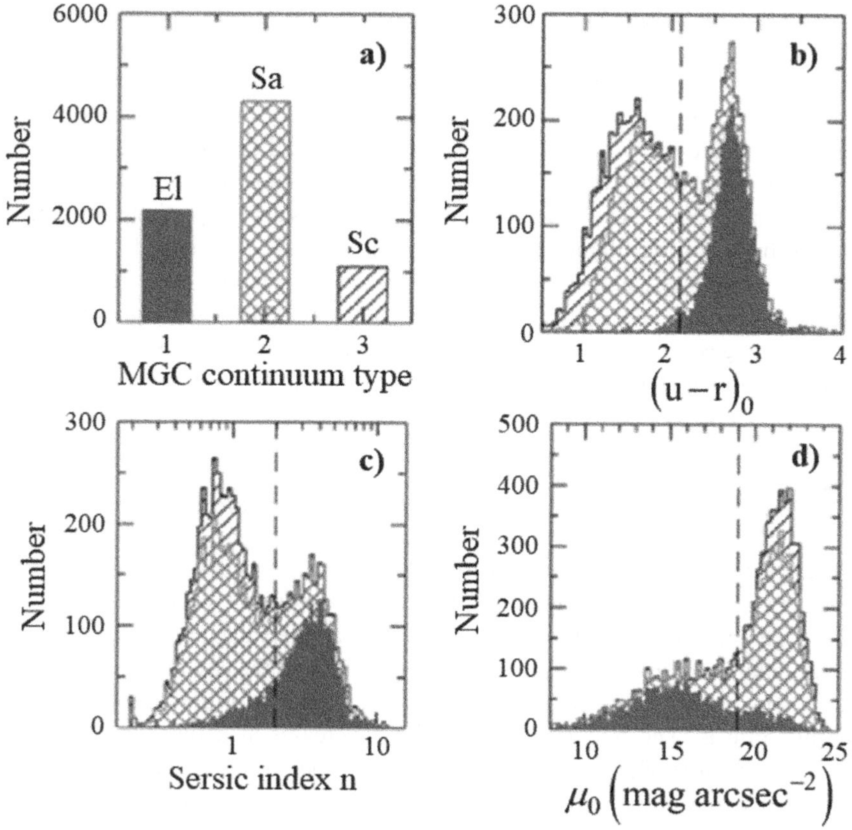

Figure 4.5. *Histograms of the number of galaxies by morphological Hubble type (a), color u-r (b), Sersic index n (c) and central surface brightness μ_0 for galaxies in the MGC catalog (reproduced from Driver* et al. *2006)*

It has been observed that, for local galaxies, the most active galaxies in terms of star formation are the least massive, namely the late-type spiral galaxies and the irregular galaxies. This trend is part of a more general phenomenon, known as "downsizing" or hierarchical inversion (Cowie *et al.* 1996): the luminosity of the most active galaxies continuously decreases over time. Indeed, according to the standard cosmological model, which is hierarchical, we expect the earliest objects that form to be the smallest; these objects then merge to form larger objects, and, today, we would expect to observe the formation of the most massive galaxies. But this is not the case at all. However, this paradox is not quite so paradoxical if we consider that the hierarchy only concerns the dark matter halos. At the beginning of the Universe, the least massive halos collapse first. By merging, they then form larger and larger halos. Overdensities correspond to groups that collapse later, and, similarly, galaxy clusters virialize at around z = 1–2. Today, most clusters continue to grow through mergers, and we frequently see multiple distinct subclusters form into a rich cluster, like the Coma Cluster. Baryons, on the other hand, have another fate. At the beginning of the Universe, gas falls into the halos, forming dwarf galaxies and, gradually, stars. The gas consumed in these stars is partially replaced by accretion from cosmic filaments, but the density of these filaments decreases as the expansion progresses, and the fraction of gas in the galaxies also decreases over time. As a result, galaxy mergers involve old stars more and more frequently, without any gas or a new generation of stars. The most massive galaxies are therefore found in the red sequence and are passive, even if they were recently involved in a merger.

4.2.2. *The parameters that determine the red sequence*

The fraction of "red" galaxies with inactive star formation depends on several parameters. We can see in Figure 4.4 that the red sequence encompasses the most massive galaxies. It turns out that there are two parameters that are strongly correlated with the color – not just the mass, but also the richness of the environment, namely the surface density Σ after projecting the neighboring galaxies. Figure 4.6 clearly shows these two effects. The fraction of red galaxies increases as a function of the stellar mass for each curve of any given color, which corresponds to a fixed surface density Σ of galaxies per Mpc^2. From these observations, we can now deduce the physical phenomena that cause quenching of star formation within a galaxy.

The angular momentum is an essential parameter in the fate of a galaxy. It is possible to measure the rotation curves of disc galaxies, as well as the rotation curves of spheroid galaxies, even though the latter rotate very little, since they are in equilibrium due to their anisotropic velocity dispersion. Given the rotation velocity at each radius, we can calculate the total angular momentum, and, more significantly, the specific momentum j per unit mass. Fall and Romanowsky (2013) showed that the specific momentum j can be used as another parameter to classify galaxies: every galaxy in the Hubble sequence is aligned along parallels with slope 2/3 in the $\log j$-$\log M_*$ diagram when only baryonic quantities are considered. The spiral galaxies of latest type are found at the top of the diagram. Many other versions of this diagram and classification have been published (e.g. Sweet *et al.* 2018). Some incorporate a third parameter, the bulge-to-disc ratio, which allows classical bulges resulting from mergers to be distinguished from pseudobulges, which result from resonance with a bar in the secular scenario.

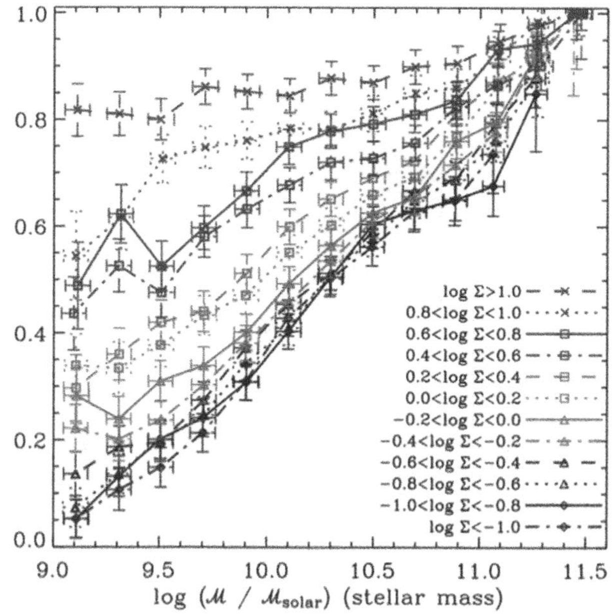

Figure 4.6. *Fraction of red galaxies as a function of the stellar mass M, ranging from 10^9 $M_{\odot}$ to 4 × 10^{11} $M_{\odot}$, for several galaxy surface density values Σ ranging from 0.1 to 10 gal Mpc^{-2} (reproduced from Baldry* et al. *2006)*

Of course, these parameters are not the only possible choices: the color of a galaxy is also a function of its gas fraction, the quantity of dust, the age of its stars, the metallicity and the morphological type. Note that the star formation history of a galaxy depends on its surface density. As shown in Figure 4.7, galaxies are not distributed randomly on the diagram of central surface density versus stellar mass. Massive galaxies have a high surface density, and spiral galaxies satisfy the relation proposed by Freeman (1970). They are all HSB (high surface brightness). By contrast, dwarf galaxies, which have a low stellar mass, are mostly LSB (low surface brightness) objects. Such objects are generally rich in atomic and possibly molecular gas. They should be capable of forming stars, but their efficiency is low. The massive (HSB) galaxies are more compact, with an older stellar population and a redder color. It is difficult to say which is cause and which is effect. If initial conditions such as angular momentum and high rotation prevent mass from concentrating, this may explain why the galaxy remains at a dwarf stage, with low surface brightness. The transition between the two categories occurs once again at the critical stellar mass of $M_* = 3 \times 10^{10}$ $M_{\odot}$, or a surface density of 3×10^8 $M_{\odot}/kpc^2$. Ultimately, the star formation history may depend more on the surface density than the mass.

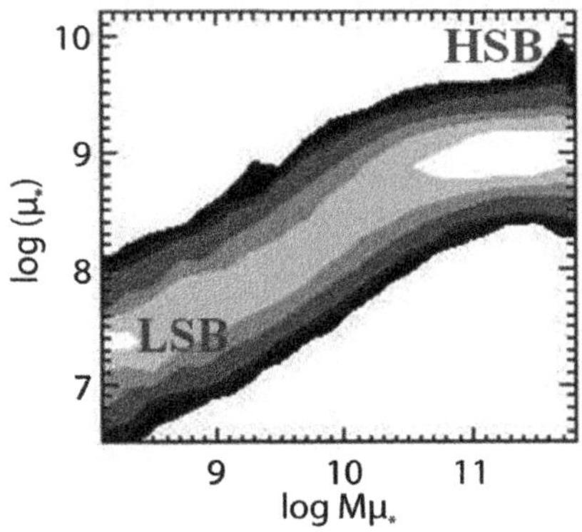

Figure 4.7. *Surface density at galaxy centers as a function of the stellar mass M*. For massive galaxies with high surface brightness (HSB), Freeman's law (1970) is satisfied with a constant value for all discs. There is also a category of low surface brightness (LSB) galaxies, which are mostly dwarf galaxies. Diagram generated with 122,000 galaxies (reproduced from Kauffmann* et al. *2003)*

The transition around the critical mass of $M_* = 3 \times 10^{10}\ M_\odot$ could reflect the importance of feedback from star formation. For shallow potential wells, the velocity of the gas ejected by supernovae, of the order of 100 km/s, is high enough to impart escape velocity to the dragged gas, and the galaxy loses its gas. By contrast, more massive galaxies retain their gas, and star formation accumulates.

4.2.3. *Mechanisms for quenching star formation*

Many mechanisms that might abruptly stop the formation of stars have been suggested. The occupied regions in the color-stellar mass diagrams of galaxies (Figure 4.4) suggest that this stopping does indeed occur suddenly. The number of galaxies between the red sequence and the blue cloud is low (green valley) and reflects the short time spent in the transition zone.

One of the first such mechanisms is suppression of the gas supply. Galaxies form in cosmic filaments consisting of dark matter and gas, and they continue to be supplied by matter that falls from these filaments over the course of their lifetime. If the gravitational potential of the galaxy is not too deep, numerical simulations show that the gas can supply the galaxy while staying relatively cold, which means that it remains available for star formation. By contrast, if the potential well exceeds a certain mass and depth, the gas experiences a shock wave that brutally heats it. Being diffuse and hot, it is no longer sufficiently dense to cool, and remains at the equilibrium of the virial theorem in the dark matter halo. Hot gas can only become dense enough to cool down again in dense clusters. In massive galaxies and groups, the gas remains very hot and is therefore not available for star formation. The critical transition mass is thought to be of the order of $M_h = 5 \times 10^{10}$ $M_\odot$ for the halo, or $M_* = 3 \times 10^{10}$ $M_\odot$ for the stellar mass (Dekel and Birnboim 2006).

This mechanism predicts that the separation between the red sequence and blue cloud in the color-stellar mass diagram will only occur at low redshifts, when the potential wells are deep enough to generate shocks. The red sequence should therefore only appear after z = 1. However, this is not consistent with observations. Even at large redshifts, there are elliptical galaxies or ETGs that are already red, no longer forming any stars. Consequently, there must be other mechanisms. Galaxies with active star formation define the so-called main sequence, since their formation rate (SFR) is nearly proportional to their stellar mass (see Figure 4.8). There are indeed bursts of star formation during interactions or mergers of galaxies, but 90% of cosmic star formation occurs on the main sequence.

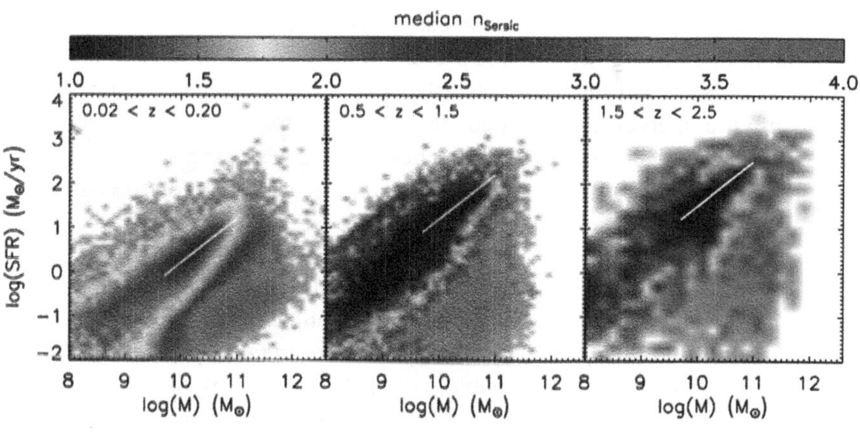

Figure 4.8. *Distribution of galaxies in the diagram of star formation rate (SFR) and stellar mass (M*$_*$*), at three redshift ranges: on the left, the local galaxies of the Sloan survey, followed by 0.5 < z < 1.5 and 1.5 < z < 2.5, both from HST deep fields. The color indicates the Sersic index n of the radial light distribution. Blue corresponds to discs (n~1-2), and red corresponds to galaxies dominated by spheroid components (n>3). The blue galaxies are located on the main sequence, where SFR is proportional to M*$_*$*. The main sequence is indicated by a white line of slope 1 and zero point that increases with the redshift. At every epoch, the galaxies that form stars are discs, whereas the passive, red galaxies are spheroids. The less common systems above the main sequence form large numbers of stars and are primarily mergers or interactions of galaxies, with a spheroid profile (reproduced from Wuyts* et al. *2011)*

Observe that, in the color-stellar mass diagram (Figure 4.4), there are also low-mass galaxies on the red sequence. These galaxies must have ceased activity due to their interaction with a more massive companion; they are satellite galaxies (as opposed to the central galaxies).

Suppressing the cold gas supply of galaxies in very massive halos is a permanent but relatively slow way to stop star formation. Indeed, the central galaxy retains its cold gas and will continue to form stars until this gas is fully exhausted. This can take of the order of 2 billion years, the average time required to consume all the gas of a spiral galaxy at its current star formation rate (Bigiel 2008). We say that they have a depletion time of 2 Gyr. There are other, much quicker ways to stop star formation: for example, stripping of gas by ram pressure when a spiral enters a galaxy-rich cluster at high velocity, or during a violent collision between two galaxies. In these scenarios, the characteristic time is of the order of 100 million years.

Another mechanism that occurs over long time scales is stabilization of the gas in the disc by forming a very massive central spheroid (Martig *et al.* 2009). This mechanism does not require the cold gas to be consumed, nor its supply to be discontinued. Even though cold gas is present in the disc, it will not form stars if the mass of the bulge is too high, provided that the gas's turbulence remains high, for example after a merger that created or increased the mass of the bulge. This mechanism is called "morphological quenching" because it only depends on the morphology. It applies to elliptical galaxies, and more generally to ETGs, or in other words precisely the galaxies found in the red sequence.

In the simulation shown in Figure 4.9, a galaxy is followed in terms of cosmological conditions for a period equal to the Hubble time: accretion of gas from cosmic filaments and minor or major mergers. In an initial phase ($t = 3.5$ Gyr), the galaxy is very rich in gas and unstable, which forms clumps, which is consistent with observations of galaxies at $z > 2$. Then, after a major merger ($t = 10$ Gyr), the galaxy experiences an inactive phase. The gas is stabilized by a massive bulge, as well as by the velocity dispersion (measured by the Toomre parameter Q). Eventually, the galaxy returns to a state with a colder gaseous disc, in particular by accretion. It experiences instabilities and begins to reform stars, like on the main sequence ($t = 13.7$ Gyr, or $z = 0$).

The Toomre parameter Q is an indicator of disc stability. It is equal to the ratio of the velocity dispersion σ and the critical dispersion $\sigma crit$ ($Q = \sigma/\sigma crit$). The critical dispersion is defined in terms of the surface density Σ of the disc and the epicyclic frequency κ, which is a function of the rotation velocity $\Omega = V/R$. Its value is $\sigma crit = 3.36\ G\Sigma/\kappa$. The more the disc self-gravitates, the more it can be expected to be stabilized by the velocity dispersion. Conversely, a massive bulge causes κ to increase, as well as Q if σ and Σ remain constant, which creates the stabilizing effect.

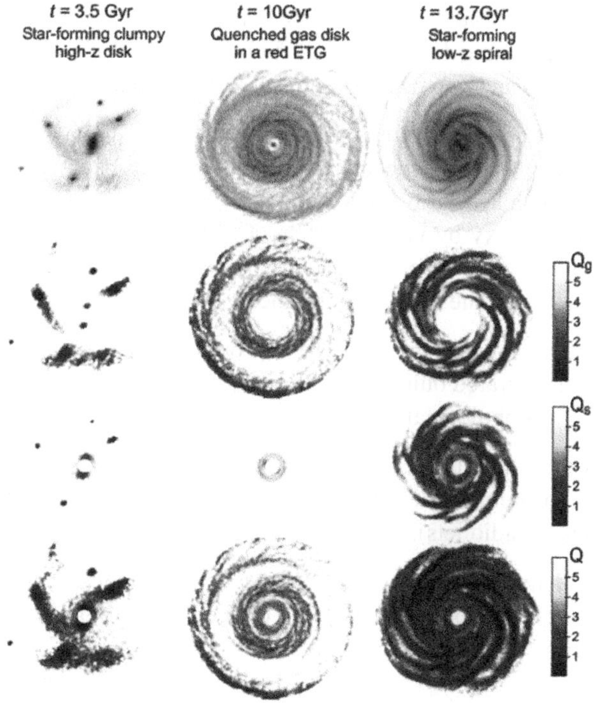

Figure 4.9. *Gas distribution in a simulation that follows the life of a galaxy for a period equal to the Hubble time. The first row shows the gas surface density (in blue). The three other rows show the values of the Toomre parameter Q for the gas (Qg), for the stars (Qs), and for a combination of both (Q). The three columns correspond to three phases of the galaxy: a clumpy morphology, an inactive phase where the Q parameter can be seen to remain constantly high and an active phase on the main sequence (reproduced from Martig* et al. *2009)*

Finally, there are internal feedback processes, caused by either star formation or the activity of the central black hole (supernova feedback and AGN feedback). These processes can rapidly eject the gas, or heat it and make it turbulent, which stops star formation. These processes are rarely permanent. In massive galaxies, the gas does not reach escape velocity and falls back into the galaxy. As a result, the effect mostly just moderates the star formation, making the gas unavailable for a certain period of time. In dwarf galaxies, where the ejection velocity is greater than the escape velocity, the effect is temporarily stronger, but the galaxy may still accrete gas from cosmic filaments at a later point.

We can summarize the various mechanisms for stopping star formation by classifying them according to their speed: two categories of processes that unfold over large time scales and two categories of rapid processes:

1) Processes that cut the cold gas supply of galaxies. The quenching time scales are larger (2–4 Gyr) because the galaxy continues to consume any gas that was already present. This may involve halos that are sufficiently massive at $z < 1$ to heat gas falling into the galaxy by violent shocks. Effects from the environment (groups and clusters) can also cause the cold gas of filaments to completely disappear. The galaxy is suffocated, and the effect is permanent.

2) Processes that stabilize the gas: These processes are slow, but not permanent. A massive and concentrated bulge forms from minor or major mergers and stabilizes the disc. This is a morphological process.

3) Processes that eject the gas present in the galaxy. These processes are rapid (<100 Myr) but not permanent. This category includes feedback from supernovae and AGNs (winds or radio jets).

4) Processes that heat the gas, rapid but very transitory. Turbulence may be generated by interactions between galaxies or feedback from supernovae. The gas dissipates its energy, and star formation resumes.

4.3. Spiral galaxies: density waves or not?

4.3.1. *The winding problem*

The morphology of disc galaxies is primarily distinguished by the existence of spiral arms. For a long time, spiral arms were a mystery, since they are usually coherent from one end of the galaxy to the other, winding around only less than once, like Messier 51 (see Figure 4.10).

Disc galaxies rotate around their center differentially. In other words, while the center completes one revolution in ~10 million years, the edges require a billion years to complete their rotation. There can be a factor of 100 between the rotation periods of the center and the edges. If the spiral arms of M51 were physical in nature, they should wind around at least 100 times, causing them to dissolve. How can this winding problem be solved?

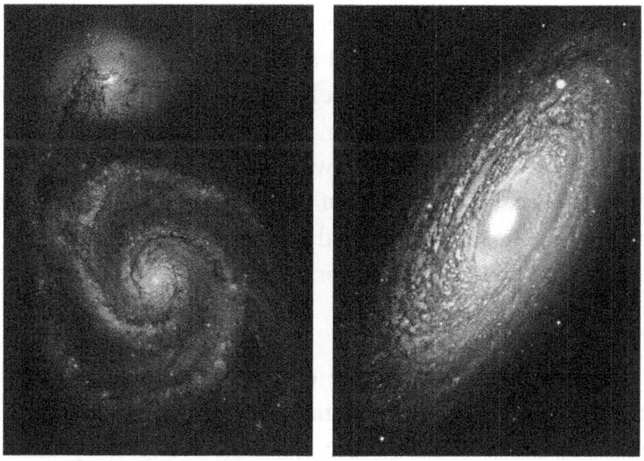

Figure 4.10. *Two examples of spiral galaxies: on the left, Messier 51, which has two arms, winding just over 180°, very contrasted. This configuration is described as "grand design". On the right, NGC2841, a flocculent spiral galaxy with just a few spiral arm fragments, very tightly wound, but lacking coherence from one end of the galaxy to the other. This configuration is called a "stochastic spiral". It can be explained solely in terms of contagious star formation and differential rotation (Gerola and Seiden 1978). Images from the HST (credit: NASA/ESA)*

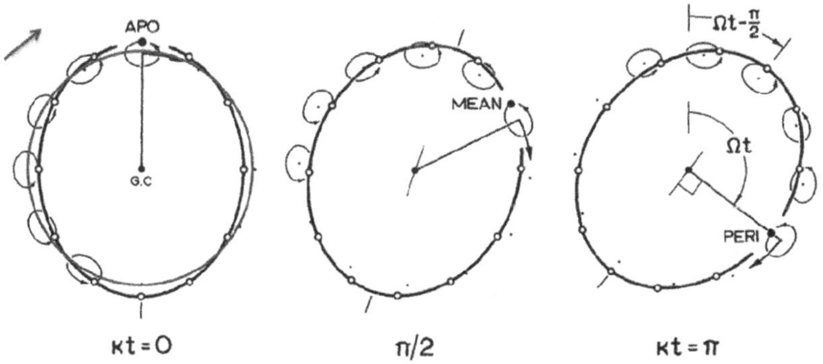

Figure 4.11. *A typical stellar orbit in the epicyclic approximation. At order zero, the star completes a circle (in red). At the first order of perturbation, a small oscillation is added, with epicyclic frequency* κ*. At any given time, the star is represented by a small empty circle that completes an epicycle around the guide center, represented by a point (still on the red circle, left). The position of the star at three epochs is marked with a solid circle (*$\kappa t = 0$*, apocenter,* $\kappa t = \pi$*, pericenter). The diagram shows the radii of a flat rotation curve (V = const.,* $\Omega \sim 1/R$*, and* $\kappa = \sqrt{2}\,\Omega$*). Note that the motion of the star can also be described as a rotation at velocity* Ω *along an ellipse with precession rate* $\Omega - \kappa/2$

The problem was solved in the 1960s to 1980s by the theory of density waves. The coherent spiral arms are not physical arms, but waves. All matter, whether stars or gas, passes through the arms during each revolution, staying proportionally longer within the spiral arms than in between them, resulting in the impression of an accumulation of matter. Bertil Lindblad was the first to suggest density waves after observing that the precession rate of the almost elliptical orbits of stars, $\Omega-\kappa/2$, remains almost constant as the radius increases, unlike the rotational angular momentum Ω, which varies like 1/R for the flat rotation curves of galaxies (see Figure 4.11).

This rapid variation of $\Omega \sim 1/R$ is the origin of the winding of the arms. If the waves rotate around the center with angular velocity Ω_p close to the precession rate $\Omega-\kappa/2$, then even a small force resulting from the self-gravity of matter will force the arms to rotate together at the same speed.

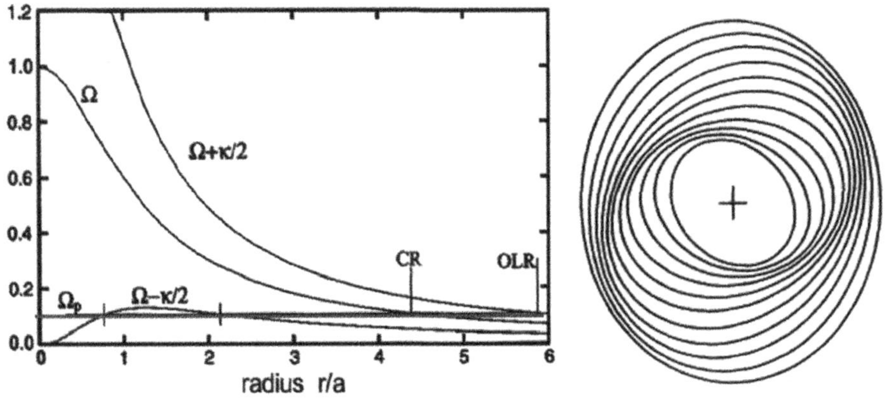

Figure 4.12. *Left: Frequencies of the rotation curve of a typical galaxy, similar to the Milky Way. The precession rate of the elliptical orbits $\Omega-\kappa/2$ is almost constant, which suggests that waves rotating at this velocity Ω_p will not wind whenever self-gravity causes them to retain their coherence. The intersection of Ω_p with Ω is the corotation (CR), and $\Omega-\kappa/2$ is the outer Lindblad resonance (OLR). Right: Distribution of a few elliptical orbits precessing by $\Omega-\kappa/2$. Initially, each of them has a major axis that is rotated with respect to the previous one by a small angle that varies logarithmically with the radius. Superimposing these orbits creates a spiral, describing the concept of a density wave. Matter orbits along the ellipses and passes through two arms on each rotation (reproduced from Kalnajs 1973)*

4.3.2. *The theory of density waves*

We therefore need to show that self-gravity can force the precession rates to remain confined around a shared value. To do this, Lin and Shu (1964) had the idea of developing the equations of tightly wound waves in the WKB ("Wenzel, Brillouin, Kramers") approximation. The wavelength λ, i.e. the space between two arms, is assumed to be small relative to the radius R of the galaxy, so that $\lambda << R$. This has the advantage of allowing a local approximation (the effects of distant waves cancel out). Without this approximation, the equations would be impossible to solve analytically.

To find the wave dispersion relation, we assume a sinusoidal perturbation in the disc surface density $\Sigma = \Sigma_0 + \Sigma_1(r)\ \exp[-im(\theta\text{-}\theta o) + i\omega t]$, where (r, θ) are the polar coordinates in the plane of the disc, m is the number of arms and ω is the temporal frequency at which the arm is crossed (ω/m is the angular velocity). For small perturbations, we linearize the dynamical equations, namely the Poisson equation, which relates the density Σ to the potential, and the collisionless Boltzmann equation, which governs the distribution function of stars in the potential. The stars of the galaxy may be assumed to be collisionless because the time required to exchange energy through two-body collisions would be 10 million times the age of the Universe.

These equations incorporate the shape of the spiral waves, and in particular the "pitch" angle, namely the angle between the spiral and the circle of radius r. This pitch angle i satisfies $\tan(i) = 1/r\ dr/d\theta_o = 1/(kr)$, where k is the wavenumber, $k = 2\pi/\lambda$.

If we assume that the spirals are very tightly wound (WKB approximation), the pitch angle is close to zero, and $kr >> 1$, or $\lambda << r$. There are two possible winding directions, with either "trailing" or "leading" waves, as shown in Figure 4.13.

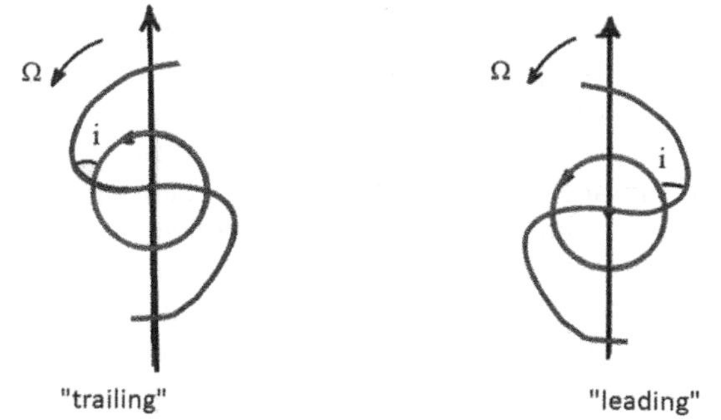

Figure 4.13. *Winding direction of the spiral arms relative to the direction of rotation of the galaxy. On the left, matter enters through the concave side of the arm, and the spiral is "trailing". On the right, matter enters through the convex side, and the spiral is "leading"*

The relation found by Lin and Shu (1964) shows the existence of two branches, short waves and long waves. The WKB approximation suggests that only the short branch is consistent with the hypotheses. Waves can only propagate everywhere in the disc, between the two Lindblad resonances (inner and outer), if the disc is not too "hot", i.e. the Toomre parameter Q is only equal to 1. Otherwise, a prohibited zone develops around the corotation that is increasingly wide at high values of Q, as shown in Figure 4.14. In the figure, the frequency ν^2 is plotted along the vertical axis as a function of the wavelength λ or wavenumber $k = 2\pi/\lambda$. The frequency of the wave is the frequency at which matter encounters a spiral arm and can be written as $\nu = m(\Omega_p - \Omega)/\kappa$ (normalized by the epicyclic frequency κ).

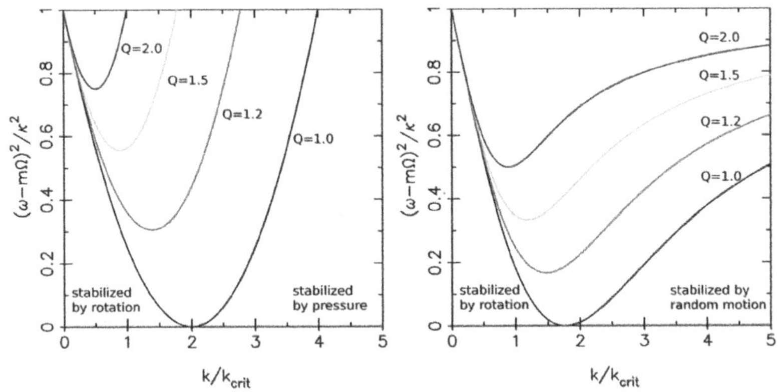

Figure 4.14. *Wave dispersion relation, revisited by Dobbs and Baba (2014). The right-hand diagram corresponds to the stellar component, without collisions, and revisits the relation calculated by Lin and Shu (1964). The left-hand diagram is the interstellar medium, fluid with dissipation. The critical wavenumber is $k_{crit} = \kappa^2/(2\pi G\Sigma_0)$. The left part of the diagram $k << k_{crit}$, where $\lambda >> \lambda_{crit}$, is the region of long waves; the right part, where $\lambda << \lambda_{crit}$, is the region of short waves. Normally, the WKB approximation imposes that only this branch should be considered. Nevertheless, the other branch gives a good idea of the solution. Note that the vertical axis corresponds to the corotation at the bottom and the Lindblad resonances at the top, either inner (ILR) or outer (OLR). For Q > 1, there is a zone around the corotation that is impossible for waves to cross in this approximation (except by the tunnel effect). The dispersion relation can tell us about the geometric shape of the waves. The wavelength is proportional to Q (in the short branch) or ~1/Q for long waves. If the disc is heated by instabilities, the spiral winds in the long branch and unwinds in the short branch*

In reality, monochromatic waves are not physical, multiple waves can coexist together, and waves travel in wave packets with group velocity $v_g = d\omega/dk$. Furthermore, waves damp relatively quickly when the response of the gas is taken into account. Interstellar gas responds strongly to excitation due to its low velocity dispersion. Even though its mass fraction is relatively low compared to the mass of stars, its response is strongly nonlinear and dissipative. There are shock waves at the entry to the spiral arms. The density contrast in the gas is much stronger, around 5–10, compared to just 1.2–1.5 for stars. The gas is compressed when it enters the arms, which triggers the formation of new stars. This makes the arms bluer and brighter, even though the density contrast of the stars is not very high. The shock waves create large velocity variations as they pass through the spiral arms – the characteristic "streaming" motion of density waves.

Ultimately, the question of the displacement of wave packets with sufficiently high velocity and the damping of waves as they cross the disc reduces the winding problem of physical spiral arms to another problem: the persistence of spiral density waves. What mechanism is at the origin of the formation of these waves, and, after being excited by an interaction or internal instability, how can the persistence of the spiral waves be explained?

Part of the solution is based on the reflection of waves, either at the center of the galaxy, or at the resonances, combined with an amplification effect. The waves either travel to the center of the galaxy, where they are reflected, or they are reflected before reaching it by a heat barrier if the Toomre parameter Q is too high in the center. The corotation plays a special role because the waves have negative energy traveling inward and positive energy traveling outward. At the corotation, a wave traveling outward can be partially transmitted and partially reflected inward. Transmitting a wave with positive energy outward amplifies the wave with negative energy that is reflected inward.

The swing phenomenon can cause a spectacular amplification that was studied and described by Toomre (1981). The amplification process occurs when a leading wave packet transforms into a trailing wave packet. It is caused by several mechanisms whose effects align and act concertedly: the differential rotation of the galaxy, self-gravity and epicyclic motion, which operate like a swing that allows self-gravity to act for longer.

The tidal forces caused by interacting with a companion were suspected to be responsible for the excitation of spiral waves as early as the 1970s (Toomre and Toomre 1972). Tidal forces have the right symmetry to form two spiral arms, and most galaxies with coherent spirals do in fact have two arms. Indeed, the azimuthal dependency of the tidal forces is a function of $\cos(2\theta)$, which explains the perturbation with $m = 2$. However, tidal forces mostly act at the edges of the galaxy, with little effect on the center. It can be shown that these forces vary like the square of the distance to the center for a distant companion. In the prototypical case of

Messier 51 and its companion NGC5195, which is located at the end of its spiral arm (see Figure 4.10), the spiral structure is very pronounced immediately from the center of the galaxy where the tidal forces are acting. And indeed, in the very simple simulations by Toomre and Toomre (1972) without self-gravity for a restricted three-body problem, the tidal spiral arms only form at the edges, and spirals are not created at the center. Self-gravity is required to reproduce the process of swing, and the characteristic amplification generated by self-gravity combines with the swing from the epicycles. The waves then propagate very quickly toward the center and are amplified so much that the tidal effects are much stronger inside the disc than at the edges, unlike in the test particle simulations.

Note that, if the orbit of the companion is inclined with respect to the plane of the first galaxy, the vertical tidal forces can distort the plane into a pancake shape. This is described as "warping". In this case, the azimuthal symmetry of these forces is m = 1.

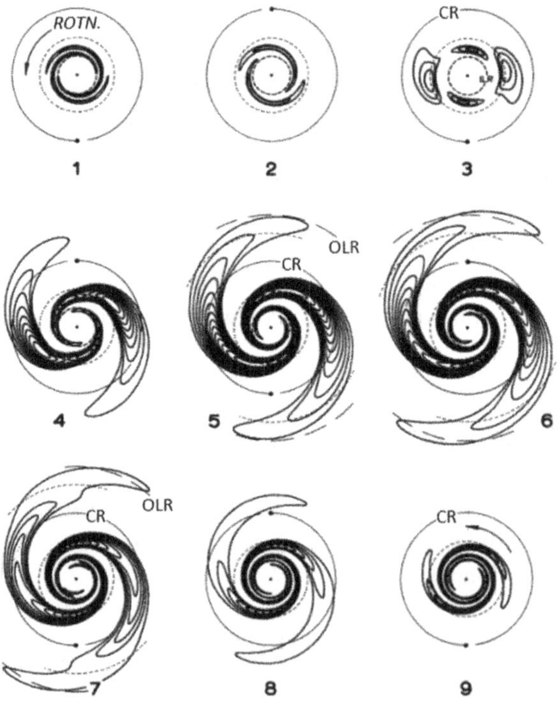

Figure 4.15. *Illustration of the amplification of the swing. In this simulation, a leading wave packet develops during the first three epochs, then unwinds and transforms into a trailing wave packet. Its amplitude increases considerably under the combined effect of self-gravity, differential rotation, and epicyclic swing. The wave packet then propagates toward the center, where it is damped (reproduced from Toomre 1981)*

4.3.3. *Role of gas and star formation*

When passing through the spiral arms, which form a gravitational potential well, the gas accelerates and is subjected to shock waves. The compression favors the transformation of the diffuse atomic gas HI into the denser molecular gas H_2. Thus, dust and molecular gas follow a very narrow spiral inside the spiral arm (see Figure 4.16).

The ionized regions (which emit the recombination line Hα) indicate young and massive stars whose lifetime is less than the time required to pass through the spiral arm. However, the images have various irregularities, and the arms appear somewhat chaotic. Star formation does not always proceed in an orderly fashion, and there are frequent exceptions, with stars forming between the arms, branches and extensions.

Figure 4.16. *Multicolor image obtained by the Hubble Space Telescope. The regions of ionized gas that emit the Balmer Hα line of atomic hydrogen are visible in pink. Only very hot stars of types O and B are capable of ionizing the surrounding gas. These massive stars consume their gas fuel very quickly, with a typical lifetime of 10 million years. They tend to be created at the entry of the spiral arms and expire before the crossing is complete. As a result, the arms are very bright and pink (credit: NASA/ESA)*

This is all caused by the nonlinearity of the physics of gas, star formation and feedback processes. Simulations have shown that the gas in the arms is subject to shearing (or Kevin–Helmholtz) instabilities that can cause extensions shaped like "spurs", as is often observed in the spiral arms of galaxies (see Figure 4.17).

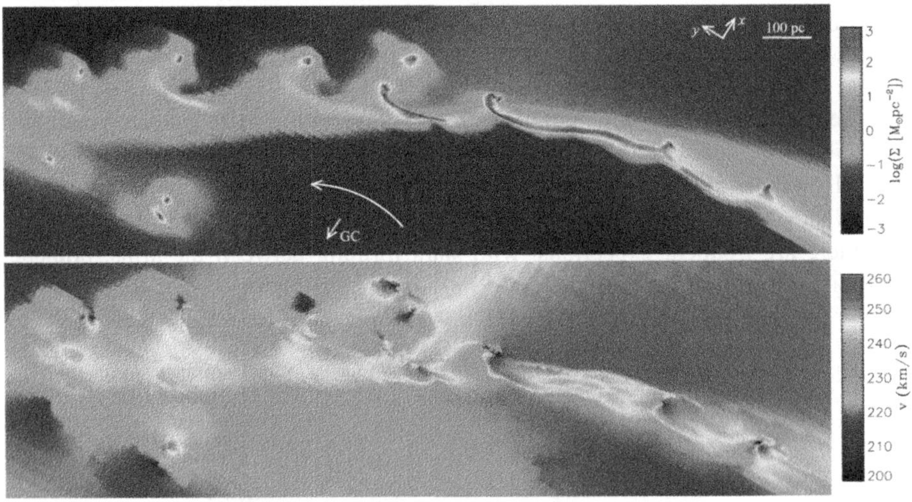

Figure 4.17. *High-resolution simulation of the gas in a spiral galaxy, with the associated star formation and feedback. The top image shows the surface density of the gas. The galaxy center is marked by a downward white arrow. Instabilities can be seen to develop, forming extensions and spurs. The bottom image shows the velocity field for the same region of the spiral arm. The very high velocity gradient in the right section produces a shear; this is the source of the Kelvin–Helmholtz instabilities developing on the left (reproduced from Renaud* et al. *2013)*

4.4. Bars: drivers of evolution

Two-thirds of spiral galaxies are barred, classified as SAB or strongly barred (SB). Of all disc galaxies, around one-third are SB and one-third are more moderately barred (SAB). Bars favor coherent spiral or "grand design" structures. Interactions between galaxies are also a determining factor, as described above. Flocculent spiral galaxies are therefore less common among barred galaxies, as shown in Figure 4.18.

The discs of galaxies are perpetually evolving; in extreme cases, they may be destroyed and reform. Whenever a galaxy evolves under its own gravity, it tends to concentrate its mass to approach the state of least energy. But rotation counteracts the concentration of mass. We must therefore determine the mechanisms to evacuate the angular momentum outward. The disc of a galaxy plays the role of an accretion disc. It is there to transfer angular momentum out of the galaxy.

Wherever there is interstellar gas, viscous torques are a possible candidate for this transfer. Dissipating the energy of this gas does indeed reduce its velocity dispersion, and the gas remains a "cold" component, but viscous torques are insufficient. The gas is very diffuse and highly fragmented into clouds. Dense molecular clouds do not behave like a fluid but like ballistic particles that are only involved in a collision once or twice per orbit around the galaxy center.

Gravity torques are therefore necessary to transfer the angular momentum. Asymmetries such as spiral arms or bars created tangential forces and torques. However, spiral arms are more transient, and their action is less impactful than bars. Bars are quasi-stationary density waves that can be viewed as a combination of leading and trailing wave packets. They are robust but still evolve secularly.

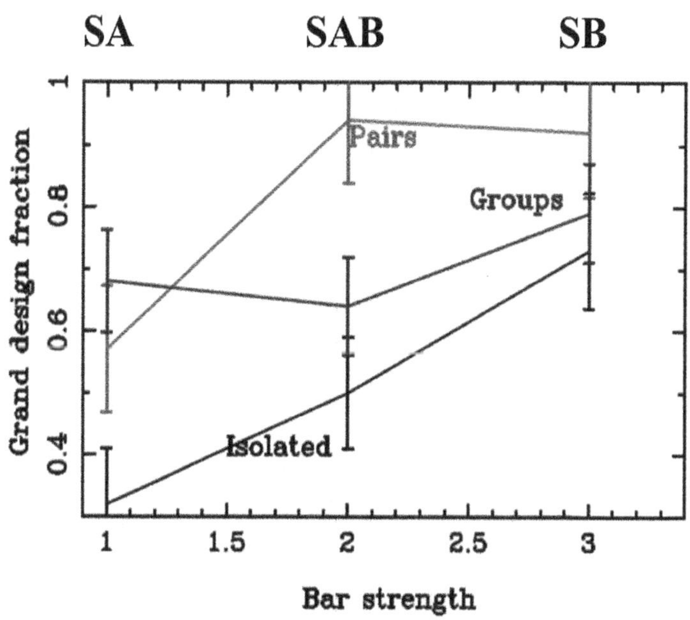

Figure 4.18. *The fraction of coherent spiral or "grand design" structures as a function of the bar strength for three types of environment: isolated galaxies, groups of galaxies, and pairs of galaxies. Barred galaxies are much more likely to have a spiral structure with density waves than a flocculent structure (reproduced from Elmegreen and Elmegreen 1983)*

4.4.1. *Formation of bars*

The goal of the first self-gravity simulations of galaxy discs in the 1970s and 1980s was to test the theory of density waves. But, although the simulated discs all started at equilibrium, satisfying the Toomre stability criterion, $Q > 1$, they were all unstable and ultimately always formed bars (Hohl 1971, 1976). The instability of bars had not been foreseen by the theory of density waves, since the latter had been working with the hypothesis of tightly wound waves, whereas bars are on the contrary very open waves.

Several solutions have been considered to explain the stability of discs, since at least one-third of spiral galaxies are not barred. The idea is to make the disc more turbulent and unsuitable for the development of waves, with a Toomre parameter that is $Q > 2$ or higher near the center and decreases away from it. Recall that the criterion $Q = 1$ stabilizes against disc fragmentation by means of velocity dispersion at small scales and differential rotation shearing forces at large scales. The first effect of increasing the value of Q is to increase the velocity dispersion. But another way to achieve stabilization is to make the disc less self-gravitating by placing it within a spherical potential from the dark matter halo. This idea was proposed by Ostriker and Peebles (1973). At the time, the rotation curves of spiral galaxies were beginning to show that invisible matter necessarily exists in the outer regions.

Ostriker and Peebles demonstrated stabilization for dark matter halos by performing self-gravitating N-body simulations with $N = 300$! They chose a spherical halo that was simulated non-coherently. An analytic potential was applied to the disc without feedback. The impact of this stability demonstration on the astronomy community was significant; it proved the utility of dark matter halos and gave evidence of their existence at a time where a consensus had not yet been reached. Later, once coherent simulations of dark matter could be performed with particles, like for stars and gas, it was observed that the dark matter interacts with the disc by exchanging angular momentum (see Athanassoula 2002). By accepting angular momentum from the disc, it enables a transfer and allows the bars to develop. Instead of suppressing bars, dark matter halos facilitate them.

Since then, various evolutions have improved the realism of simulations. First, interstellar gas was incorporated, a dissipative component that behaves differently than stars, which are a collissionless medium (see Figure 4.19). Later, incorporating star formation from gas and its feedback allowed various stellar populations in the disc to be accounted for, as well as stochastic effects in the morphology of the spiral arms. Resonance effects allowed the formation of rings to be explained (Buta and Combes 1996, see Chapter 1), and radial migration (Sellwood 2014) shed light on metallicity gradients.

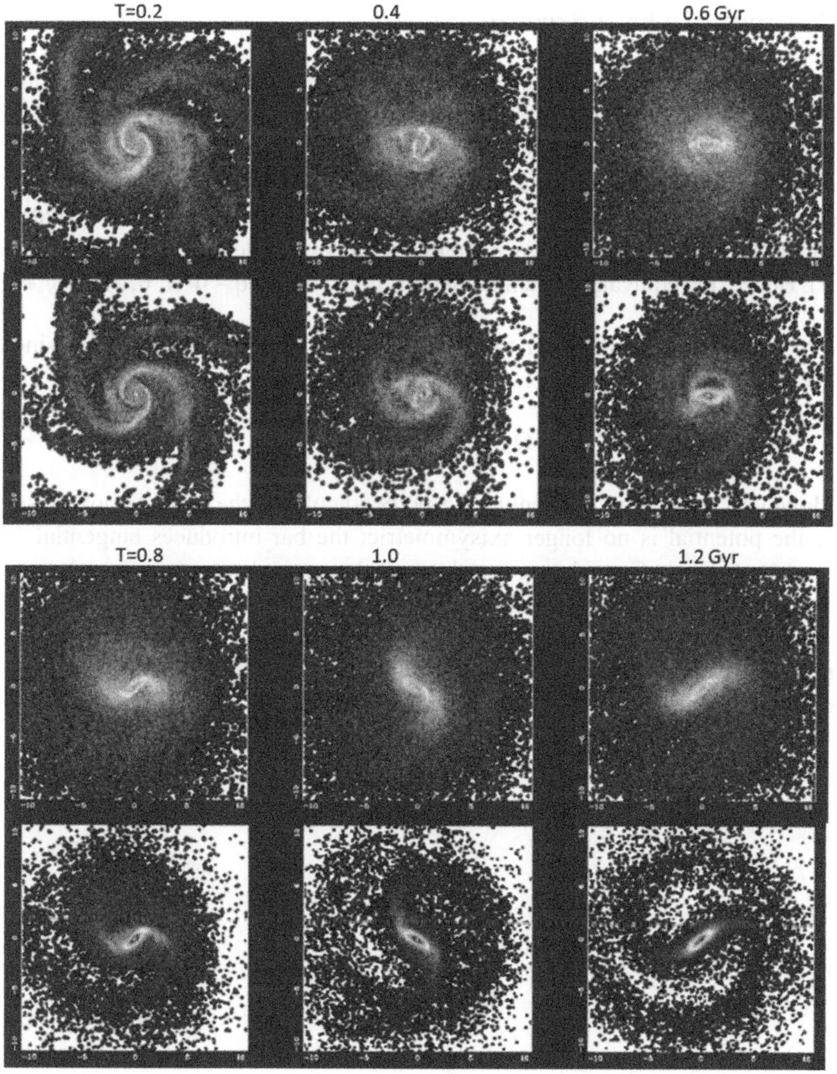

Figure 4.19. *(a) Simulations of a self-gravitating disc, containing stars (top panel) and interstellar gas (bottom panel). Both components are subject to the same self-gravitating potential well, but they behave very differently, since the gas is dissipative and is constantly cooled by radiation, whereas there is nothing to counteract the heating of the stars. Each epoch lasts for 200 million years. At first, a transient spiral forms to evacuate the angular momentum. (b) Continuation of the simulation of the formation of a bar. A barred wave develops in the stars, while continuing to generate spiral waves in the gas. Note the resonance rings in the gas at the inner Lindblad resonance. The ring has an elliptical shape that is elongated parallel to the bar*

4.4.2. *Orbits in a barred galaxy*

To better understand the dynamical phenomena associated with bars, studying the orbits of stars in a barred potential, in particular in a frame of reference that rotates with the bars, is essential. These orbits were calculated and precisely described by Contopoulos and Papayannopoulos (1980). In the frame rotating with the bar at angular velocity Ω_b, the inertial term $\frac{1}{2}\,\Omega_b^2\, r^2$ is added to the bisymmetric gravitational potential Φ (m = 2) of the bar (since the inertial force that derives from this potential is $\Omega_b^2\, r$). Hence, the equivalent potential Φ eq is given by $\Phi\ \text{eq} = \Phi\,(r, \theta, z) - \Omega_b^2\, r^2/2$.

The energy in this frame, or the Jacobian EJ, is an integral of motion, since the potential is independent of time:

$$EJ = v^2/2 + \Phi\,(r, \theta, z) - \Omega_b^2\, r^2/2.$$

However, the kinetic momentum Lz perpendicular to the plane is not conserved, since the potential is no longer axisymmetric; the bar introduces tangential forces and gravity torques. Considering the shape of the equivalent potential in the rotating frame is instructive because it determines the shape of the orbits. Figure 4.20 shows the equipotentials. The bar is horizonal, and the Lagrange points are the stationary points: L4 and L5 are maxima, L1 and L2 are saddle points, all around the corotation with the bar.

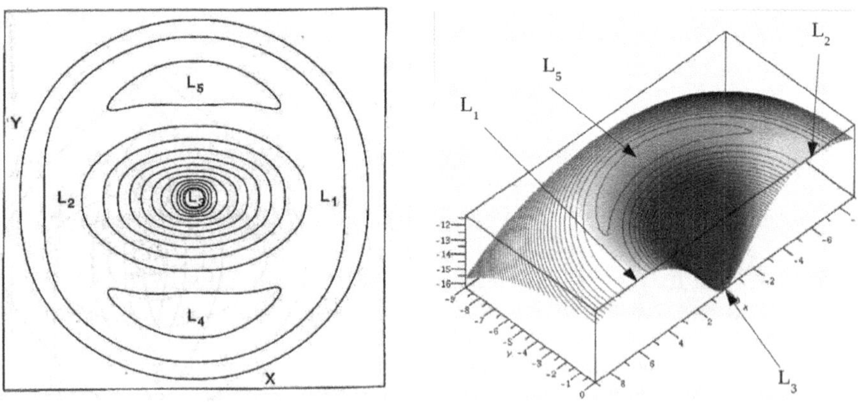

Figure 4.20. *Left: Shape of the equipotential lines in the frame that rotates with the bar, for a bar along the Ox axis. There are five Lagrange points: L3 at the center of the potential, L1, L2, L4 and L5 at the highest points, located at the corotation of the bar along the axes Ox and Oy. Right: Perspective and cross-section to better illustrate the potential in three dimensions (credit: D. Pfenniger)*

The shape of the equipotentials already suggests that the principal orbits will align with the bar inside the corotation; outside, they will be perpendicular to it. There will be stable orbits that only revolve around L4 and L5. The periodic orbits form a skeleton that defines the orbits of every star in the galaxy. They attract and trap the other orbits, the majority of which are not periodic. There are also chaotic orbits that are not trapped around periodic orbits.

More precisely, the orbits are divided into the following families:

1) very close to the center, and up to the corotation if there is no inner Lindblad resonance (ILR), the orbits are parallel to the bar and support it. This family is called x1;

2) between the two ILRs, if they exist, there are x2 orbits that are perpendicular to the bar, direct and stable. The x2 family disappears if the bar strength is too high (which also removes the ILRs);

3) if there are 2 ILRs, the x1 orbits parallel to the bar resume between the second ILR and the corotation;

4) at the corotation, there are stable orbits around the Lagrange points L4 and L5 that do not rotate, in the reference frame of the bar;

5) after the corotation, the orientation of the orbits changes again, and they become perpendicular to the bar and do not support it. The bar must therefore end at its corotation (see Figure 4.21).

There is a simple way to summarize the orientation of the orbits: they are either parallel or perpendicular to the bar, and the orientation of the orbits changes (by 90°) at each resonance.

By knowing the families of orbits, we can understand the evolution of the bars better. When bars first form, simulations show that they have relatively high angular momentum, and no ILR. The orbits in the x1 family support the bar, which grows as it traps more and more stars with a low precession rate, slowing down. One ILR can potentially develop, followed by a second, and orbits in the x2 family can weaken the bar, possibly even leading to its destruction. This process can be viewed as a form of self-regulation of the bar strength.

Furthermore, the orbits no longer support the bar past the corotation (CR), becoming perpendicular to it. This explains why bars generally end just inside their corotation, sometimes with a 4:1 resonance ring not far from the CR. This is an excellent diagnostic for determining the angular momentum of the bar Ω_b.

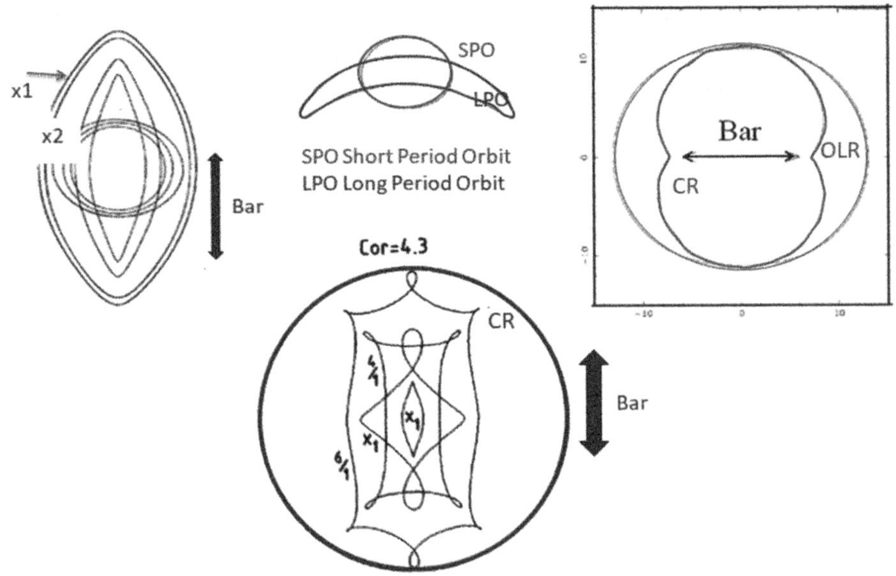

Figure 4.21. *Periodic orbits characteristic of a barred galaxy in the rotating frame of reference. Left: The x1 and x2 families at the center of the bar. Center: The stable orbits around the L4 and L5 Lagrange points. Right: Periodic orbits perpendicular to the bar, outside of the corotation, up to the outer Lindblad resonance (OLR). Bottom: Higher resonances, very close to the corotation (reproduced from Contopoulos and Papayannopoulos 1980)*

4.4.3. *Response of gas to a barred potential*

In the disc of a barred galaxy, the gas tends to follow the same periodic orbits as the stars; however, stars are collisionless, whereas interstellar gas is not. The gas clouds collide, which prevents the orbits or streamlines from crossing in practice. Instead of rotating abruptly by 90° at each resonance, the gas streamlines turn gradually to finally reach 90° at each resonance. The gas response therefore has a spiral structure, whereas this spiral disappears with stars (see Figure 4.22).

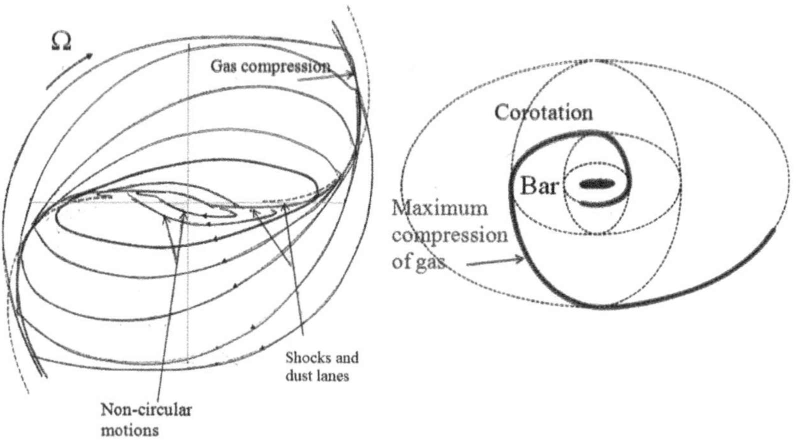

Figure 4.22. *Due to collisions between gas clouds, the streamlines of the interstellar gas gradually rotate by 90° at each resonance, rather than abruptly like for stars. The orbits resemble Kalnajs kinematic ellipses (see Figure 4.12), and they produce spiral arms as they tighten. If there are three resonances in the disc, the spiral winds by 270° (see right diagram)*

Once the gas has distributed itself into a spiral structure, it experiences tangential forces from the bar, and hence torques, which causes a transfer of angular momentum. This does not happen if the gas is distributed axisymmetrically (e.g. in a ring) or symmetrically with respect to the bar (e.g. aligned with the bar). In this case, the tangential forces and associated torques cancel out on average over each orbit. Figure 4.23 gives a geometric representation showing the direction in which the torques act. The bar defines four quadrants in the space of the disc: two opposite quadrants (NE and SW) where the torque is positive, and two other quadrants (NW and SE) where the torque is negative relative to the direction of rotation of the galaxy. The torques change sign at each resonance. Outside of the corotation, the gas receives angular momentum and is driven outward to the outer Lindblad resonance (OLR), where it forms a somewhat elongated ring (perpendicular to the bar just before the OLR, in the shape of a figure of eight; or elongated parallel to the bar just outside of the OLR; see Figure 4.21). Arranged as a symmetric ring relative to the bar, it no longer receives torque and is stable. Inside of the corotation, the gas loses angular momentum and spirals toward the center. If there is an ILR, the gas stops there and forms a nuclear ring. Inside the ILR and up to the center, the spiral continues in the "trailing" direction in opposite quadrants, and the torque is positive, carrying the gas toward the ILR, where it accumulates and forms stars. If there are two ILRs, and especially if the gas falls within the sphere of influence of the central black hole, transient leading spirals can form rapidly in regions where the precession rate $\Omega - \kappa/2$ increases as the radius decreases. In these cases, the sign of the torque is once again inverted, and the gas can supply the nucleus (see Buta and Combes 1996).

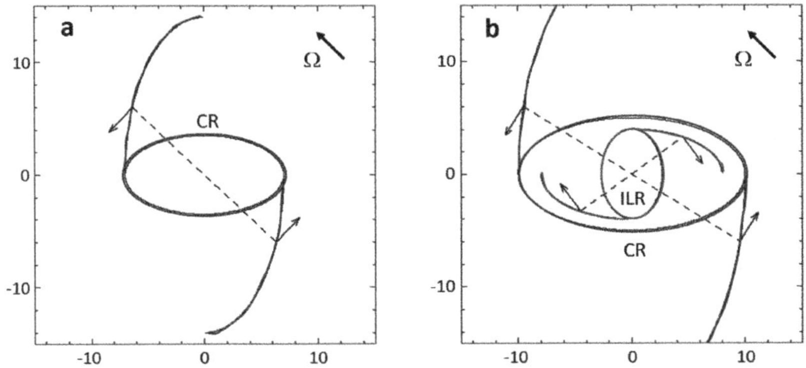

Figure 4.23. *Diagram showing the direction of the gravity torques exerted by the bar on the gas distributed in a spiral structure: (a) outside of the corotation, the torque is positive, and the gas is driven out toward the outer Lindblad resonance (OLR); (b) inside of the corotation, the torque is negative, and the gas will spiral toward the center*

After the gas loses its angular momentum, it accumulates toward the center and forms stars. The mass concentration gradually increases, which also increases the precession frequency $\Omega - \kappa/2$ and generates two ILRs. The bar is weakened by the perpendicular orbits, which causes a secondary or nuclear bar to decouple within the last ILR. This bar, affected by the high precession rate at the center, rotates with a higher angular velocity than the primary bar. In numerical simulations, it is often possible to observe a nuclear bar that shares resonances with the primary bar in order to minimize chaotic orbits. For example, the ILR of the primary bar is the corotation of the secondary bar (Friedli and Martinet 1993; Garcia-Burillo *et al.* 1998).

The nuclear bars extend the action and gravity torques of the primary bar to the center of the galaxy. They supply the active galaxy nuclei (e.g. Shlosman *et al.* 1989). There may even be a whole cascade of nested waves, as has been found in simulations (Hopkins and Quataert 2010).

4.4.4. *Vertical resonances and peanuts*

Viewing simulations of barred galaxies edge-on reveals a peanut structure, similar to a bulge but caused by the stars in the disc (Combes and Sanders 1981; Combes *et al.* 1990; Raha *et al.* 1991). These structures had been identified in the observations of the earliest catalogs but had previously not been interpreted. The peanut galaxy NGC 128, shown in Figure 4.24, is a prototype for them.

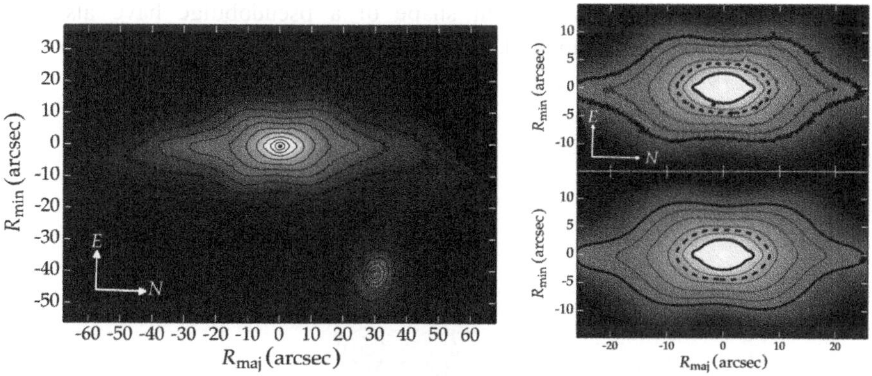

Figure 4.24. *Example of a peanut structure: the galaxy NGC 128, viewed edge-on. On the left, red band image from the SDSS. On the right, near infrared image from the Hubble Space Telescope (top) and model with ISOFIT/CMODEL (bottom) (eproduced from Ciambur and Graham (2016)). NGC 128 is in the process of interacting with its companion NGC 127 to the north-west (here, bottom-right)*

Simulations have shown that this structure is caused by the vertical elevation of the stars of the disc, which resonate with the bar at the vertical Lindblad resonance radius when $\Omega_b = \Omega - \nu z/2$, where νz is the vertical oscillation frequency (Combes *et al.* 1990). This resonance is not far from the ILR in the plane, since the frequency νz is close to the epicyclic frequency κ. The peanut shape is one way to recognize the existence of a bar in galaxies viewed edge-on (Bureau and Freeman 1999). The peanut thickens the plane, which causes a three-dimensional dilution of the bar and weakens it. Nevertheless, the bar can still evolve by exchanging angular momentum with the dark matter halo. The bar strength may still increase, and even its angular velocity Ω_b may still decrease. Resonance then occurs at a larger radius, and a second peanut with a larger radius develops (Martinez-Valpuesta *et al.* 2006).

The peanut shape is especially apparent when the bar is seen along its major axis in a galaxy viewed edge-on. In other projections, it mostly appears as a box shape. In any case, it is a bulge, called a "pseudo-bulge" as opposed to a classical bulge. The key differences between these two types of bulge are: their rotation – pseudo-bulges originate from the disc and are still strongly rotating, unlike classical bulges, which originate from minor mergers; their light profile – pseudobulges have a Sersic index close to n = 1–2 rather than >4 like classical bulges; and their flattening – pseudo-bulges are flatter, whereas classical bulges are more spheroidal (Fisher and Drory 2010).

The bulge of our own galaxy, the Milky Way, is a pseudobulge caused by the vertical resonance of its bar in the center (e.g. Di Matteo *et al.* 2015).

The orbits supporting the peanut shape of a pseudobulge have also been identified. They are either shaped like a "pretzel" (Portail *et al.* 2015) or a "banana" (Combes *et al.* 1990; Figure 4.25).

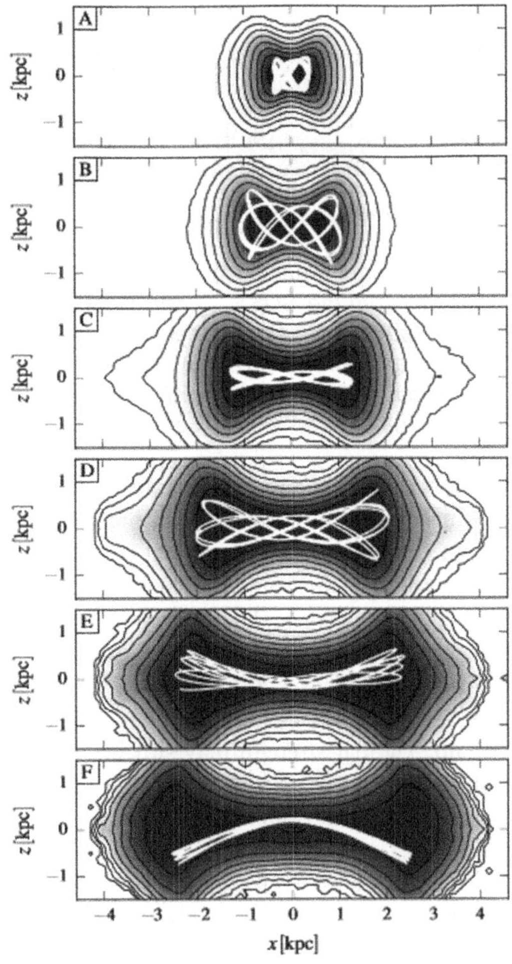

Figure 4.25. *Families of orbits in the projection where the galaxy is viewed edge-on and the bar is viewed along its major axis. The orbit categories A to F make up the morphology of the peanut bulge. In particular, the category C contains "pretzels" and the category F contains "bananas" (reproduced from Portail* et al. *2015)*

4.4.5. *Dark matter and bars*

In section 4.4.1, we saw that dark matter halos facilitate the formation of bars by exchanging angular momentum. The presence of a bar cannot be stabilized by the dark matter halo, as predicted by Ostriker and Peebles (1973). Nevertheless, the bar can constrain the quantity of dark matter at the center of the galaxy, within the radius of the bar, by dynamical friction effects.

The standard cosmological model of dark matter (CDM) predicts a central concentration of dark matter, corresponding to a central peak in the radial density distribution, described as a cusp. This is namely the NFW profile (Navarro *et al.* 1997), where the density $\rho(r) \propto 1/r$, meaning that it satisfies a power law with slope 1. This cusp is not observed in rotation curves. Instead, the observed density is compatible with a core, i.e. a flat density function $\rho(r)$. The existence of a bar could further constrain the quantity of dark matter in the center. Indeed, as the bar rotates, it is slowed by dynamical friction on the dark matter halo, losing all of its speed if the halo density is too high. The corotation should therefore be pushed out of the disc, but this is not observed in practice. The relevant calculations were performed by Debattista and Sellwood (2000). They show that, to be compatible with the observations of rapid bars, the rotation curve must be dominated by the visible mass within the radius of these bars. The definition of a rapid bar can be specified in terms of the ratio of the corotation radius and the bar radius. The observed value of this ratio is RCR/Rbar ~1.2. The resulting constraint on dark matter is already significant, but it can be further reinforced by considering bars in galaxies with low mass or dwarf galaxies, such as the large Magellanic Cloud. However, it remains difficult to find the pattern speed of the bar in these galaxies, which are irregular, very rich in gas, and do not have any resonance rings.

Another possibility might be to completely change the model by adopting modified gravity without a dark matter halo, for example the MOND model proposed by Milgrom (1983). A study of the formation and evolution of bars was performed by Tiret and Combes (2007, 2008). Without a halo of particles, dynamical friction is essentially inexistent for the bar, which does not slow down, as shown in Figure 4.26.

Note that the observed frequency of bars is also a very critical test for any dynamical model of galaxy formation, as well as any dark matter model. In a modified gravity model without a dark matter halo, the disc is completely self-gravitating and more unstable given the same observed velocity dispersion. We might therefore expect to find more unstable discs, either fragmented or forming more bars. But there are multiple effects combined: in the MOND model, the bars arise earlier and weaken faster by forming peanut bulges. Once they weaken, they can no longer be amplified, unlike the bars of the CDM model, where the halo can retake some angular momentum from the stars of the disc and amplify a second bar. In the modified gravity model, angular momentum is exchanged with stars in the disc, and the outer regions are stretched out. After the first bar weakens, these disc particles are no longer available to amplify a second bar. If we summarize and count the bars of both models, we find that they are slightly more numerous for MOND at a given strength, which seems to fit the observations better.

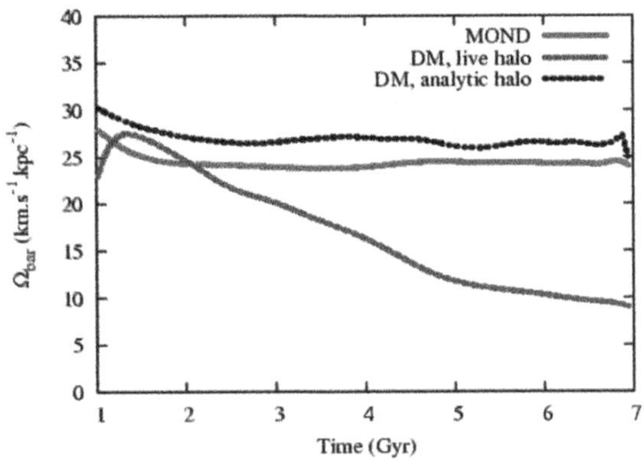

Figure 4.26. *Evolution of the pattern speed of the bar over 7 billion years for three simulated models. In blue, a dark matter model (CDM) with a rigid dark matter halo that only imposes an analytic potential, without dynamical friction. In green, a realistic, live CDM model with particles, which slows the bar down by dynamical friction. In red, the MOND modified gravity model, without a halo of particles, with dynamical friction only from stars in the disc (reproduced from Tiret and Combes 2007)*

Even in the standard model, the frequency of bars and strong bars is problematic (see Block *et al.* 2002; Eskridge *et al.* 2002). For a sample of around 200 galaxies observed in the near infrared, the histogram of galaxies as a function of their bar strength is deficient for unbarred galaxies relative to simulated predictions. In simulations that account for gas, the bars lead the gas toward the center via gravity torque. This accumulation of matter in the center weakens the bar, especially since it accepted angular momentum from the gas by exerting torque upon it. This causes the bar to weaken even further, since it is a wave with negative momentum. To resolve this discrepancy between models and observations, we must consider that spiral galaxies are not isolated throughout their lifetimes but are enriched by accretion of matter originating from cosmic filaments. In particular, galaxies accrete gas that refills the disc and cools it down due to its low velocity dispersion. The disc then becomes unstable once again to form a second bar or reinforce the first bar (Bournaud and Combes 2002; Bournaud *et al.* 2005). In this way, by considering several barred episodes in spiral galaxies together, we can reproduce the near infrared observations of the frequency of bars in such galaxies more closely. For a good level of agreement, galactic discs would need to double their mass in 7–10 Gyr, i.e. less than the Hubble time. In the modified gravity model, the accretion of gas must also be taken into consideration, and the bar frequency is also compatible with observations (Tiret and Combes 2008).

4.5. Environment of spiral galaxies

4.5.1. *Morphological segregation*

As a population, spiral galaxies are found in different environments than the other categories of ETGs, namely lenticular and elliptical galaxies. Dressler observed as early as 1980 that the fraction of the various populations varies highly significantly with the surface density of galaxies projected onto the sky (see Figure 4.27). A recent version of this diagram was compiled by Cappellari (2016) with more galaxies. Here, the main effect of the environment is not just to change the morphology of galaxies, but also to stop their star formation, transforming galaxies that are still active into inactive red galaxies. At the beginning of this chapter, we cited various mechanisms that can cause this phenomenon: stripping of gas by ram pressure, tidal interactions and harassment, and most importantly suffocation, by suppressing the accretion of cold gas from cosmic filaments. However, although it is conceivable to imagine that gas stripping could transform a spiral galaxy into a lenticular galaxy, the increased frequency of ellipticals due to mergers within galaxy clusters cannot be explained in this way, since the relative velocities within the clusters are far too high, with hyperbolic trajectories; galaxy pairs rarely form in rich environments. Within groups, merging pairs can still form, since the relative velocity is sufficiently low. It therefore seems that galaxy mergers resulting in elliptical galaxies take place at a lower level; they are pre-evolved outside of their clusters, so to speak, within a group or filament. These groups then merge within richer environments, namely clusters, supplying the latter with elliptical galaxies and explaining why such galaxies accumulate in denser regions.

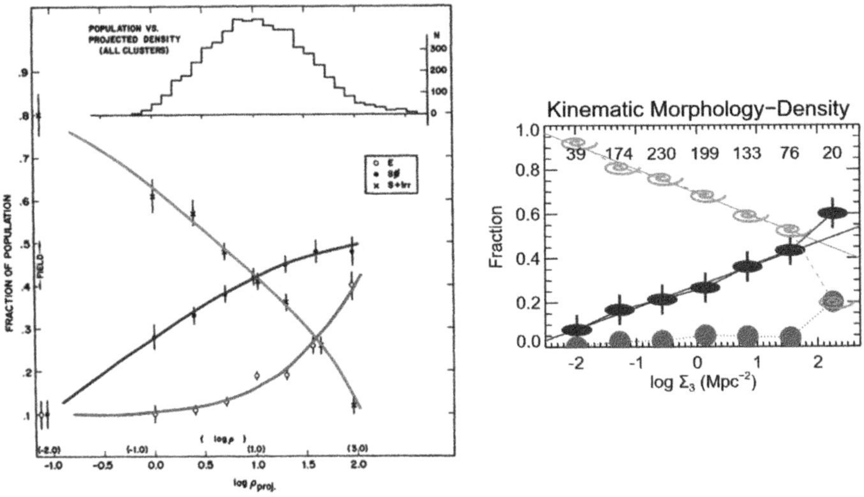

Figure 4.27. *Fraction of galaxies for three different populations: spiral galaxies (green), lenticular galaxies (blue) and elliptical galaxies (red) as a function of the local environment, i.e. the surface density in galaxies Σ or ρ-proj. On the left, the original diagram by Dressler (1980), with the number of affected galaxies shown above. On the right, a more recent version of this diagram (Cappellari 2016)*

4.5.2. *The problem of bulgeless galaxies*

Rich galaxy environments are fairly rare, and, locally, most galaxies are spiral galaxies (around two-thirds). But the vast majority of them only have a very weak bulge, or no bulge at all. This observation is problematic for the standard cosmological model, which is hierarchical.

Locally, around two-thirds of bright spiral galaxies have very little bulge or no bulge at all (Kormendy and Fisher 2008; Weinzirl *et al.* 2009; Kormendy *et al.* 2010). The remaining one-third consists of galaxies with both a classical bulge and a pseudobulge, sometimes also with clusters of stars in the nucleus (Böker *et al.* 2002). Galaxies viewed edge-on and extremely thin galaxies are common (Kautsch *et al.* 2006). According to the statistics of more than 100,000 galaxies from the Sloan survey, more than 20% of bright spiral galaxies have no bulge at all up to a redshift of z = 0.03 (Barazza *et al.* 2008). How can these observations be reconciled with the hierarchical scenario?

Classical bulges, which are spheroidal with little flattening, no rotation, and a Sersic profile with index n greater than 3, are known to form in galaxy mergers, both minor and major, which causes them to lose their angular momentum and concentrate their mass. Conversely, pseudobulges are the result of the secular evolution of discs, with vertical resonance caused by bars. However, at large redshifts, in the first half of the history of the Universe ($z > 1$), every galaxy experienced a phase where the gas fraction dominated. In these galaxies, where the gas fraction is larger than 50%, the disc is very unstable and fragments into 3-4 clumps with a baryonic mass of 10^8–10^9 Mo, giving distant galaxies a very piecewise appearance, often described as "clumpy" (see Figure 4.28).

Simulations reveal discs of fragments so massive that their dynamical friction on the dark matter halo slows them down very quickly. In less than a billion years, they spiral toward the center, forming a bulge (Noguchi 1999; Bournaud *et al.* 2007).

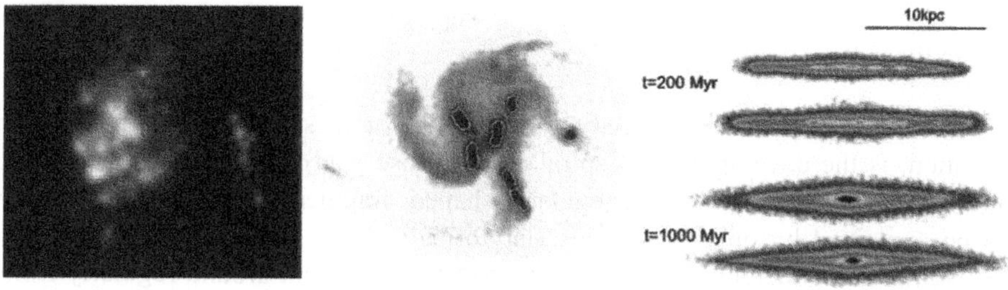

Figure 4.28. *Galaxies in the first half of the history of the Universe are dominated by gas, which makes them very unstable and fragmented into "clumpy" objects. On the left, observations of a typical galaxy in the ultra-deep field (UDF) of the HST, at z = 1.4. In the center and on the right, a simulation, viewed face-on and edge-on, showing the formation of a bulge over a billion years (reproduced from Bournaud* et al. *2007)*

This problem could be resolved by assuming that every spiral galaxy without a bulge or with a classical bulge of very low mass (less than 20% of the total visible mass) has not experienced a major merger since a redshift of $z = 4$ (during around the last 12 billion years). But there is still a classical bulge from the fragmented phase of the first half of the history of the Universe in every galaxy, which does not allow us to account for extremely thin galaxies without any bulge. We would therefore need to invoke a modified gravity model to avoid dynamical friction of the fragments on dark matter halos (Combes 2014). In such a case, the fragments would have time to be destroyed by feedback from star formation and supernovae before being assembled into a bulge.

4.6. Conclusion

Recent large-scale sky surveys have made it possible to study hundreds of thousands of galaxies statistically. This has given rise to a different perspective, one that complements the Hubble sequence classification of galaxies, which remains very useful. Galaxies appear to be separated into two sequences: red and blue, with a relatively unpopulated transition between them, known as the green valley. This bimodality suggests that the transition from the active star formation phase and the inactive red phase occurs rapidly. This rapid transition is closely linked to both the stellar mass of galaxies and their environments.

Various mechanisms have been suggested to explain why star formation is quenched, and numerical simulations have been very helpful. Internal feedback processes originating from either star formation or active galaxy nuclei play a role. But the environment is the most essential factor, as it can permanently suffocate a galaxy by stripping its interstellar gas or suppressing its cold gas supply from cosmic filaments.

One of the driving evolutionary forces, both for mass concentration and star formation, is the development of spiral arms in disc galaxies. There are flocculent spiral galaxies whose arms are irregularly shaped, fragmented and stochastic; they can be explained by only contagious star formation and differential rotation. But most spirals are coherent, typically with two arms that wind around regularly from the center to the edges. These arms are in fact density waves, generated by the tidal interaction of a companion galaxy, then amplified by swing.

Spirals are sometimes transient, whereas bars are a more robust form of density wave. They allow angular momentum to be exchanged between the inner and outer regions of the galaxy, facilitating the concentration of its mass. They are the driving force of secular evolution, forming rings at the Lindblad resonances, pseudobulges by vertical resonance, and spiral structures in the gas.

Spiral galaxies are dominant in the field but give way to lenticular and elliptical galaxies in richer environments such as groups and clusters. These processes likely unfolded gradually, with mergers between galaxies within small groups before these groups then merged within clusters. Many observations are still poorly understood, and there remain dynamical processes to be discovered. Over the next few years, large surveys such as those being conducted with the Euclid satellite, or the LSST on the ground, will further expand our statistical knowledge of galaxies, from 1 million to 10 billion. Giant telescopes both on the ground (both optical with extremely large telescopes and radio with the square kilometer array) and in space

(James Webb Space Telescope) will allow a fine study of galaxies with high sensitivity and spatial resolution extending up to the confines of our horizon.

4.7. References

Athanassoula, E. (2002). *ApJ*, 569, L83.

Baldry, I.K., Glazebrook, K., Brinkmann, J., *et al.* (2004). *ApJ*, 600, 681.

Baldry, I.K., Balogh, M.L., Bower, R.G., *et al.* (2006). *MNRAS*, 373, 469.

Barazza, F.D., Jogee, S., Marinova, I. (2008). *ApJ*, 675, 1194.

Baugh, C.M. (2006). *RPPh*, 69, 3101.

Bigiel, F., Leroy, A., Walter, F., *et al.* (2008). *AJ*, 136, 2845.

Block, D.L., Bournaud, F., Combes, F., *et al.* (2002). *A and A*, 394, L35.

Böker, T., Laine, S., van der Marel, R.P., *et al.* (2002). *AJ*, 123, 1389.

Bournaud, F., Combes, F. (2002). *A and A*, 392, 83.

Bournaud, F., Combes, F., Semelin, B (2005). *MNRAS*, 364, L18.

Bournaud, F., Elmegreen, B.G., Elmegreen, D.M. (2017). *ApJ*, 670, 237.

Bureau, M., Freeman, K.C. (1999). *AJ*, 118, 126.

Buta, R., Combes, F. (1996). *Fund. Cosmic Phys.*, 17, 95.

Cappellari, M. (2016). *ARA and A*, 54, 597.

Ciambur, B.C., Graham, A.W. (2016). *MNRAS*, 459, 1276.

Combes, F., Sanders, R.H. (1981). *A and A*, 96, 164.

Combes, F., Debbasch, F., Friedli, D., *et al.* (1990). *A and A*, 233, 82.

Combes, F. (2014). *A and A*, 571, A82.

Contopoulos, G., Papayannopoulos, T. (1980). *A and A*, 92, 33.

Cowie, L.L., Songaila, A., Hu, E.M., *et al.* (1996). *AJ*, 112, 839.

Croton, D.J., Springel, V., White, S.D.M., *et al* (2006). *MNRAS*, 365, 11.

Debattista, V., Sellwood, J.A. (2000). *ApJ*, 543, 704.

Dekel, A, Silk, J (1986). *ApJ*, 303, 39.

Dekel, A, Birnboim, Y. (2006). *MNRAS*, 368, 2.

Di Matteo, P., Gomez, A., Haywood, M., *et al.* (2015). *A and A*, 577, A1.

Dobbs, C., Baba, J. (2014). *PASA*, 31, 35.

Dominguez Sanchez, H., Huertas-Company, M., Bernardi, M., *et al.* (2018). *MNRAS*, 476, 3661.

Dressler, A. (1980). *ApJ*, 236, 351.

Driver, S.P., Allen, P.D., Graham, A.W., *et al.* (2006). *MNRAS*, 368, 414.

Eggen, O.J., Lynden-Bell, D., Sandage, A.R. (1962). *ApJ*, 136, 748.

Eke, V.R., Baugh, C.M., Cole, S., *et al* (2006). *MNRAS*, 370, 1147.

Elmegreen, B.G., Elmegreen, D.M. (1983). *ApJ*, 267, 31.

Eskridge, P.B., Frogel, J.A., Pogge, R.W., *et al.* (2002). *ApJS*, 143, 73.

Fall, S.M., Romanowsky, A.J. (2013). *ApJ*, 769, L26.

Fisher, D.B., Drory, N. (2010). *ApJ*, 716, 942.

Freeman, K.C. (1970). *ApJ*, 160, 811.

Friedli, D., Martinet, L. (1993). *A and A*, 277, 27.

Garcia-Burillo, S., Sempere, M.J., Combes, F., *et al.* (1998). *A and A*, 333, 864.

Gerola, H., Seiden, P.E. (1978). *ApJ*, 223, 129.

Helmi, A., White, S.D.M (1999). *MNRAS*, 307, 495.

Hohl, F. (1971). *ApJ*, 168, 343; (1976). *AJ*, 81, 30.

Hopkins, P.F., Quataert, E. (2010). *MNRAS*, 407, 1529.

Jenkins, A., Frenk, C.S., White, S.D.M. *et al* (2001). *MNRAS*, 321, 372.

Kalnajs, A.J. (1973). *PAS*, Au 2, 174.

Kauffmann, G., Heckman, T.M., White, S.D.M., *et al.* (2003). *MNRAS*, 341, 54.

Kautsch, S.J., Grebel, E.K., Barazza, F.D., *et al.* (2006). *A and A*, 445, 765 and 451, 1171.

Kormendy, J., Fisher, D.B. (2008). *ASPC*, 396, 297.

Kormendy, J., Drory, N., Bender, R., *et al.* (2010). *ApJ*, 723, 54.

Lin, C.C., Shu, F.H. (1964). *ApJ*, 140, 646.

Lintott, C.J., Schawinski, K., Slosar, A., *et al.* (2008). *MNRAS*, 389, 1179.

Madau, P., Dickinson, M. (2014). *ARA and A*, 52, 415.

Martig, M., Bournaud, F., Teyssier, R., *et al.* (2009). *ApJ*, 707, 250.

Martinez-Valpuesta, I., Shlosman, I., Heller, C. (2006). *ApJ*, 637, 214.

Milgrom, M. (1983). *ApJ*, 270, 365.

Navarro, J.F., Frenk, C.S, White, S.D.M. (1997). *ApJ*, 490, 493.

Noguchi, M. (1999). *ApJ*, 514, 77.

Ostriker, J.P., Peebles, P.J.E. (1973). *ApJ*, 186, 467.

Peng, Y., Lilly, S.J., Kovac, K., *et al* (2010). *ApJ*, 721, 193.

Portail, M., Wegg, C., Gerhard, O. (2015). *MNRAS*, 450, L66.

Raha, N., Sellwood, J.A., James, R.A., *et al.* (1991). *Nature*, 352, 411.

Renaud, F., Bournaud, F., Emsellem, E., *et al.* (2013). *MNRAS*, 436, 1836.

Schawinski, K., Urry, C.M., Simmons, B.D., *et al.* (2014). *MNRAS*, 440, 889.

Searle, L., Zinn, R. (1978). *ApJ*, 225, 357.

Sellwood, J.A. (2014). *RvMP*, 86, 1.

Shlosman, I., Frank, J., Begelman, M.C. (1989). *Nature*, 338, 45.

Sweet, S.M., Fisher, D., Glazebrook, K., *et al.* (2018). *ApJ*, 860, 37.

Tiret, O., Combes, F. (2007). *A and A*, 464, 517.

Tiret, O., Combes, F. (2008). *A and A*, 483, 719.

Toomre, A. (1981). *Proceedings of the Advanced Study Institute*, Cambridge University Press, 111–136.

Toomre, A, Toomre, J. (1972). *ApJ*, 178, 623.

Weinzirl, T., Jogee, S., Khochfar, S., *et al.* (2009). *ApJ*, 696, 411.

Wuyts, S., Förster Schreiber, N.M., van der Wel, A., *et al.* (2011). *ApJ*, 742, 96.

5

Galaxy Mergers and Interactions through Cosmic Time

Christopher J. CONSELICE
Astrophysics, The University of Nottingham, UK

This chapter reviews our understanding of galaxy mergers and interactions including how both processes are important methods within galaxy formation and evolution as well as for triggering star formation and central galaxy black hole build-up. First a background is given to the subject, including the various definitions used to define and measure the merger and interaction history of galaxies. Then observations are described that show that the merger fraction and rate increase steeply up to $z = 6$ – over 13 billion years of cosmic time. This information can be used to determine the number of galaxy mergers a typical galaxy experiences over time, as well as the quantitative role of mergers in forming the masses of galaxies through measurements of the merger mass accretion history. Another issue discussed in this chapter is the correlation of galaxy mergers and interactions and the triggering of star formation and AGN activity. In conclusion, future applications of measuring mergers are discussed, such as correlations with the properties of dark energy and dark matter, and also future observational efforts with telescopes such as James Webb Space Telescope (JWST) and Laser Interferometer Space Antenna (LISA).

For color versions of the figures in this chapter, see www.iste.co.uk/combes/galaxies.zip.

Galaxies,

5.1. Introduction

Galaxy studies typically focus on normal galaxies, with unusual ones considered rare and often simply curiosities that can be more or less ignored. After galaxies were first studied in the 18th and 19th Centuries, very few of the ones examined in those early days of extragalactic astronomy look like what we would today call peculiars. However, peculiar features in both galaxy structure and spectra have proven to be major clues for the formation methods of galaxies, and have driven the discovery of new astronomical objects such as active galactic nuclei (AGN) and central massive black holes. Likewise, galaxy structural peculiarities hold important clues to the formation of galaxies as these features are often evidence that galaxy mergers are a process responsible for some aspects of galaxy formation and evolution (Figure 5.1).

Examples of peculiar features in galaxies that turned out to be important new phenomenon include quasars that appeared as galaxies with very bright nuclei (e.g. Schmidt 1963), and active galactic nuceli that were originally discovered as having peculiar spectra of otherwise normal looking galaxies (Seyfert 1943). Even though galaxies with these properties are rare, these observational features reveal unique aspects of galaxy formation and evolution. Fundamentally, we also want to know the duty cycle and history of these processes to fully understand their role in assembling galaxies, their host stars and black holes.

To a large degree, the systematic study of galaxies began with the Messier catalog created by Charles Messier, working at the Hôtel de Cluny in Paris, who compiled a list of 110 nebulous non-cometary objects. This was later expanded by the more systematic catalogs of John and William Herschel in the southern and northern hemispheres. Most of the bright nearby galaxies in which Messier and the Herschels classified as "nebula" were actually what we would today call galaxies. Later, astronomers with large aperture telescopes were able to discover the spiral pattern within some of these systems, most famously by William Parsons, 3rd Earl of Rosse in Ireland who examined, among other galaxies, the Whirlpool nebula, what is often called M51.

Even after photography was applied to galaxies in the early 20th Century few peculiar galaxies were known for some time. In fact, most galaxies imaged with early photography turned out to be spiral galaxies, the most numerous type of bright galaxy in the nearby universe. The first galaxy classification schemes developed by Wolf (1908), Lundmark (1926), Hubble (1926) and Sandage (1961) typically did not include unusual or peculiar galaxies in their systems. This is due to the fact that nearby peculiar galaxies are very rare – they make up about 1% of the nearby population and they tend to be among the fainter lower mass systems.

However, as photography developed and expanded to include fainter and more representative galaxies it became clear that there were systems that indeed were peculiar in appearance and did not fit into any existing classification scheme. These galaxies were thought to be very unusual and their origin that was not initially thought to be due to merging. It was in fact thought that these galaxies were in some throes of early formation, or that we are witnessing the formation of galaxies through magnetic fields, or that the structures were dominated by some type of physics besides that of gravitational attraction that combines two galaxies into one.

Due to their rarity, it may thus come as a surprise that peculiar galaxies hold important clues for the formation of all galaxies. At one time or another nearly all galaxies, or their ancestors, would have been classified as peculiar. In the now famous Hubble (1926) tuning fork diagram, there is no reference to peculiar galaxies, but it did include irregular galaxies.

These irregulars are now known to be low mass star forming galaxies that typically are not produced by galaxy interactions or mergers, but by dynamical quiescent star formation. In fact, Hubble's system is dominated by ellipticals and spiral galaxies, which are the most common galaxy types in the Local Universe (e.g. Conselice 2006).

Only in the 1960s did it become generally recognized that some galaxies are merging. No doubt the very rarity of mergers made it seems unlikely that galaxies would go through this process when they were first found and cataloged. We now know that galaxy mergers and interactions are one of the dominant ways in which galaxies form. Mergers also induce processes that drive the formation of stars, planets and black holes during the history of the universe. However, we are just now learning about the full observational role of galaxy mergers, and how they influence the formation of galaxies themselves.

Galaxy mergers are the result of the hierarchical methods in which galaxies are forming. This hierarchical formation of structure formation is the cornerstone of the dominant ideas for how galaxy formation and cosmology operate in the universe today. The idea here is basically that because the dark matter particle is likely cold, that is moving slowly, this gives us a cosmology dominated by cold dark matter (CDM). The result of this is that the first structures to form in the universe would have been in small low-mass systems that later merged to produce more massive systems as cosmic time goes on. In fact, it is in principle possible to use the merger history to determine the cosmological history of the universe, as we discuss later in this chapter.

That the dark matter particle has not been found yet is still controversial to some extent, however theoretical predictions and models based on CDM for structure formation clearly show a good agreement with observations on large scales, such as the formation of super-clusters, walls and void. On the scale of galaxies, however, there are still problems with CDM being able to match the observations. Galaxy mergers may be one way in which this model can be tested and probed in more detail in the future.

Not long after mergers were "discovered" in the 1960s, it became clear that one of the ways in which galaxies, especially merging galaxies, can be studied is through computer simulations of how dynamical structures change with time in different scenarios. It was in fact Toomre and Toomre (1972) who first showed that the peculiar structure seen in Arp's (1966) atlas could indeed be produced by two systems undergoing a collision and merger. Their computer simulation showed the effects that two merging galaxies can have on the structure of spiral galaxies merging with each other, transforming the new resulting system into a new single elliptical galaxy. This has long remained the nominal conjecture for how massive galaxies that are elliptical have formed in the universe. It has however not been without some controversy, which we explain in this chapter.

Figure 5.1. *Optical images of galaxies taken by the Hubble space telescope in various stages of merging in the nearby universe. Some of these mergers are just starting and can be characterized as galaxy interactions. Other systems are in various postmerger phases where two galaxies have merged to become a single system. Some of these galaxies were originally classified by Arp (1966)*

There was a significant boost of interest in merging galaxies after the Hubble space telescope was launched in 1990. This was particularly the case when Hubble took the first deep pictures of the distant universe. These early deep images of distant galaxies taken by Hubble clearly showed that galaxies in the distant universe were quite different from those in which we see today. Some of this early work included Dressler *et al.* (1994) who demonstrated that indeed many distant systems appear to be merging galaxies. More deep HST surveys have shown quantitatively that this is indeed the case, in particular the Hubble Deep Fields North and South, the Hubble

Ultra-Deep Field, and later surveys such as GOODS, COSMOS and CANDELS. All of these deep HST surveys have demonstrated the commonality of merging systems in the distant universe that increases steeply with look-back time. Other studies have demonstrated how much these mergers contribute to the formation of galaxies throughout the history of the universe, which we review in this chapter.

In summary, in this chapter we review our understanding of galaxy interactions and mergers throughout cosmic time, and how these mergers have influenced the evolution of galaxies including the formation of stars and central massive black holes. We also discussed the theoretical attempts to explain how galaxy mergers are driving the formation of galaxies. When observational measurements are quoted and discussed, we use a cosmology such that $\Omega_\lambda = 0.7$ and $\Omega_m = 0.3$ with $H_0 = 70$ km s^{-1} Mpc^{-3}.

5.2. The physics of merging

The merger process is fairly straightforward to understand with just basic physics, although the actual act of merging is a nonlinear process that cannot necessarily be directly traced through the assembly of matter through cosmic time, but depends on factors such as the dynamical friction time scale for galaxies in pairs (Figure 5.2).

In general, the idea for how galaxies form is that they assemble early in the history of the universe once matter begins to condense after the big-bang. The way to examine this early process of galaxy formation is through the dominate ideas of cosmology in the Λ-cold dark matter (ΛCDM) cosmology, which is the canonical cosmological model today. The Λ term describes the existence of dark energy and the CDM aspect the temperature of the dark matter. Specifically, the relative abundance of dark matter and dark energy is prescribed through recent measurements such as the Planck cosmic background satellite (Planck Collaboration 2013).

In the ΛCDM cosmology, the first galaxies to form would have been small low mass ones, as the dark matter particle is thought to be a slow moving one, given that the dark matter temperature is "cold". Because of this gravity cannot act over large distances before the first objects form, which gives us many lower mass objects. Within this paradigm, the larger objects then form from the mergers of these smaller objects, in a process that is often described as a hierarchical assembly of matter. This is also the way in which galaxy clusters are formed by the accumulation of galaxies.

It is also thought that galaxy mergers can only occur effectively within a dark matter halo that dominates the mass of a system such as a galaxy group. Galaxies are often found in groups, such as the Milky Way within the Local Group, but groups are one of the most common environments in which galaxies are found and are likely the necessary location for producing merging.

The majority of galaxies in the nearby universe are in fact found in groups, and galaxies at all redshifts up to $z \sim 6$ cluster on the scale of nearby groups (~ 1 Mpc). Simple dynamical friction arguments however show that galaxy mergers should be relatively common, and a dominant process for driving evolution.

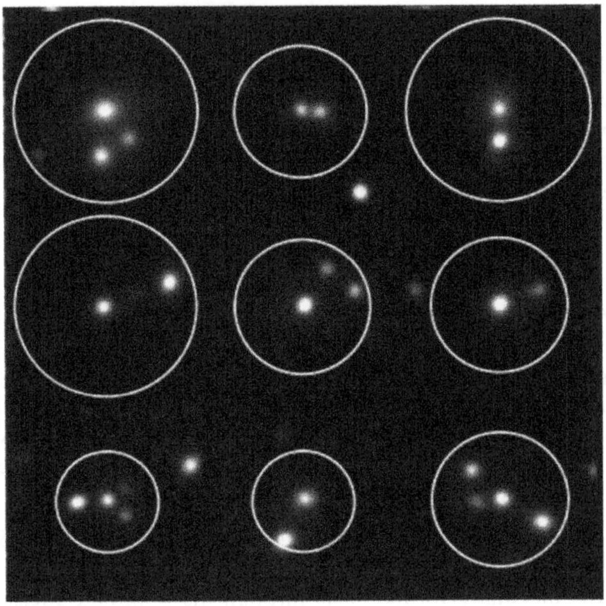

Figure 5.2. *Examples of galaxy pairs as seen in the distant universe. Each circle contains at least two different galaxies as separate systems, but which are close enough to merge within a few billion years. These are examples of major mergers such that the mass of the smaller galaxy is at least 25% of the mass of the larger system*

In general, galaxy evolution occurs via both internal and external drivers. The environments of galaxies likely have a strong effect on how galaxies form and evolve. For example, if galaxies are surrounded by other galaxies, then gravitational interactions and mergers induce the formation of new stars, remove stars/gas from galaxies due to tides, increase the masses of galaxies through accretion of satellites, decrease the number of galaxies from mergers and change the structures and morphologies of galaxies. If the groups these galaxies are within also contain an intragroup medium, that is if there is gas in the space between the galaxies in a group, then gas within the group galaxies can be removed through processes such as ram-pressure stripping and exhaustion whereby the gas in a system cannot be replenished and all the gas is used up in star formation.

The importance and strength of these various processes can be characterized by the distribution of velocities within a group which is denoted by the velocity dispersion of its members, σ, which gives some idea of the range of velocities within a group. The removal of gas through interacting with the intragroup medium, known as ram pressure stripping, results in a force $\sim \rho\sigma^2$, where ρ is the density of the intragroup medium. As the velocity dispersion of groups are typically $\sim 250 - 400$ km s^{-1}, the efficiency of this process in groups is not large, at least compared with clusters of galaxies where both ρ and σ are both much higher. Likewise high-speed galaxy interactions, such as galaxy harassment (e.g. Moore *et al.* 1996), are unlikely to produce significant effects in groups due to the relatively low velocities of member galaxies. This type of process is most efficient in rich clusters of galaxies where the velocity dispersion is high.

Galaxy Mergers Notes

- Mergers should be common in groups - dynamical friction time-scale evolves as (group σ^3)
- ~250 km/s upper limit for dragging galaxies into center of a group over a Hubble time in a group
- Low redshift merger rate expected to be low, around 2% of galaxies in groups merge per Gyr
- At higher-redshift the mass density increases as H ~ $(1+z)^3$. Resulting in a higher merger rate of $\sqrt{(\rho)}$~$(1+z)^{1.5}$

Figure 5.3. *Summary of properties of merging galaxies in their most likely environment – galaxy groups*

Galaxy exhaustion, whereby a galaxy uses all of its gas to form stars without being replenished, is also not a fast process and has a time scale of roughly $\sim$ Gyr in clusters. It also requires a removal mechanism, which is usually an intracluster medium, that prevents further gas accretion for member galaxies. We know these mechanisms cannot be extremely effective in groups of galaxies as we observe star formation in galaxies in groups.

Galaxy mergers, however, should be relatively common in groups of galaxies compared to rich clusters or field galaxies. This is because of the low velocity dispersion of galaxies in groups, and the close proximity of members. Various types of nearby groups, such as the compact groups and fossil groups, are possibly the result of the galaxy merger process. One way to see this is through dynamical friction calculations that show that the merger time scale varies roughly as $\tau_{\rm merge} \sim \sigma^3$.

Groups with lower velocity dispersions therefore have a shorter (quicker) time scale for mergers than galaxies in clusters. Basic calculations show that groups with velocity dispersions > 300 km s^{-1}, which is a limit that contains all galaxy clusters, will not have a significant number of mergers over a Hubble time. This is one reason why galaxy mergers, almost by definition, must occur in groups of galaxies.

Simulations and analytical calculations of the dynamical friction process show that galaxies in groups become more centrally concentrated by about a factor of two since $z \sim 1$, but galaxies within these groups do not necessarily merge. This effect is strongest for groups with lower velocity dispersions, < 150 km s^{-1}. Merging in groups at $z \sim 0.4$ should be low, about 2% per Gyr per group.

The merger rate for galaxies in groups, and therefore galaxies in general, should increase significantly at earlier times, that is at $z > 1.5$. The reason is simply because the universe was denser and galaxies were physically closer together, producing more merging. Within a Λ-dominated universe, the mass density increases as the Hubble parameter squared, or $\mathrm{H}^2 \sim (1 + z)^3$, with a merger rate given by $\sim \sqrt{(}H^2) \sim (1+z)^{1.5}$ (see Figure 5.3 for a summary of this). This merger rate may explain why we see a mix of galaxy types in groups, such as early-types and spirals with bulges. The spheroidal components in these group galaxies were formed when systems underwent mergers early in the universe, with disk/bulge systems forming when spheroids re-acquire a disk via gas accretion from the intergalactic medium.

5.3. The merger history of galaxies

5.3.1. *Defining mergers and merging*

Before we describe the background of galaxy mergers and how they have evolved throughout the universe it is first important to discuss definitions of merging, and what we mean when we refer to galaxy mergers. There is significant confusion within the literature about merging, and apparently conflicting results on merger evolution that are the result of inconsistent definitions being used.

First, it is important to get a visual idea of what we mean by the merger process for two galaxies. In this discussion, we have to assume that galaxies have already formed at least some mass before they can merge. It is thus quite clear that the very first galaxies could not form from mergers of preexisting galaxies, by definition. The ways in which the very first galaxies may form is largely beyond the scope of this chapter and is currently unconstrained observationally, but will be addressed by the JWST scheduled to be launched in 2021.

The basic idea is that when galaxies have formed their first seeds of mass, whatever these might be, because the universe is so dense their stars will likely collide with another galaxy's star through gravity. Because of this gravitational attraction, these systems eventually collide and coalesce into a new galaxy. This process can then continue with this new galaxy smashing into and merging with another galaxy, and then another, and so on through cosmic time.

There are two main ways of finding interacting and merging galaxies. The first are the classical mergers where the system looks peculiar (Figure 5.1) and the second are those systems that are close together on the sky in "pairs" (see Figure 5.2). In the nearby universe, these merging systems are fairly rare, making up just 1% or so of all galaxies.

Merging galaxies will obviously be more common when and where galaxies are closer together, as in the group environment. Galaxy groups, or gravitationally bound systems with at most a few dozen members, have a velocity dispersion that is no greater than a few hundred kilometers per second. At higher velocities major mergers are rare as the escape velocity of a pair is smaller than the typical velocity in clusters of galaxies for instance. This however does not preclude galaxy interactions within clusters, but these are high speed and simple inject energy into the systems (e.g. Moore *et al.* 1996). Because of the very large velocities of cluster galaxies, it is very difficult for two galaxies to merge within these environments. This however does not mean that the galaxies which we see in clusters today were not found by the merger process in the distant past. This is certainly possible if, for example, the environment in which those systems formed were quite different then from the environment in which we find them today, which is extremely likely.

That is, galaxies form in mergers when they are in a group environment, whereby these groups later merge together and form a larger σ dispersion system, thereby creating a less effective environment for direct mergers.

The most common type of mergers studied to date are the so-called major mergers. Major mergers are usually defined as merging systems where the mass ratio varies between equal mass (1:1) mergers, down to either a third or fourth of the more massive galaxy, relative to the lower mass system (1:3 or 1:4). These major mergers dominate the study of galaxies simply because they are the easiest to find and study. The reason is because typically when a major merger occurs we are examining systems of nearly equal mass, and therefore the effects on the galaxy are easiest to see. Minor mergers, where the mass ratio is less than 1:3, often do not produce morphological or structural changes. Furthermore, it is more difficult to find minor mergers given that the secondary galaxy is usually much fainter than the primary and pairs with small mass ratios are difficult to detect (e.g. Duncan *et al.* 2019).

The other type of merger, which is very difficult to study are the so-called micromergers that are very low mass galaxies and systems of about 1/100th or less of the mass of the central galaxy. These type of mergers probably coincide with the creation of extensive amounts of cold gas, which can furthermore increase the star formation rate of the galaxy beyond what would be expected from the amount of stars that have been produced in the system when it is observed.

The selection of mergers is also important. If you select mergers based on a stellar mass cut or a luminosity cut, then you will likely get different fractions and rates (Man *et al.* 2016). This does not imply that the merger fraction is inconsistent, just that by selecting in different ways we are selecting the underlying galaxy population differently. Because the merger fraction can vary significantly with galaxy property (e.g. Casteels *et al.* 2014), then the choice of how to find mergers is critical for having consistently measured merger histories.

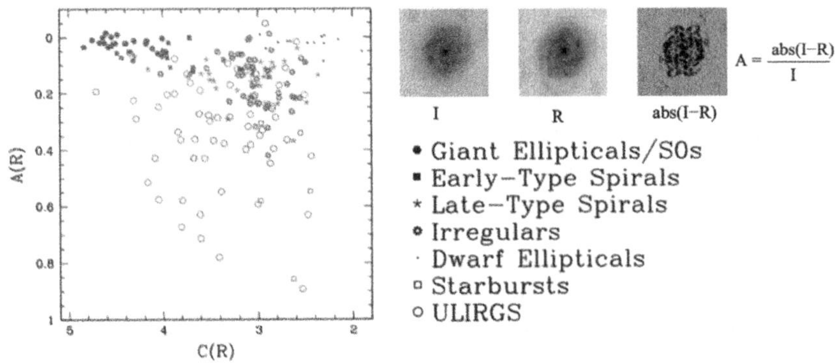

Figure 5.4. *Plots showing the use of quantifying galaxy structures to find mergers. The computation of the asymmetry index is shown in the upper right, whereby an example galaxy has been rotated (R) and subtracted from its original (I). The absolute value of the residuals are then summed and divided by the original flux of the galaxy to measure the asymmetry parameter. The plot of the left shows where galaxy types seen in the legend fall in the asymmetry-concentration diagram. As can be seen, ULIRGs, which are often major mergers, have a uniquely high asymmetry value*

It is important to realize that galaxy merger fractions and rates cannot be compared to each other unless they are measured with exactly the same galaxy selection (luminosity magnitude, mass/luminosity ratios), otherwise any comparisons are invalid.

5.3.2. *Merger rates in the nearby universe*

The merger fraction and rate of galaxies in the Local Universe is a long-term problem, which has been wondered about for many decades. There are many ways in which this nearby merger rate can be measured. This includes looking for galaxies in pairs, as well as looking for galaxies that are morphologically distorted, which are an indication that a merger has recently occurred. The two methods give similar results, which is that the local merger fraction is about $\sim 1\%$.

In De Propris *et al.* (2007), 3,184 galaxies with $-21 < M_B < -18$ mag. and $0.010 < z < 0.123$ were selected from the Millennium Galaxy Catalogue to determine the merger fraction within this sample. What they found was a 2% merger fraction and a merger rate of $(5.2 \pm 1.0) \times 10^{-4}$ Mpc^{-3} Gyr^{-1}. An alternative way to determine the nearby galaxy merger rate is to use galaxies, which are asymmetric and thereby likely having undergone a recent merger. When carrying out this analysis using nearby systems, De Propris *et al.* (2007) calculated that the nearby galaxy major merger rate over the mass range $10.8 < \log M_* < 11.5$ is $(1.2 \pm 0.5) \times 10^{-3}$ Mpc^{-3} Gyr^{-1}. There is therefore good consistency between the merger rate measured in the nearby universe between these two techniques. Later studies such as Casteels *et al.* (2014) have found similar results.

5.3.3. *Galaxy merger fraction evolution to* $z = 6$

A major part of this chapter is about the formation of galaxies through the merger process. We describe later, based on these observations, how galaxy mergers are influencing the assembly of galaxies, which consist of already formed stars in two separate systems that merge into a new larger system. That is, we are interested in how much new mass is brought into a galaxy by mergers and for the most part, due to technical limitations, ignoring the amount of gas that might be present within the merging systems. This gas in the distant universe could be quite significant and substantial and will be the focus of future work with radio telescopes such as the square kilometer array (SKA).

To understand the role of mergers, we first need to investigate how common mergers are within the universe, and how their presence changes with cosmic time. The first step is to find mergers and to measure the merger fraction – the fraction of galaxies in a given population – usually defined by a mass and redshift range – over cosmic time. Often this quantity is measured as the "merger fraction", which is the number of mergers or pairs of galaxies divided by the total number of galaxies in a system. The fraction of galaxies merging and the merger fraction can therefore differ by a factor of ~ 2 (Conselice 2006).

In general, there are two main ways in which distant galaxies can be studied for understanding the merger process. These two can be used together as complementary probes of how galaxies are forming through the merger process. These two methods are known as the morphological or structural method, and the other is the pair method. Both of these were introduced in section 5.2. How they are used to determine the presence of merging is very simple to explain.

Ideally, the measured merger history using both would give very similar results when comparing the derived merger history. In fact, this is a requirement if we are to have confidence that we are able to measure this process accurately. It is also the case however that we are making assumptions about the dark matter content of these systems, as in a universe with Modified Newtonian Dynamics (MOND) the different fractions would be up to a factor of 10 times different (e.g. Tiret and Combes 2007).

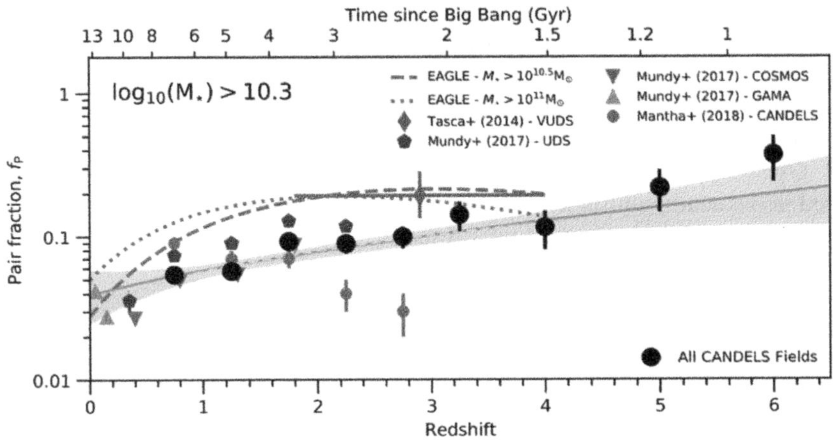

Figure 5.5. *The measured pair fraction of galaxies up to* $z \sim 6$ *taken from Duncan* et al. *(2019). This plot shows our latest estimate of how the merger history changes as a function of time. The black points are from deep Hubble space telescope imaging while the other colored points are from large ground based surveys, including Mundy* et al. *(2017). As can be seen, the merger fraction continues to increase at higher redshifts up until* $z \sim 6$*, where it reaches a value of* ~*50%. Also shown on this figure are model comparisons from the Eagle simulation, which generally find a higher pair fraction than what is observed*

Using the background information in section 5.2 describing the basic idea behind galaxy mergers, we can see how the pair method and the structural method work. The pair method is done by searching for two galaxies of a given mass ratio that are separated by a certain distance (see Figure 5.2). This apparent separation is usually 20 kpc or 30 kpc at maximum. The assumption here, which we will revisit later, is that when two galaxies are close together, there is some probability that they will merge within a given dynamical friction time scale. If we know that these two galaxies have a probability of merging within that time scale, or that we know for sure they will merge, then this allows us to calculate the fraction of galaxies that are undergoing a merger at any given epoch.

The other method for finding mergers involves the later part of the merger process when what were two separate galaxies are becoming a single system through dynamical violent relaxation (e.g. Lynden-Bell 1968). These are often seen as peculiar galaxies that we discussed earlier in this chapter. Many of these but perhaps not all are involved in some kind of merger process. Examples of these systems are shown in Figure 5.1.

Both of these methods for finding mergers have their disadvantages and distinct advantages. For the pairs we do not know for sure whether the two galaxies are actually in the same physical aspect of space, although that can be alleviated by photometric, or in unique cases spectroscopic redshifts (e.g. Tasca *et al.* 2014; Mundy *et al.* 2017). For the galaxies that are undergoing a merger and are peculiar in appearance we want to be able to distinguish a merger from systems, which are undergoing significant star formation that can make a galaxy appear to be peculiar in appearance.

Historically, the earliest studies of mergers was largely based on finding galaxies in pairs that came from the goal of deriving cosmological parameters from galaxy mergers in Patton *et al.* (1994). Later these studies were extended out to slightly higher redshifts by LeFevre *et al.* (2000). Pairs studies can now be carried out to $z \sim 6$ (e.g. Duncan *et al.* 2019).

The morphological method for finding mergers really began in the mid-1990s when the Hubble Deep Fields programs were initiated. It is not, however, just simply enough to find galaxies that are peculiar to identify unambiguously a merger, but to use quantitative ways to define and locate merging systems. From this, we can statistically determine whether or not the system is undergoing a merger. Very early studies using visual morphologies to determine whether or not a galaxy was undergoing a merger was performed in LeFevre *et al.* (2000).

However, one of the largest problems with using structure to find galaxies that are merging in the distant universe is that when we take images of distant galaxies we see them at wavelengths, which are often much bluer than the typical wavelengths used to measure galaxy structure in the nearby universe that is typically rest-frame optical light. This morphological k-correction must be dealt with as it can be a significant contribution to the quantitative structure of the system (Mager-Taylor *et al.* 2018), as we are seeing younger stars when we probe bluer light.

When the Hubble space telescope first examined distant galaxies, it used the Wide-Field Planetary Camera (WFPC) and WFC2 cameras, which are optical camera whose reddest filter goes only to about 900 nm. That is, when we used these cameras to study galaxies at redshifts at $z = 1.5$ and higher we can only observe the rest-frame ultraviolet morphologies of these distant galaxies. However, in 1998, the Hubble space telescope was refurbished by a shuttle servicing mission to install a new camera called NICMOS, which has an infrared camera that allowed us to be probe into the rest-frame optical for galaxies up to $z \sim 2.5$.

The NICMOS camera was the first that allowed us to probe the rest frame optical light of distant galaxies at high resolution. Because young stars can dominate the morphologies of galaxies in the ultraviolet, it was not understood whether the peculiar galaxies seen with the optical cameras on Hubble were simply just due to the fact that we were looking at the star formation in these systems or to the underlying disturbance of the structure due to a merger.

Soon after the NICMOS camera was installed, an infrared version of the Hubble Deep Field was taken (Dickinson *et al.* 2000). These new data clearly showed that the structures and morphologies of distant galaxies were actually quite similar in the rest frame UV compared with rest frame optical seen through the NICMOS observations (Popovich *et al.* 2003, 2005; Conselice *et al.* 2005). This is often the case when a system is dominated by star formation – its morphology is similar at all wavelengths.

There is also the issue of how we determine whether or not the structure of a galaxy is produced by a merger rather than other physical effects such as dust or star formation. The key to this is to examine galaxy structures through a quantitative process. Quantitative morphologies of galaxies is however a huge topic and we only discussed in this chapter its relevance to finding merging galaxies.

The first use of quantitative indices for studying the structures of distant galaxies was produced by Shade *et al.* (1994). This paper introduced the asymmetry index, which is a way to quantify how asymmetric a galaxy is. This was later used for distant galaxies by Abraham *et al.* (1996). However, it was not quite understood at the time how this index worked for finding distant mergers and whether it was finding systems in a unique way.

The first calibration of the symmetry index was performed by Conselice (1997). In this paper, the asymmetry is measured for 43 nearby galaxies to determine how the asymmetry of these galaxies changes with other galaxy properties such as color and star formation rate.

The asymmetry index was later defined with various improvements to make it among the most robust non-parametric structural parameter. The formula for calculating the asymmetry index (A) is given by (Conselice 2003):

$$A = \min\left(\frac{\Sigma|I_0 - I_{180}|}{\Sigma|I_0|}\right) - \min\left(\frac{\Sigma|B_0 - B_{180}|}{\Sigma|I_0|}\right), \qquad [5.1]$$

where the value I_0 is the original galaxy image and I_{180} is the image after rotating it 180° from its center. There is an iterative approach toward finding the center that gives the minimum asymmetry value, which is then used to calculated the asymmetry value (e.g. Conselice *et al.* 2000). The values for the background B is computed in a similar way as for the galaxy itself and this term is subtracted from the rotational and subtracted value.

This type of work was later expanded to the concentration index by Bershady *et al.* (2000). Furthermore, merging galaxies were included in the study by Conselice *et al.* (2000), which for the first time showed that nearby merging systems deviate from normal galaxies such that they have a high asymmetry index. This was later expanded in Conselice (2003) who further showed that a new index call the clumpiness or smoothness index (S) can be used to find galaxies undergoing intense star formation.

It is possible to use the clumpiness and the asymmetry index together to find systems, which deviate from a tight linear relationship by having a high symmetry but a modest clumpiness, or concentration, similar to the mergers (Figure 5.4). The basic idea here is that if a galaxy has a high clumpiness, it means that there is a significant fraction of light coming from small-scale structures in the galaxy. Galaxy mergers do not produce small-scale structures, but produce bulk large asymmetric ones. Therefore, while an intense starforming galaxy produces a high asymmetry and clumpiness, a merging system can only produce a high asymmetry, especially if the system is observed in the infrared.

This methodology works well for finding mergers in the nearby universe as shown by the systematic deviation of ultra luminous infrared galaxies (ULIRGs) from normal galaxy such as spirals and ellipticals. In general, what is found to be the case for mergers is that the value of the asymmetry A must be greater than $A > 0.35$ for a major merger. To avoid conflicts with galaxies undergoing intense star formation, the criteria that $A > S$ must also hold (Conselice 2003). Thus, for a merger the following criteria must hold:

$$A > 0.35 \,\&\, A > S \qquad [5.2]$$

When this methodology is applied to distant galaxies what we find is that the derived merger fraction increases significantly when we go to higher redshifts. The merger fraction was studied in this way in Conselice *et al.* (2003) who showed that the peak merger fraction at $z = 2.5$ was about 40%. At the same time, this merger fraction increase was parameterized as a power law that goes as $(1 + z)^m$, where m is the power lower index for the increase in the fraction of mergers with time.

This formula for quantifying the merger history was first introduced by Carlberg (1991). Carlberg (1991) was interested in measuring the increase in the merger history as a way to determine the cosmology of the universe. The merger history is usually parameterized by an increase in a power law given by:

$$f_{\mathrm{m}} = f_0 \times (1 + z)^m \qquad [5.3]$$

where f_0 is the merger fraction at $z = 0$ and m is the merger power-law index. The value of this index depends highly on the value for the $z = 0$ merger fraction value. A study using the COSMOS field also shows that the merger fraction measured with morphology increases up to $z = 1$ in accordance with what is found in the deeper but smaller area data (Conselice *et al.* 2009). The latest version of the merger fraction history is shown in Figure 5.5 up to $z \sim 6$.

Later studies such as within the Hubble Ultra Deep Field has found similar results with a slightly lower peak fraction. Furthermore, as shown in Conselice *et al.* (2008), the merger fraction might in fact even start to decline at redshifts higher than $z > 2$.

A good representation of this behavior is that the merger fraction can be parameterized by a combined power law and exponential, a form of the Euler gamma function (e.g. Conselice *et al.* 2008; Duncan *et al.* 2019) whose form is given by:

$$f_{\rm m} = \alpha \times (1+z)^m \exp[\beta(1+z)]. \qquad [5.4]$$

The merger fraction gives us some idea of the number of mergers happening within a given galaxy population as a function of time. In itself this does not tell us how mergers are affecting the formation of galaxies however because the time scale under which these mergers occur is unknown from observations. We need to understand these time scales, as without this information we do not know how long a galaxy is in the merger phase. The merger fraction is then simply just an observation without a proper context. When the merger fraction is interpreted with a time scale, it gives us the merger rate, the topic of the next section of this chapter.

5.3.4. *Galaxy merger rates*

Ultimately, it is the galaxy merger rate that is what we are interested in for learning about how mergers influenced the formation of galaxies and their constituents. The galaxy merger rate is with some exceptions a theoretically derived quantity based on the time scales for galaxies to merge.

The times scales for galaxy mergers are quite different for galaxies in pairs versus the galaxies which are morphologically disturbed. For those galaxies that are separated by some apparent distance in a pair, typically what is done is to calculate the dynamical friction time scale for two systems to merge. This time scale depends on the masses of the systems such that the more massive the galaxy, the shorter the time period for the merger to happen.

When calculating the merger time scale ($\tau_{\rm m}$) using the dynamical of friction formula for galaxies with typical separations of $\sim 2 - 30$ kpc, the time scale is $\tau_{\rm m} =$

0.5–1 Gyr. This however assumes a couple of things. The first is that these distant galaxies are as dominated by dark matter as local systems are. While it is not entirely clear yet whether or not galaxies in the distant universe have a dark matter to seller mass ratio similar to what we find. In local galaxies today, it does appear that they do (e.g. Conselice *et al.* 2018). From this, we can tentatively conclude that the dynamical friction formula is applicable in the distant universe.

The merger rate is then defined as:

$$\Gamma(M_*, z) = \frac{\tau_m}{f_m}. \quad [5.5]$$

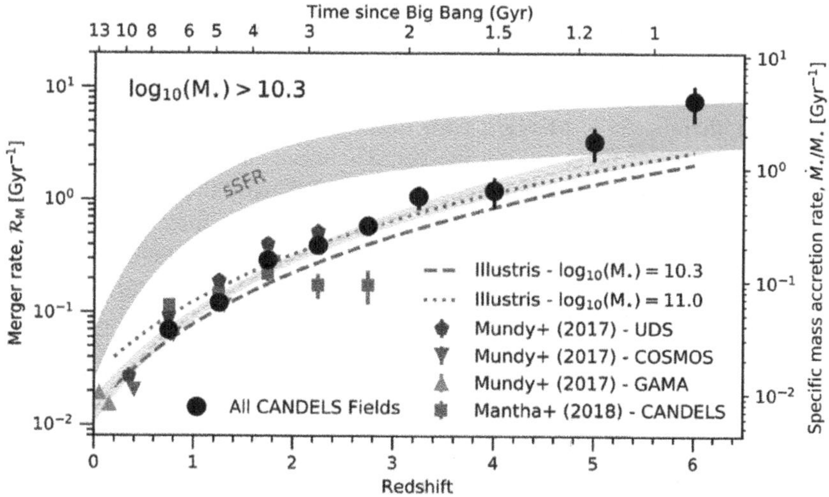

Figure 5.6. *The merger rate (number of mergers occurring per galaxy per unit time) as a function of redshift (left-hand side). The right-hand side shows the specific mass accretion rate, the amount of stellar mass brought into a galaxy due to the merger process. The dashed and dotted lines show the predictions for these merger rates within the Illustris simulation as described in Duncan* et al. *(2019). The shaded green curve shows the corresponding specific star formation rate (sSFR) for galaxies at the same mass selection as for the mergers as a function of redshift. Results from previous work, including Mundy* et al. *(2017) and Mantha* et al. *(2018), are also shown*

We use galaxy structures such as with CAS to locate which distorted and peculiars systems are mergers to derive merger fractions. However, we have to use other methods to derive the time scales for mergers for the simple reason that we are not able to wait a billion years to determine how long a merger occurs. Dynamical fiction time scales are not suitable for this calculation simply because the two galaxies are already merged and are dynamically relaxing at this stage. The way in which these time scales are calculated is through using simulations of the merger process. In this way, we get some idea of the different types of mergers that produce asymmetric structures, which can then can be converted into a time scale.

This approach was first carried out by Conselice (2006) who examined pairs of galaxies merging in N-body models. This first analysis did not include galaxies undergoing starformation, or other structures that can produce asymmetries such as dust lanes. Since then more detailed simulations incorporating starformation and other effects have been employed (e.g. Lotz *et al.* 2010), but the main results found by Conselice (2006) still hold.

Conselice (2006) showed that the merger time scales for asymmetric galaxy are around 0.5–1 Gyr, which is similar to what is found for galaxies with a pair separation of about 30 kpc. Furthermore, it can be shown through simulations that high values of the asymmetry index locates systems that have merge with a merger ratio greater than $\sim 1/4$. Mergers with lower mass ratios do not produce a high enough asymmetry to be seen as a merger. Therefore, the mergers measured with CAS are those that are undergoing major mergers. Likewise the viewing angle of the emerging system is found to have little effect on the resulting asymmetries. This shows that the asymmetry index is a robust way of finding major merging galaxies, as opposed to other indices such as Gini index, whereby an unknown mix of minor and major mergers are identified (e.g. Lotz *et al.* 2010).

Resulting major merger rates from these calculations to shown in Figure 5.6 using the pair methodology up to $z \sim 6$. As can be seen, the derived merger rate increases quite rapidly up until the highest redshifts in which we can currently probe this value. In fact, we find that at $z \sim 6$ the merger rate is 10 per Gyr, which means that there is on average a merger every 100 Myr at these redshifts. This is very often, and implies that merging is an extremely important process during this time. Over the entire range of $z \sim 0 - 6$, there are in total on average a few major mergers for galaxies with masses $M_* > 10^{11}\ M_\odot$. This merger history is very similar if instead of selecting with stellar mass a co-moving volume selected sample is used (e.g. Duncan *et al.* 2019).

The right-hand axis in Figure 5.6 shows the amount of mass accreted into galaxies due to the merger process, in terms of the specific mass accretion rate that is defined as $\dot{M}/M$, where $\dot{M}$ is the amount of mass brought in through mergers. The form of this tracks the merger rate itself, and it can be seen when comparing to the specific star formation rate for the same galaxies (sSFR) that the star formation rate is much more important at lower redshifts, but the specific mass accretion rate is similar, or higher at $z > 5$. This implies that at least major mergers are a dominate way in which galaxies are forming their mass at very early times in the universe's history. This is however not necessarily the case at lower redshifts were only up to 40% of the mass in galaxies is established through major and minor mergers (e.g. Ownsworth *et al.* 2014).

5.4. The added value of mergers

5.4.1. *Galaxy and black hole assembly through interactions and mergers*

Now that we have some idea of the galaxy merger rate, and hence the number of mergers that have occurred for galaxies as a function of time we can now ask more detail questions. One of these is how much mass is added to galaxies as a function of time through the merger process? This question of course has some subquestions including how much of this mass is added due to major mergers as opposed to minor mergers.

Before we describe this process in which mass is added to galaxies due to the merger process, it is important to understand and review the basics for how we think galaxies are assembling themselves. There are two ways in which we can look at this question. The first is through theoretical models that predict a slew of different histories for galaxy assembly depending on the merger history and the gas content of galaxies as a function of time, with the role of mergers strongly model dependent (e.g. Jogee *et al.* 2009). However, we do have a good observational idea of the ongoing star formation rate for galaxies as a function of time. As shown in many different papers, the star formation rate increases when we look back in time to peak at about z = 2 (Madau and Dickinson 2014), which roughly agrees with the evolution of the merger rate (Conselice 2014).

Star formation rates are typically measured by using light from galaxies – either the ultraviolet or the far-infrared that directly traced star formation rates by looking at young massive stars or dust emission (Kennicutt 1994). It is fairly easy to show that we can identify most of the stellar mass forming in the universe through this star formation rate, which we can now probe up to redshifts of $z = 7$ (e.g. Madau and Dickinson 2014). However, it is important to realize a few things about these measured star formation rates.

The first is that all stars have to be created in starformation events, by definition. Thus, it makes sense that the star formation rate should follow the assembly of stellar mass. However, the measurement of star formation rates does not reveal how that star formation is triggered. It is also not the case that just because star formation happens in a galaxy that those stars stay in that galaxy as an isolated system throughout time. When we discuss galaxy interactions and mergers what we are referring to is the assembly of matter through the merger process, which necessarily means that this later assembly of galaxies occurred after the stars were already formed.

It is also not understood or well known yet what is the role of interactions and mergers in triggering starformation. Galaxy mergers and interactions are often observed in nearby groups, which may be a common way to induce star formation in these systems. In general, we know that star formation happens within galaxies when gas densities become high and the gas is cold. This is a natural outcome of galaxies that are interacting and merging. The reason being is that interactions will alter the density of gas due to gravitational fields that can lead them into colliding, becoming denser, and then forming new stars.

Do galaxy interactions and mergers trigger star formation in galaxies? This is a long-standing question with however some ambiguous results. This can be examined both in the Local Universe, and also at higher redshifts. What we find is that the star formation rate in interacting galaxies separated by a few 10s kpc is about 80% higher than in galaxies, which are not interacting (Robaina *et al.* 2009). However, starbursting galaxies are often found in merger events, showing that the most intense star formation is often triggered by mergers. This is a recurring theme such that mergers do not seem to generally increase on average the star formation rate a large amount, but that the most star forming galaxies are indeed those undergoing mergers.

It is found that the average enhancement of star formation in merging galaxies is indeed modest, with most of the star formation occurring at $z < 3$ due to gas accretion rather than merging activity (e.g. Knapen *et al.* 2015; Lofthouse *et al.* 2017). However, it is important to understand that the triggering of star formation is just one way in which mergers are influencing the formation of galaxies. By far, the process of mass assembly of already existing galaxies merging together is the dominate way in which mergers influence galaxy formation. This can be determined based on the merger rates calculated earlier. As discussed already, the prevalence of mergers over star formation only occurs at $z > 5$ and this is where the location of intense starbursts triggered by mergers is likely to be found.

The other idea which we can investigate is how, or even if, the formation and fueling of AGN is triggered by merging or not. The idea is simple – two galaxies merge, presenting an unstable situation in which gas is flowing through a galaxy. Some of that gas in principle could be collated or collected toward the center, some of which will enter the accretion disk of an AGN and thereby later fuel it, producing active galaxies such as quasars and other types of AGN.

Whether or not this is happening is a longstanding question with a mix of answers depending on what type of source is looked at. It is clear that quasars, which are the most luminous AGN, are often in the centers of galaxies undergoing mergers (Urrutia *et al.* 2008). This is now in fact also seen in deep ALMA imaging of extremely high redshift sources at $z > 7$ (Bañados *et al.* 2019). However, there is not a universal agreement on this with some finding that dusty Type II quasars having a merger fraction of $\sim$ 25% (Schawinski *et al.* 2012), which is higher than that of field galaxies but certainly not a large fraction of these types of quasars. This is also found for typical X-ray selected AGN in deep HST field imaging, such that only a small fraction of systems are undergoing mergers (e.g. Kocevski *et al.* 2012).

However, it is likely that a big part of these conflicting results is that the merger fraction for AGN depends strongly on the luminosity of the sources and the underlying masses of the host galaxies. The most luminous sources are inevitably often mergers, while lower luminous sources can be triggered in other ways including through gas accretion. Thus, similar to star formation, mergers seem ubiquitous for the most luminous and intense systems, but less important for more common systems.

5.4.2. *Cosmological relevance*

The dominant model for galaxy formation, as well as the cosmology the universe, is one in which the universe is hierarchical. To really understand galaxies and cosmology together as two parts of a single system, it is important to fully appreciate that galaxies are the constituents of the universe, and therefore galaxies in their forms and evolution are strongly dependent on the underlying cosmology.

The basic idea in a physical sense relates to the dark energy and dark matter balance in the universe, and thus how much the universe is expanding and accelerating, and how this will influence the formation of structure, including galaxies.

Galaxy mergers could potentially play a key role in testing these ideas and also refining our knowledge of cosmology. It is not just the fact the galaxy mergers are producing galaxy evolution, and also triggering a significant amount of galaxy star formation and AGN activity. It is also the case that different cosmological parameters will produce a different merger history of the universe.

The idea is fairly simple. The universe's history and the galaxies within it are dominated by the cosmological features of the matter density and the density of dark energy, which is often parameterized by the parameter Ω_Λ. If the universe is dominated by dark energy, then the universe is accelerating faster and the expansion of space would have happened earlier in the universe than what has appears to have happened within our own (Riess and Turner 2003). This would naturally lead to a universe where galaxies become more separate from each other earlier, driving down the merger rate (e.g. Conselice *et al.* 2014b).

From basic kinematic arguments, it can be shown using a simple formalism that the acceleration of the universe must have started at redshifts of $z \sim 0.7$. This is due to the relative contribution of matter and dark energy. If the universe was more dark matter dominated and had no dark energy, then the effect will be such that mergers will be much more common, and galaxies today would be much more massive than they actually are and would have experienced more mergers.

Thus, by examining the contribution of dark matter and dark energy one gets a unique value of the merger history. This was examined in detail in Conselice *et al.*

(2014) who examine how the merger history of dark matter halos would vary between different universe models. While it was found that different relative amounts of dark energy and dark matter produce different merger histories, the differences are small between realistic possible cosmologies. This will however be a possible test to make once merger fractions error shrink when large area surveys are carried out by the Euclid Space Telescope and LSST.

For example, a reliable difference between w = -1 and w = -0.9 would require that the errors on the merger fractions be close to 0.01. This is in principle possible if we can measure accurate errors on measuring merger fractions and also have enough galaxies in the sample to beat down the statistical fluctuations.

These upcoming surveys such as Euclid, LSST and WFIRST will give us some idea of how to do this. Whether or not this will be a competitive measurement of cosmology at the same level as galaxy clustering, weak lensing, or the cosmic background radiation remains to be determined. However, this is a alternative way of measuring cosmology that does not depend on any definitive properties of the physics that dominates these other methodologies. Since this method is unique and complementary to other well-established ones, it could be used as a test on our ideas of cosmology to make sure they are consistent with galaxy formation itself.

There are alternative ways in which we can study the cosmological properties of the universe with galaxy mergers and interactions. Another one is that the dark matter particle and temperature can determine how mergers occur. In the $\Lambda\,CDM$ universe, it turns out that galaxy mergers are fairly common because in this cosmology there are many small mass galaxies and halos in the early universe that later merge to form larger and larger galaxies as time goes on.

However, if the dark matter particle was warm or hot, it would change dramatically this merger history. As shown in Conselice *et al.* (2014), the merger history can be quite different for galaxies that are merging within a universe made up of warm dark matter (WDM). WDM is an alternative idea to CDM such that the power spectrum is not as steep at the low mass end, thereby producing fewer low-mass galaxies per giant galaxy. From this, the merger history is less than what you get with CDM, which is what we find when you compare to models of WDM. This remains an open issue, and more simulations with different forms of dark matter are needed to determine the relative role of mergers compared to other dark matter temperature cosmologies.

5.4.3. *Future uses and methods*

The future is very promising for using galaxy mergers for tracing the galaxy formation process, and the use of it for deriving properties of galaxy evolution and cosmology. There is still much about galaxy mergers, including their basic measurements, that we do not know or have measured to a high accuracy. Also, using mergers as a cosmological tool has yet to be exploited to any significant way. Different methods for finding mergers are also in their infancy, as is studying mergers in the context of gravitational wave experiments.

Observationally, we still do not have a firm understanding of the merger history as a function of time to an accuracy similar to the star formation rate. This is mostly due to the relative low level of research activity to date studying mergers as compared to star formation. Major progress can therefore still be made. There are still many uncertainties due to the lack of known spectroscopic pairs of galaxies at different redshifts. This is the due to the fact that most galaxies surveys are done over large areas such that galaxies close together are very hard to obtain spectroscopy for in the same exposure, as these surveys are not optimized to find small-scale galaxy structures. Therefore, we have to use techniques such as photometric redshift or morphologies, both of which have their downsides as explained earlier in this chapter. Obtaining spectroscopy for a large number of apparent pairs will go a long way toward obtaining precise measurements of the galaxy merger history through cosmic time, as well as its effect on star formation and AGN triggering.

New large surveys with new upcoming telescopes such as LSST, Euclid and WFIRST will revolutionize the study of galaxy mergers. This will be supplemented by spectroscopic surveys with 4MOST and MOONS among others. Using these surveys mergers can be found through morphological as well as pair methods that can be utilized to a large extent over all types of galaxy properties and environments. Currently, we are very limited to what types of pairs and mergers can be studied. For example, the Euclid survey will cover 15,000 square degrees with HST-like resolution. LSST will not have the same resolution as Euclid, but will be deeper and therefore complementary for studying galaxies undergoing mergers in the more distant universe using pairs.

It is also clear that new techniques in machine learning and deep learning will greatly advance our understanding of mergers and how to find them. An exciting prospect with this is the ability to find different phases of a merger as well as to use machine learning trained on a simulation and then applied to data. This will greatly advance our understanding of merging simply because we would be able to study them in much more detail and discover how the phases of merging are correlating with different aspects of star formation and AGN triggering.

5.5. Summary

In this chapter, we have explained the basic processes involving galaxy mergers and how mergers are responsible for building up a significant fraction of the stellar, dark and gaseous mass in a galaxy. We have carefully explained the definition for mergers, and how mergers can be seen in all aspects of galaxy formation. Even our own Milky Way is undergoing merging with lower mass dwarf companions, and thus this is a ubiquitous process that has been ongoing through all of cosmic history to some degree or another.

When taking into account all of this evidence, it is clear that up to 50% of the stellar mass in galaxies is built up from stars that were born in other galaxies and deposited into new galaxies through the merger process. Results so far suggest that the methods for implementing these mergers are equally divided between major mergers, whereby the systems merging are of similar mass, and minor mergers where the masses of the galaxies merging differs by more than a factor of three.

Overall, the evidence is clear that mergers are much more common at higher redshifts than they are today. The merger rate is up a factor of 1,000 times higher at the highest redshifts we can probe ($z \sim 6$) compared with galaxies today. This increase in merger rates can be expressed as a power law of the form $(1 + z)^m$, where $m \sim 2.5$.

In addition to being a primary driver for assembling galaxies and their stars, there is evidence that this process of merging can trigger star formation within the galaxies where this process occurs, as well as build up the masses of central massive black holes. Merging is therefore a major method for assembling and creating the stars and black holes found in galaxies.

In the future, gravitational wave experiments will find and study galaxy mergers up to $z > 10$ with instruments such as LISA. Experiments such as Euclid and LSST will also allow us to determine the merger history for galaxies at all masses up to $z \sim 3$ including the elusive lower mass systems, whereby galaxy mergers may help to find the black holes in LIGO/Virgo sources (e.g. Conselice *et al.* 2020). At higher masses JWST will probe the first galaxies and their merger histories. From all this data in the next decade, we will have a full picture of galaxy evolution throughout all of cosmic history and the role of mergers within this process.

5.6. Acknowledgments

I thank my collaborators for their leadership and assistance on many of these topics, especially Ken Duncan and Carl Mundy. Some of the research presented in this chapter was funded by NASA, the National Science Foundation in the United States, and the STFC, and the Leverhulme Trust in the UK.

5.7. References

Arp, H. (1966). *ApJS*, 14, 1.

Bañados, E., Novak, M., Neeleman, M., *et al.* (2019). *ApJ*, 881, L23.

Barnes, J.E., Hernquist, L.E. (1991). *ApJ*, 370, 65.

Beckwith, S.V.W., Stiavelli, M., Koekemoer, A.M., *et al.* (2006). *AJ*, 132, 1729.

Bershady, M.A., Jangren, A., Conselice, C.J. (2000). *AJ*, 119, 2645.

Bertone, S., Conselice, C.J. (2009). MNRAS, 396, 2345.

Bluck, A.F.L., Conselice, C.J., Buitrago, F., *et al.* (2012). *ApJ*, 747, 34.

Bluck, A.F.L., Conselice, C.J., Almaini, O. *et al.* (2011). *MNRAS*, 410, 1174.

Bluck, A.F.L., Conselice, C.J., Bouwens, R.J., *et al.* (2009). *MNRAS*, 394, 51L.

Buitrago, F., Trujillo, I., Conselice, C.J., *et al.* (2008). *ApJ*, 687, 61L.

Buitrago, F., Trujillo, I., Conselice, C.J., *et al.* (2013). *MNRAS*, 428, 1460.

Bridge, C., Appleton, P.N., Conselice, C.J., *et al.* (2007). *ApJ*, 659, 931.

Carlberg, R.G. (1991). *ApJ*, 375, 429.

Casteels, K.R.V., Conselice, C.J., Bamford, S.P., *et al.* (2014). *MNRAS*, 445, 1157.

Conselice, C.J., Bershady, M.A., Jangren, A. (2000). *ApJ*, 529, 886.

Conselice, C.J. (2003). *ApJS*, 147, 1.

Conselice, C.J., Bershady, M.A., Dickinson, M., *et al.* (2003a). *AJ*, 126, 1183.

Conselice, C.J., Blackburne, J., Papovich, C. (2005). *ApJ*, 620, 564.

Conselice, C.J. (2006a). *MNRAS*, 373, 1389.

Conselice, C.J. (2006b). *ApJ*, 638, 686.

Conselice, C.J, Rajgor, S., Myers, R. (2008). *MNRAS*, 386, 909.

Conselice, C.J., Yang, C., Bluck, A.F.L. (2009). *MNRAS*, 394, 1956.

Conselice, C.J., Bluck, A.F.L., Ravindranath, S., *et al.* (2011). *MNRAS*, 417, 2770.

Conselice, C.J., Morlock, A., Bluck, A.F.L., *et al.* (2013). *MNRAS*, 430, 1051.

Conselice, C.J. (2014). ARAA, 52, 291.

Conselice, C.J., Bhatawdekar, R., Palmese, A., *et al.* (2020). *ApJ*, 890, 8.

De Propris, R., Conselice, C.J., Liske, J., *et al.* (2007). *ApJ*, 666, 212.

Dickinson, M., Hanley, C., Elston, R., *et al.* (2000). *ApJ*, 531, 624.

Dressler, A., Oemler, A., Butcher, H.R., *et al.* (1994). *ApJ*, 430, 107.

Driver, S.P., Windhorst, R.A., Ostrander, E.J., *et al.* (1995). *ApJ*, 449, 23L.

Duncan, K., Conselice, C.J., Mundy, C., *et al.* (2019). *ApJ*, 876, 110.

Förster Schreiber, N.M., Genzel, R., Bouche, N., *et al.* (2009). *ApJ*, 706, 1364.

Giavalisco, M., Ferguson, H.C., Koekemoer, A.M., *et al.* (2004). *ApJ*, 600, 93L.

Grogin, N.A., Conselice, C.J., Chatzichristou, E., *et al.* (2005). *ApJ*, 627, 97L.

Hubble, E.P. (1926). *ApJ*, 64, 321.

Jogee, S., Miller, S.H., Penner, K., *et al.* (2009). *ApJ*, 697, 1971.

Kocevski, D.D., Faber, S.M., Mozena, M., *et al.* (2012). *ApJ*, 744, 148.

Knapen, J.H., Cisternas, M., Querejeta, M. (2015). *MNRAS*, 454, 1742.

Kocevski, D.D. (2012). *ApJ*, 744, 148.

Le Févre, O., Abraham, R., Lilly, S.J., *et al.* (2000). *MNRAS*, 311, 565.

Lofthouse, E.K., Kaviraj, S., Conselice, C.J., *et al.* (2017). *MNRAS*, 465, 2895.

Lopez-Sanjuan, C., Balcells, M., Perez-Gonzalez, P.G., *et al.* (2010). *ApJ*, 710, 1170.

Lopez-Sanjuan, C., Le Fevre, O., de Ravel, L., *et al.* (2011). *A&A*, 530, 20.

Lopez-Sanjuan, C., Le Fevre, O., Ilbert, O., *et al.* (2012). *A&A*, 548, 7.

Lotz, J.M., Primack, J., Madau, P. (2004). *AJ*, 128, 163.

Lotz, J.M., Davis, M., Faber, S.M., *et al.* (2008). *ApJ*, 672, 177L.

Lotz, J.M., Jonsson, P., Cox, T.J., Primack, J.R. (2010). *MNRAS*, 404, 590.

Lundmark, K. 1926, Arkiv for Matematik, Astronomi och Fysik, 19(8), 1–5.

Lynden-Bell, D. (1967). *MNRAS*, 136, 101.

Madau, P., Dickinson, M. (2014). *ARAA*, 52, 415.

Man, A.W.S., Zirm, A.W., Toft, S. (2016). *ApJ*, 830, 89.

Mantha, K.B., McIntosh, D.H., Brennan, R., *et al.* (2018). *MNRAS*, 475, 1549.

Moore, B., Katz, N., Lake, G., *et al.* (1996). *Nature*, 379, 613.

Mortlock, A., Conselice, C.J., Hartley, W., *et al.* (2013). *MNRAS*, 433, 1185.

Ownsworth, J.R., Conselice, C.J., Mortlock, A., *et al.* (2014). *MNRAS*, 445, 2198.

Papovich, C., Giavalisco, M., Dickinson, M., *et al.* (2003). *ApJ*, 598, 827.

Papovich, C., Dickinson, M., Giavalisco, M., *et al.* (2005). *ApJ*, 631, 101.

Patton, D.R., Pritchet, C.J., Yee, H.K.C., *et al.* (1997). *ApJ*, 475, 29.

Planck Collaboration, Ade, P.A.R., Aghanim, N., *et al.* (2014). *A&A*, 571, A16.

Robaina, A.R., Bell, E.F., Skelton, R.E., *et al.* (2009). *ApJ*, 704, 324.

Sandage, A. (1961). *The Hubble Atlas of Galaxies*. Carnegie Institution, Washington DC.

Schawinski, K., Simmons, B.D., Urry, C.M., *et al.* (2012). *MNRAS*, 425, L61.

Seyfert, C.K. (1943). *ApJ*, 97, 28.

Schmidt, M. (1963). *Nature*, 197, 1040.

Tasca, L.A.M., Le Fèvre, O., López-Sanjuan, C., *et al.* (2014). *A&A*, 565, A10.

Toomre, A., Toomre, J. (1972). *ApJ*, 178, 623.

Tiret, O., Combes, F. (2008). Proceedings of "Formation and Evolution of Galaxy Disks", Funes, J.G., Corsini, E.M. (eds), 259. arXiv e-prints, arXiv:0712.1459.

Trujillo, I., Conselice, C.J., Bundy, K., *et al.* (2007). *MNRAS*, 382, 109.

Urrutia, T., Lacy, M., Becker, R.H. (2008). *ApJ*, 674, 80.

Williams, R.E., Blacker, B., Dickinson, M., *et al.* (1996). *AJ*, 112, 1335.

Wolf, M. (1908). *Publikationen des Astrophysikalischen Instituts Koenigstuhl-Heidelberg*, 3, 109.

6

Cosmic Evolution of Galaxies

David ELBAZ and Emeric Le FLOC'H
CEA, Saclay, France

Barely a few years after discovering the existence of galaxies, we learned by studying their dynamics that the Universe has a history, following on from the Big Bang and the expansion of the Universe. Despite this, many years passed before we realized that galaxies are constantly evolving in terms of their shape, color, star formation activity and the growth of their galactic black holes. In the late 1970s, some clues were discovered indicating that galaxies were different in the past; galaxies belonging to clusters were bluer – the Butcher-Oemler effect (Butcher and Oemler 1978) – and so were field galaxies (Tyson *et al.* 1979). But it still was not until the mid-1990s, after the bitter failure of the quest for primordial galaxies since the 1980s, that we realized that galaxy formation is a truly continuous process. The history of the cosmic evolution of galaxies closely reflects the progress in our techniques for observing the Universe, such as the arrival of the new class of 8–10 m telescopes, the Hubble Space Telescope and the multiwavelength approach, which reveals a different facet of galaxies at every frequency in the electromagnetic spectrum.

6.1. Introduction

The study of cosmic evolution encompasses (i) the assembly of galaxies into gravitationally bound (virialized) units, (ii) the conversion of interstellar gas into stars, (iii) the dissipation of the energy generated by gravity (e.g. *turbulence in the*

For color versions of the figures in this chapter, see www.iste.co.uk/combes/galaxies.zip.

Galaxies,
coordinated by Françoise COMBES.

interstellar medium generated by differential rotation, radiation of the accretion disc around black holes, jets from active nuclei) and nuclear reactions at the heart of stars (*stellar nucleosynthesis*), (iv) the influence of this energy on the efficiency of forming new generations of stars, (v) the evolution of the shape of galaxies (galaxy morphogenesis), with a particular focus on the formation history of discs and spheroids, and (vi) the role of the environment on the history of galaxies, and vice versa.

Research in this field aims to determine at which epochs these mechanisms occurred, and which processes control the efficiency of the conversion of gas into stars: whether they are **internal** (*in situ*, e.g. turbulence, stellar feedback generated by radiation pressure – or supernova explosions, dynamical instabilities responsible for bars or the creation of kpc-sized clumps) or **external** (*ex situ*, e.g. tidal effects created by the crossing of other galaxies, galaxy mergers, extraction of part of the gas reservoir of a galaxy due to pressure from intergalactic gas within a galaxy cluster or stripping caused by ram pressure).

Thus, the history of galaxies is at the interface between structure formation in the Universe at very large scales and the radiation mechanisms of molecules, atoms and interstellar dust grains at the very tiniest scales.

Gravity is the key actor, or at least the initiator, of all of these mechanisms. In the current cosmological paradigm, gravity is largely dominated by the presence of 5.3 times more dark matter than classical (so-called baryonic) matter in the Universe. The effect of dark energy, whose role is to accelerate the rhythm of expansion of the Universe, on the cosmic evolution of galaxies is marginal, except when calculating from the redshift the cosmic age and the distance. Its effects are primarily felt at the scale of galaxy clusters, which are measured in millions of light-years. Today, the formation of these galaxy clusters is slowed by the accelerating expansion of the Universe. If we disregard the alternative theory of modified gravity instead of dark matter, a paradigm that is not widespread in the scientific community today, the role of dark matter is to generate gravitational wells, the most massive of which range up to the mass of galaxy clusters of around 10^{14}–10^{15} solar masses. Thus, each galaxy resides at the heart of a gravitational well that is globally dominated by dark matter, forming a so-called "dark matter halo". Over the past few years, observations have developed tools that have allowed the cosmic evolution of galaxies to be linked to that of dark matter halos, leading to one of the field's biggest paradoxes: the content of galaxies in luminous baryons, whether the latter take the form of gas, dust or stars, is well below the universal proportion of 16% baryons and 84% dark matter. The proportion of luminous baryons – i.e. baryons observable at any wavelength accessible from the ground or space – in a dark matter halo is at best six times lower than than the universal proportion of 16%, as observed in galaxies of around 5 × 10^{10} Msun, and can be up to 50 times lower at the two extreme ends of the spectrum, in galaxies with masses much higher or lower than this preferential value. Were the baryons ejected from their galaxies by galactic winds? Or were they left upstream, for example due to heating generated by an active nucleus or a gravitational shock generated by intergalactic gas falling into the galaxy? Perhaps some combination of both effects or even another process that remains to be discovered? Today, this question lies at the heart of the problematic of the cosmic evolution of galaxies.

For a long time, it was thought that dark matter primarily acted on the cosmic evolution of galaxies through mergers of dark matter halos and the galaxies at their centers, transforming spiral galaxies into elliptical galaxies and triggering starbursts. Various observations, which are described below, later raised doubts about this view, suggesting that it is too simplistic. Findings from observations and numerical simulations including hydrodynamic effects in the history of baryons within dark matter halos have initiated a true paradigm shift by suggesting that the cosmic evolution of galaxies is secular, i.e. continuous, without jolts or bumps, rather than highly variable, with streaks of starbursts caused by galaxy mergers. The supply of diffuse matter in the form of filaments of intergalactic gas, known as cold flows, has become a key ingredient of the cosmic evolution of galaxies and may in fact be the primary source of mass growth for galaxies (Dekel *et al.* 2009). In this new paradigm, galaxy mergers might prove to play only a minor role in controlling star formation, and we may find that their influence on the morphogenesis of galaxies has also been exaggerated in light of certain galaxies that formed many of their stars *in situ* with a very high concentration of gas and interstellar dust at their centers.

6.2. Characteristics of galaxies used to define their cosmic evolution

6.2.1. *Decoding multi-wavelength radiation*

From X-rays to radio waves, galaxies have very diverse spectral energy distributions (SEDs) that depend not only on the characteristics of their stellar populations (e.g. age, metallicity), but also on their gas content (neutral and ionized), recent star formation activity, extinction effects from dust in the interstellar medium and whether or not an active nucleus is present in their central region. Analyzing and decomposing the SEDs of galaxies therefore gives us a wealth of information about their physical properties.

6.2.1.1. *Visible domain*

At wavelengths ranging from ultraviolet (UV, ~0.1 μm) to near-infrared (~2 μm) bands, the spectrum of a galaxy is primarily determined by the contributions of its stars, whose "black-body" radiation varies as a function of their surface temperature and hence mass. Thus, the stellar populations of a galaxy collectively emit continuum radiation whose shape tells us directly about the type of stars that dominate the galaxy's spectrum. For example, galaxies characterized by strong star formation activity typically have a very blue continuum that originates from young, hot ($T > 10{,}000$ K) and massive stars, which radiate at short wavelengths, for example in the ultraviolet band. Conversely, the spectra of galaxies where star formation has stopped tend to be dominated by the presence of old, cold (~3,000 K), and low-mass stars like our Sun, which primarily radiate at red and infrared wavelengths (Bruzual and Charlot 2003).

In the visible domain, the spectral distribution of galaxies is also characterized by the presence of chemical emission and/or absorption lines, which are added to the stellar radiation continuum. Emission lines primarily originate from gas in the interstellar medium that has been ionized by UV radiation from massive stars. The most commonly observed lines are the Balmer series (Hα, Hβ, Hγ, etc.) and oxygen lines ([OII] and [OIII]), as well as the lines of heavier elements such as sulfur, nitrogen and neon. Absorption lines, on the other hand, originate from the outer layer of evolved stars, which tells us about their chemical composition, and in particular their metallicity.

Finally, we can distinguish the presence of two characteristic discontinuities in the UV/visible spectra of galaxies. The Lyman discontinuity at 912 Angstroms (also known as the Lyman break) is caused by the absorption of shorter-wavelength photons by neutral hydrogen in galaxies; consequently, it occurs at the limiting wavelength of the Lyman series. The Balmer break, which occurs toward 4,000 Angstroms, is only observed in the spectra of evolved star populations. As we will see later, these discontinuities play an essential role in measuring the distance of various galaxies in the distant Universe, since the wavelength at which they are observed allows us to estimate the redshift of their source of emission, and hence their recession velocity and distance.

6.2.1.2. *Infrared and submillimeter domain*

The dust found in the interstellar medium has a considerable impact on the SED of galaxies. It absorbs very efficiently the photons at short wavelengths emitted by massive stars and re-emits this energy in the infrared domain ($\lambda > 5$ μm). Extinction therefore shapes the global SED of galaxies in its own way, not just by attenuating emissions in the blue and UV bands (described as spectrum reddening), but also by generating larger-wavelength thermal radiation whose spectral shape depends on the composition of dust grains and the intensity of the radiation field that heats them (Désert *et al.* 1990; Galliano *et al.* 2018). In some galaxies, the extinction rate can be so high that infrared luminosity (typically integrated between 8 and 1,000 μm) accounts for more than 95% of the bolometric luminosity of an object (Sanders and Mirabel 1996), which is why being able to accurately estimate this part of the spectrum of galaxies is so important.

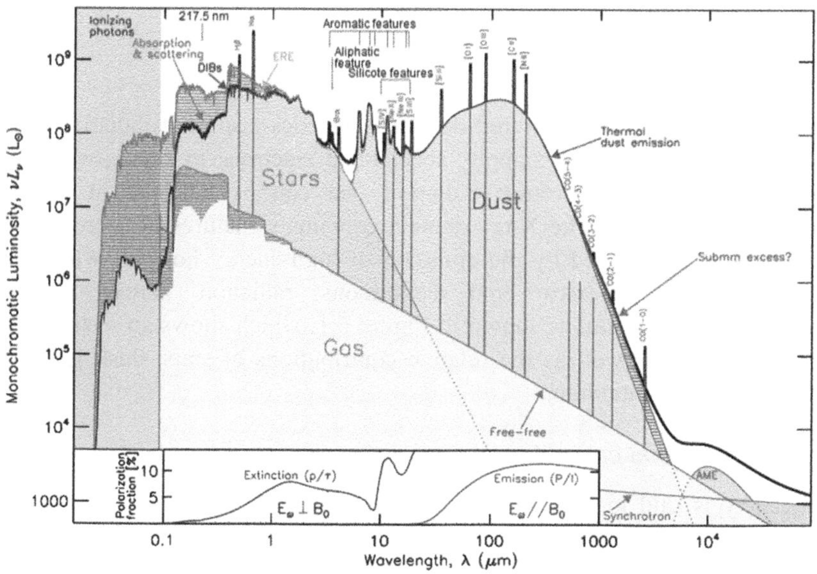

Figure 6.1. *Example of the spectral energy distribution of a galaxy, from UV to radio wavelengths, showing the contributions of stars in the visible domain (in blue) and dust in the infrared domain (in red), as well as the contribution of gas and radio synchrotron radiation (in green). Note also the presence of numerous emission lines, which originate from neutral gas and ionized gas (Galliano* et al. *2018)*

As well as the thermal continuum produced by dust in the interstellar medium, there is also a spectrum of relatively wide bands, primarily observed between 3 μm and 12 μm, that can be attributed to aromatic composites (polycyclic aromatic hydrocarbons [PAHs]). These particles are found in photodissociation regions, and their radiation tells us about recent star formation activity (Draine and Li 2007). Fine structure emission lines, such as neon, sulfur, silicon and argon lines in the mid-infrared region, also allow us to characterize the ionization conditions and the hardness of the radiation field in HII regions (ionized regions), as well as the local environments of active nuclei. Finally, in the far-infrared and submillimeter domains, other lines, such as [OIII]-88 μm, [CII]-158 μm and the transitions of the carbon monoxide CO line, allow us to constrain the properties of cold gas (e.g. Carilli and Walter 2013). In particular, the luminosity of the fundamental transition CO(1-0) is often used to estimate the total mass of molecular gas present in a galaxy. Dihydrogen H_2, the primary component of this molecular gas, is not bright enough to be directly observable. Instead, it is typically estimated from the mass of CO by assuming a conversion factor whose value depends on the type of environment and in particular its metallicity (Bolatto *et al.* 2013).

6.2.1.3. *Other wavelength domains*

Beyond the visible and infrared domains, galaxies also emit radiation at X-ray and radio frequencies. For example, accelerated electrons in supernova remnants generate continuum synchrotron radiation that can be detected at meter and centimeter wavelengths. The X-ray domain contains signatures of accreting binary star systems, characterized by the emission of high-energy hot gas, which is also detectable. Thus, galaxies emit continuous radiation across the entire electromagnetic spectrum, as shown in Figure 6.1, which shows an example of the SED of a galaxy, as well as the relative contributions of stars, dust, gas and the synchrotron radio continuum.

6.2.1.4. *The special case of active galaxy nuclei*

A galaxy is said to have an "active galaxy nucleus" (AGN) when significant matter accretion activity supplies and enlarges the supermassive black hole at its heart. During these phases of intense activity, the viscosity forces in the accretion disc heat the surrounding matter, and the disc radiates an especially energetic continuum, most notably in the X-ray and ultraviolet domains. This continuum, whose shape is also modulated by extinction from dust, is added on top of the radiation of the galaxy itself, which complicates the interpretation and analysis of the observed total SED.

6.2.1.5. *Estimation of physical parameters*

As we saw just above, analyzing the SED of a galaxy gives us valuable information about the nature of the galaxy's various components, and thus about the physical parameters that characterize the galaxy. For example, a key parameter for understanding the cosmic evolution of galaxies is the star formation rate (SFR), which can be estimated in several ways (Kennicutt 1998). The SFR can be determined from the monochromatic radio luminosity or the total infrared luminosity (integrated between 8 μm and 1,000 μm), which exhibits one of the most robust correlations observed in the extragalactic Universe, and additionally have the advantage of being unaffected by extinction from dust (Condon 1992). If detection in the radio or infrared domains is not possible, the SFR can also be determined from the UV luminosity produced by young stars, or the lines of hydrogen ionized by star formation (Hα, Hβ, etc.), provided that extinction from dust is properly accounted for. Finally, if we have measurements of both the observed UV luminosity (not corrected for extinction) and the IR luminosity, we can calculate the balance of energy produced intrinsically by young stars even more precisely, and hence quantify the SFR even more accurately.

Note however that these various observables are only sensitive to radiation that is characteristic of massive stars ($M^* > 8\ M_{\odot}$), whose lifetimes never exceed a few dozen millions of years. The calibrations established between these quantities and the SFR therefore implicitly assume an "Initial Mass Function" (IMF) that describes the mass distribution of the stars created by each new episode of star formation – right down to the very smallest of them – which must also be specified when taking measurements.

The luminosity of the observed emission lines also gives us precious insight into the physics of galaxies. The lines of neutral gas enable us to measure the total mass of the gas supplies (atomic and molecular) that are potentially available for the formation of new generations of stars. The emission lines of ionized gas can also be used to measure the metallicity of the gas, estimate the SFR, or even quantify the ionization state of the interstellar medium or diagnose whether or not an active nucleus is present. The equivalent width of some of these lines (i.e. the ratio between the flux of the line and the flux of the underlying continuum) can also be used to estimate the relative extent of ionized gas emissions from young stars compared to the continuum produced by evolved stars, and hence quantify the average age of a stellar population.

Finally, by fitting so-called "stellar population synthesis" models to the spectrum of observed galaxies in the UV to near-infrared bands, we can reconstruct the global history of their star formation over time (Bruzual and Charlot 2003). With this approach, we can therefore also estimate physical quantities that are crucial for understanding the evolution of these galaxies, such as the total mass of stars that have already formed (M^*, expressed in units of solar mass), their metallicity, their average age, or the extent of extinction from dust. These models can also constrain the SFR, which is often averaged over the last 10–100 million years of the history of the galaxy, and which can then be compared to the SFR determined on the basis of the observables described above.

6.2.2. *Populations of galaxies*

6.2.2.1. *Blue galaxies and red galaxies*

The color distribution of galaxies in the local Universe has an especially remarkable bimodal character. Galaxies can largely be divided into two distinct classes: star-forming galaxies, which have blue colors, and red galaxies, where star formation has stopped. Even today, understanding the physical processes that lead to this bimodality remains one of the major questions of extragalactic astrophysics. Of course, to find the answer, we need to be able to identify the family of a given galaxy as unambiguously as possible. In the past, many studies relied on simple "color-magnitude" diagrams, where blue and red galaxies are separated by their *U-V* color (i.e. the magnitude difference between two bands), expressed in terms of an absolute magnitude or the stellar mass (e.g. Baldry *et al.* 2004). However, this approach causes star-forming galaxies that have been reddened by extinction to be classified together with red objects where star formation has ceased, skewing the estimate of the number of these passive galaxies. To remove the degeneracy between reddening caused by extinction and the red color of old stellar populations, we can combine the *U-V* color with another criterion, such as the *V-J* color (giving a so-called "*UVJ*" diagram), to separate the two populations more distinctly (Figure 6.2).

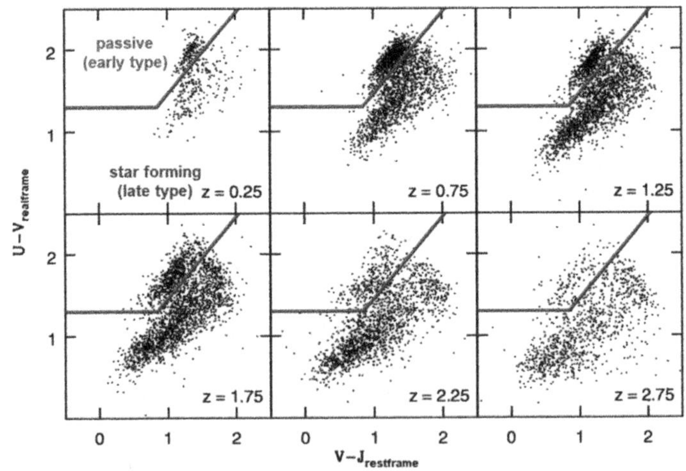

Figure 6.2. *Diagrams that allow "blue" star-forming galaxies (active galaxies) to be separated from "red" galaxies without star formation (passive galaxies) on the basis of the U-V and V-J colors at rest for different redshift intervals (van der Wel* et al. *2014). The blue galaxies are distributed along a diagonal line whose direction is given by the extinction vector in the UVJ plane. This clearly separates the red and passive galaxies in the top-left part of the diagram*

Since red and passive galaxies primarily have an elliptical appearance, we can also use morphological classification criteria to separate them from blue galaxies, which generally have discs with spiral arms or irregular morphologies. We can also separate the two classes using the specific star formation rate (SFR/M*), which measures the intensity of recent star formation relative to the mass of already formed stars. However, when studying distant galaxies, estimating these physical parameters and/or the morphology is not always straightforward. In such cases, the UVJ diagram is a very effective diagnostic tool, provided that we can accurately estimate the relevant colors in the rest-frame domain.

6.2.2.2. *Luminosity function*

Besides identifying different classes of galaxies, understanding their evolution at the scale of cosmic history also requires us to accurately quantify how each family is distributed per unit of volume, for example as a function of their luminosity. This is the role of the *luminosity function* of galaxies, which is generally accurately described by a so-called "Schechter" function (see Figure 6.3). Globally, the Schechter function behaves like a power law weighted by an exponential cut-off above some characteristic luminosity L*. It can be expressed as follows:

$$\phi(L)\, dL = \phi^* \, (L/L^*)^{\alpha} \, \exp\,(-L/L^*)\; dL/L^*.$$

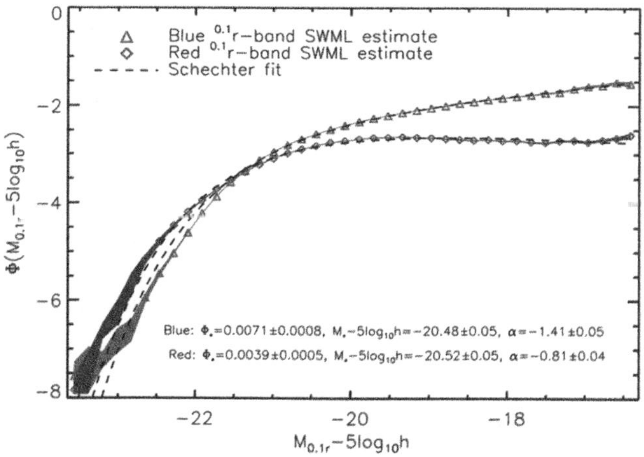

Figure 6.3. *Luminosity functions of galaxies in the local Universe obtained in the r band (0.6 µm) by the SDSS (Sloan Digital Sky Survey; Montero-Dorta and Prada 2009), expressed as a function of the absolute magnitude (M = –2.5 log$_{10}$(L) + constant, where L is the luminosity). The red and blue curves represent the luminosity functions of red and blue galaxies. The values of Φ* and M* (and hence L*) are shown in the figure*

In this expression, the term Φ^* is the normalization factor. At low luminosities, the term of the exponential form tends towards 1, and the luminosity function behaves like a power law characterized by a slope α. At high luminosities (i.e. above L^*), the exponential term dominates, and the density of galaxies decays towards zero very rapidly.

In the Local Universe, but also at larger distances, it has been observed that the density of galaxies increases as their luminosity decreases, which means that the integral of the Schechter function diverges if we extrapolate it beyond the limit of the weakest galaxies accessible to observations. However, if we are interested in the luminosity density rather than the numerical density of galaxies, we need to weight this function by the luminosity L. In this case, it can be shown that the integral of $L \times \phi(L)$ converges for the typical values measured for the slope α. Moreover, this new function attains its maximum at around $L \sim L^*$. We can therefore consider that the value L* corresponds to the luminosity of the most typical galaxies in terms of energy balance (i.e. the galaxies that contribute the most to the total luminosity density) and that their numerical density is directly given by the normalization factor 6 Φ^*.

6.3. Starbursts, secular evolution and universality of star formation

Over the course of its history, a galaxy may intermittently form stars at a highly accelerated rhythm compared to its average SFR. This is known as a starburst. The next section discusses the following questions: is there a universal and measurable definition of the starburst phase of a galaxy, i.e. can we unambiguously distinguish starbursts from galaxies that form their stars "normally" or secularly? What role did starbursts play in the cosmic history of galaxies; in other words, what proportion of the stars of a given galaxy formed during a starburst? Finally, what can starbursts teach us about the mechanisms affecting the star-forming efficiency of galaxies? Can they help us to identify the internal and external causes of the ignition of star formation?

6.3.1. *Definition of a starburst*

Some galaxies in the Local Universe, or galaxies with very low redshift, meaning that they are relatively local, formed their stars at rhythms 10–100 higher than the Milky Way. If they had continuously formed stars at this rate since their birth, these galaxies would now have a mass much larger than their current stellar mass. Therefore, this star-forming phase must necessarily be brief by comparison with the age of these galaxies, as shown in Figure 6.4. If we write <SFR> for the average

SFR of a galaxy, where <SFR>=stellar mass/age of the galaxy, then any galaxy in a starburst phase will have a "birthrate parameter" or "Scalo parameter" b well above 1, so that b = SFR / <SFR> >> 1. In practice, we speak of a starburst whenever b exceeds a value of 2–3 (Heckman 2005).

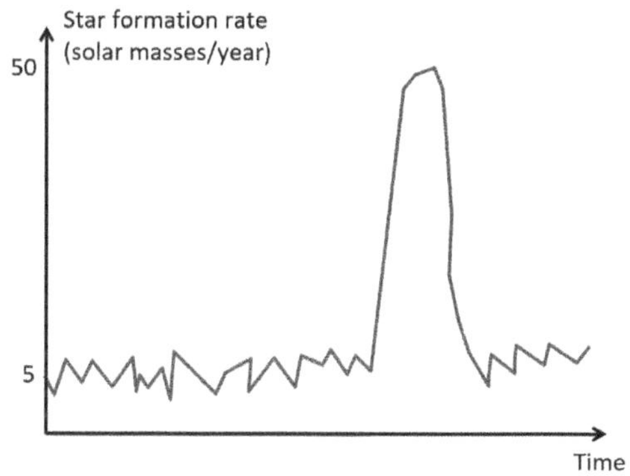

Figure 6.4. *Star formation history of a galaxy experiencing a starburst*

This initial definition is problematic when we attempt to apply it to galaxies in the distant Universe. Indeed, observations show that distant galaxies experienced a rapid growth in star formation since they formed, which implies that the b parameter is systematically higher than 1 throughout the early life of a galaxy. This undermines the view that a starburst is an exceptional phase; some other definition is required.

Another widely used definition of a starburst is based on the depletion time (t_{dep}) of the gas reservoir of a galaxy, either compared to the age of the Universe when the galaxy is observed (possibly approximated by the Hubble time) or compared to the age of the galaxy itself. Given that determining the age of a galaxy is difficult in general, the definition $t_{dep} << t_{Hubble}$ is commonly used. This definition assumes that it is possible to measure the diffuse gas reservoir of a galaxy. For the most extreme starbursts in the local Universe, we generally assume that the gas density is sufficiently high that it essentially consists of molecular gas H_2, with a negligible proportion of atomic hydrogen. Interestingly, the two prototypical starbursts, the Antennae Galaxy and the Super-Antennae Galaxy, have approximately the same depletion time of around 200 million years, even though the latter forms stars at rate of around 200 solar masses per year, whereas the SFR of the former is 10 times lower. This is because the Super-Antennae Galaxy contains 10 times more molecular gas (approximately $M(H_2) = 3 \times 10^{10}\ M_\odot$). The depletion time is equal to the inverse of the star formation efficiency: $SFE = 1/\ t_{dep} = SFR/M_{gas}$, so saying that a galaxy forms stars with high efficiency is equivalent to saying that it has a short depletion time. This definition is more robust than the first. Nevertheless, it also became problematic once we realized that galaxies in the distant Universe, at redshifts of z~2, are made of approximately 50% of interstellar gas and 50% of stars (Tacconi *et al.* 2010, 2018; Magdis *et al.* 2012). This proportion may seem high, but these galaxies were forming new stars at very high rates. The same holds true for every galaxy at these epochs. Consequently, the depletion time of these galaxies is around 600 million years. In other words, every galaxy that forms stars at z = 2.2 should seemingly have already depleted its gas reservoir by z = 1.8, which is clearly not the case. This observation is perhaps the most convincing argument that galaxies are not isolated systems but are constantly replenished with diffuse matter.

This last observation leads us to the third definition of a starburst, which requires us to introduce the concept of "star-formation main sequence" (SFMS).

6.3.2. *The SFMS and the secular evolution of galaxies*

Just above, we claimed that galaxies form their stars with a depletion rate of around 600 million years, as if star formation were in some way universal. This idea, proposed in 2007, triggered a shift in our conception of the history of galaxies. Today, starbursts are believed to play a less important role than was long thought to be the case. Sky surveys allowed us to collect complete samples of galaxies in well-defined zones of the sky. Of course, we still have not found every galaxy in every given volume of the Universe. Even nearby, we are still discovering dwarf galaxies or galaxies with low surface brightness. Nevertheless, it can be shown that it is possible to obtain a complete sample of galaxies, i.e. a sample that does not overlook any galaxy more massive than a certain critical mass within a specific volume of the Universe – for a given redshift domain and area on the sky. Sky surveys aiming to establish a complete overview of specific luminosity domains, generally in the ultraviolet band, were initiated as early as the 1990s with the Canada-France Redshift Survey (Lilly *et al.* 1996), followed by the Hubble Deep Field (HDF; Madau *et al.* 1996; Williams *et al.* 1996). Later, we will see how these surveys were used to establish a cosmic history of star formation. With the advent of deep images in the near infrared captured by ground or space telescopes equipped with the new WFC3 camera, which is capable of recording observations at 1.6 μm (H band), followed by images in the 3.6 and 4.5 μm bands from the IRAC camera of the Spitzer satellite, it became possible to measure visible radiation in the rest frame of galaxies and hence determine the stellar mass of galaxies to a constantly improving level of precision. Because of the catalogs of galaxies extracted from images obtained at multiple wavelengths, we can now find all galaxies more massive than a certain threshold, typically equal to a few 10^{10} $M_\odot$, in specific well-defined volumes of the Universe. In the local Universe, many galaxies were successfully analyzed thanks to the contributions of the Sloan Digital Sky Survey (SDSS). This allowed us to compare the star formation rhythms of galaxies at specific epochs of the Universe by restricting attention to galaxies with the same stellar mass, for example $M_\star = 4–5 \times 10^{10}\ M_\odot$. This led us to realize that, if we exclude "dead" galaxies that no longer form stars (red galaxies), then the star-forming galaxies (blue galaxies) form stars at a nearly universal rate. Regardless of the epoch, galaxies with the same mass form stars at the same rate, and thus have the same SFR, to within a factor of 2. More precisely, galaxies follow a log-normal distribution (the logarithms of the values are normally distributed) with a standard deviation of 0.3 dex, i.e. 68% of galaxies have the same SFR to within ±0.3 dex (to within a factor of 2), as summarized in Figure 6.5.

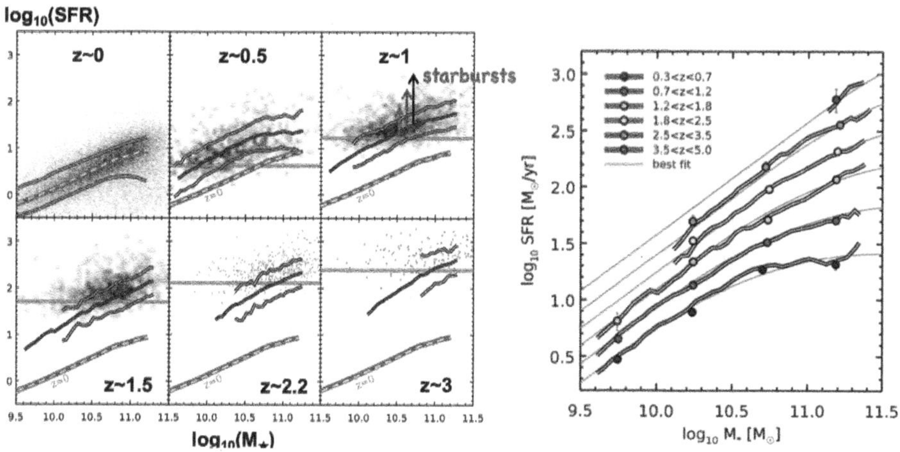

Figure 6.5. *Left: The star-formation main sequence or SFR – $M_\star$ relation, also called the SFMS, observed at six different epochs and hence redshifts, and compared to the local value (red and yellow line). Right: Evolution of the SFMS as a function of the redshift (figured reproduced from Schreiber* et al. *2015). The arrows at z~1 show the positions of starbursts with an intensity 3x, 5x, and 10x higher than the average of the main sequence*

The SFR – $M_\star$ correlation, also known as the SFMS relation, implies that most stars in the present Universe formed at some universal SFR determined by the mass of their galaxies. This is described as secular star formation.

6.3.3. *Origin of starbursts*

Starbursts are now defined according to a third approach based on the first two definitions described in the previous section: a starburst is a galaxy located above the SFMS, i.e. a galaxy that forms stars at a rate higher than the typical secular value. It is understood that the intensity of a starburst varies according to its distance from the SFMS. This intensity is measured by the $R_{MS} = SFR/SFR_{MS}$ index, i.e. the ratio of the observed SFR of a galaxy and the SFR that it would have had if it were located on the SFMS.

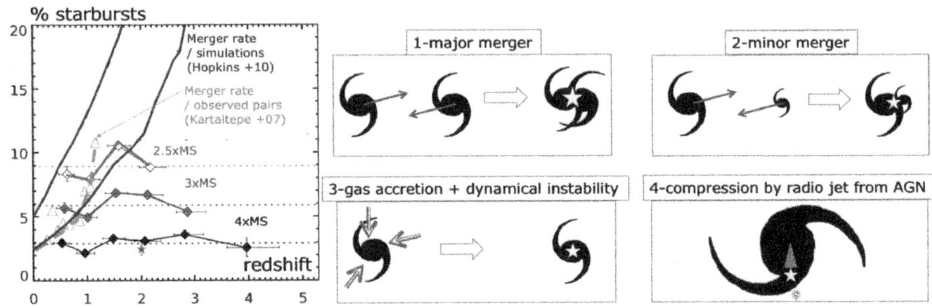

Figure 6.6. *Left: Evolution of the fraction of starbursts as a function of the redshift (figure reproduced from Schreiber* et al. *2015). Right: Diagram summarizing the four possible causes of a starburst*

The origin of the SFMS is still under debate, but it is thought to reflect the fact that galaxies form stars secularly, i.e. with low variation, due to a continuous supply of intergalactic matter that feeds them; this would explain why they continue to form stars despite having a low depletion time.

The origin of starbursts, i.e. why a galaxy might migrate to a position above the SFMS, is also debated. Four primary causes have been suggested in the literature, as illustrated in Figure 6.6: (1) a major merger between two galaxies of similar mass to within a factor of 3-4, which causes the gas reservoirs to concentrate, and accelerates star formation. This at least seems to explain the starbursts in the local Universe, the vast majority of which exhibit signatures of major mergers, such as tidal tails; (2) a minor merger between a massive galaxy and another galaxy at least four times less massive; (3) intense accretion of intergalactic gas, which generates a dynamical instability followed by a collapse of the central gas mass (e.g. Dekel *et al.* 2009; Dekel and Burkert 2014; Barro *et al.* 2017); (4) the impact of an active nucleus, which can compress the interstellar gas and trigger *positive feedback* that accelerates star formation instead of slowing it down (Elbaz *et al.* 2009; Silk 2013; Harrison 2017; Molnar *et al.* 2017). These mechanisms should be able to explain a troubling observation illustrated in Figure 6.6 (left): the proportion of starbursts does not depend on the cosmic time, i.e. the redshift, regardless of whether we define starbursts to be a factor of 2.5, 3, or 4 above the SFMS, and yet galaxy mergers were more common in the past. This topic is discussed in section 6.5.4.

6.3.4. *Secular evolution and the influence of the galactic environment*

If we separate the passive (red) and active (star-forming, blue) galaxies of the Local Universe, we observe that the dense regions that form groups and galaxy clusters are primarily populated by red galaxies, whereas blue galaxies are distributed relatively homogenously throughout space (see Figure 6.7). Denser environments are thought to accelerate the evolution of galaxies, transforming them into red galaxies more quickly and causing them to become more massive at an earlier stage, since red galaxies are also the most massive. Galaxy mergers should occur at an accelerated rhythm in dense regions.

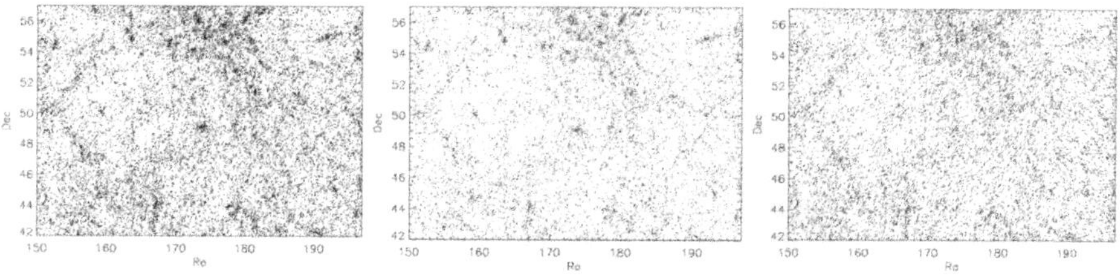

Figure 6.7. *Spatial distribution of galaxies at 0.005 ≤ z ≤ 0.1 in the SDSS (left), red galaxies only (center), blue galaxies only (right) (reproduced from Elbaz* et al. *2007)*

If we once again consider blue, star-forming galaxies in order to understand what exerts the greatest influence on the intensity of their star formation, we realize that the environment does not in fact have much of an effect on their SFR, contrary to the stellar mass (in accordance with the SFR – M* relation), as shown in Figure 6.8. Thus, the environment has indeed influenced the formation of galaxies, since most galaxies in groups and clusters are red and passive; however, it does not affect star formation activity in the Local Universe. As mentioned earlier, it therefore seems reasonable to propose that the influence of the environment on star formation might have taken place in the past and, based on our observations of local starbursts, to argue that its effect might have manifested as more frequent galaxy mergers. But there is a parameter of local galaxies that undermines this hypothesis. Studies of galaxies from the SDSS have shown that, if we consider how stars are spatially distributed throughout galaxies, their compactness does not depend on the environment (e.g. Hogg 2005; Kauffmann *et al.* 2017). For example, if we restrict attention to red galaxies, the compactness index (measured by the Sersic index) does not vary between low-density regions and high-density regions. The same is true for blue galaxies. This observation prompted David Hogg to end his article from 2005 by stating an enigma: "What physical processes can make the galaxy star-formation rate depend on its environment, while its morphology is not affected? I think that the conclusion has to be that the processes that set the morphology are somehow internal to the galaxies. That's going to be hard to reconcile with CDM cosmogony" (Hogg 2005). Since then, the ALMA radio telescope has allowed us to study star formation distributions inside distant galaxies. The findings of this study confirm D. Hogg's intuition: most star formation unfolds in a very concentrated manner inside galaxies, which implies that there must be some physical process that conveys interstellar gas and dust to the center of the galaxy very efficiently. The nature of this process is still debated to this day, and hydrodynamical simulations have struggled to reproduce it.

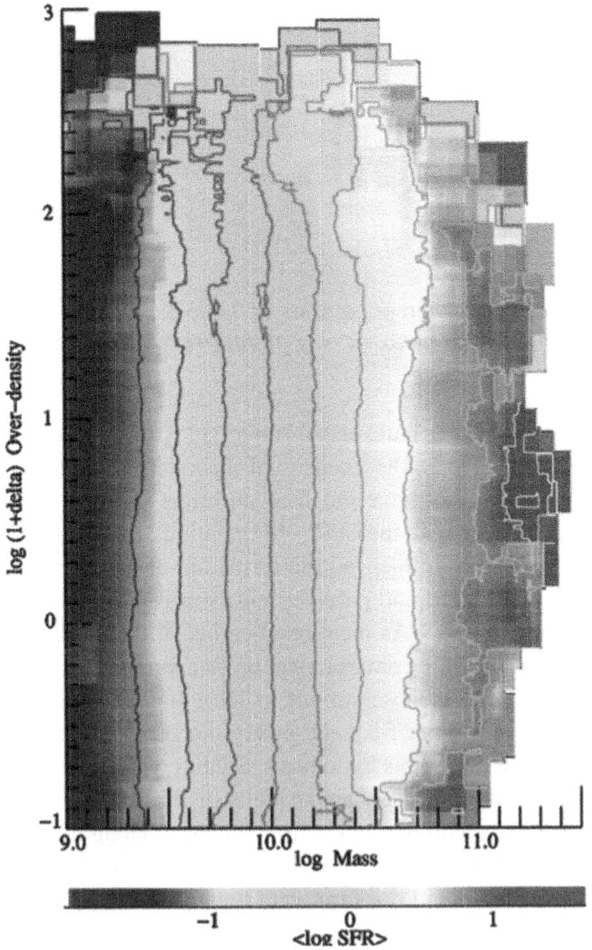

Figure 6.8. *Dependency of the star formation rate (SFR) on the galaxy density (Y-axis) and the stellar mass (X-axis)*

6.4. Detection of distant galaxies

The SED of a galaxy reflects the combined black-body spectrum of all of its stars, modulated by the presence of atoms/ions in the stellar photospheres, which are responsible for the absorption lines, and atoms/ions in the interstellar medium, which are responsible for emission lines. In theory, it should be possible to interpret this information about the present composition of a galaxy in terms of its star formation history by using the fact that the massive stars responsible for most ultraviolet radiation have a shorter lifetime than the low-mass stars that dominate in the near infrared and which store the majority of the galaxy's stellar mass. But there are major degeneracies, for example between the age and the metallicity of stars, and even low uncertainties when measuring magnitudes can have significant ramifications on the age of a stellar population, as can uncertainties in the convection within stars, the mix of elements associated with rotation, or the loss of mass. This imposes restrictions on the precision that we can hope to achieve for the formation and evolution history of a galaxy by measuring the SED alone (see, for example, Charlot *et al.* 1996). Despite such limitations, this approach – which could be described as galactic archeology – can tell us the global trends, averaged over the entire Universe, of the relative contributions of galaxies of a certain stellar mass to star formation, integrated at the scale of the Universe. Indeed, we can consider a given unit volume of the Universe, measured in millions of comoving parsecs (accounting for the expansion of the Universe) or Mpc, and determine the number of solar masses that form per unit of time, e.g. per year. This frees us from having to consider specific types of galaxy, as well as from the uncertainties associated with a history derived from a particular SED, providing us with a global history at the scale of the Universe. Historically, this global history was originally established after the first deep surveys of the Universe, as we will see below. But it can also be approximated by galactic archeology, as illustrated by Figure 6.9, which shows that the most massive galaxies of the Universe formed most of their stars in the distant past, whereas in the least massive galaxies star formation was spread out over time. This trend has been baptized "downsizing" (Cowie *et al.* 1996) to reflect the fact that, if we plot the mass of the galaxies where star formation is predominantly occurring as a function of time, we find that it decreases over the course of cosmic history.

This history seems to contradict the mainstream cosmological model, the hierarchical model of galaxy formation favored by so-called ΛCDM cosmology, which states that, due to the cold (non-relativistic) nature of dark matter (CDM stands for cold dark matter), less massive structures such as galaxies form first, and more massive structures gradually form later. This apparent contradiction is an opportunity to discuss one of the major limitations of deep sky observations, namely a bias that depends on the mass of galaxies: the most massive galaxies form in the densest regions of the Universe, where galaxy mergers occur at an accelerated rhythm. Because of this, the observation that massive galaxies formed quickly at remote epochs does not necessarily translate to a contradiction of the cosmological model; instead, it merely illustrates the importance of the environment in accelerating the evolution of galaxies.

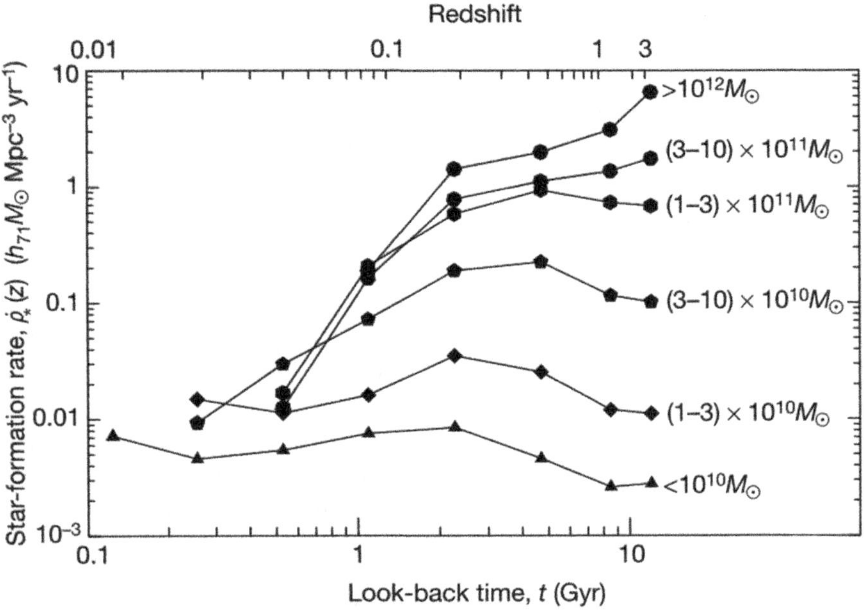

Figure 6.9. *History of the star formation rate in the Universe as a function of the look-back time (and redshift) (figure reproduced from Heavens* et al. *2004)*

As we saw above, the most massive galaxies today are red galaxies (with old stellar populations) found in especially dense regions of the Universe, such as galaxy clusters in the most extreme cases. Only a few percent of local galaxies are found in galaxy clusters, but this percentage increases with the galaxy mass; the trend becomes increasingly pronounced the further we look back into the past: every massive galaxy observed in the very distant Universe occupies a region that will become a galaxy cluster today. This observation helps us to appreciate the complexity of exploring the distant Universe. To resolve the degeneracies associated with translating the SEDs of galaxies into a star formation history, we can use:

1) the fact that, the more distant a galaxy, the more we see it as it was in the distant past;

2) the cosmological principle that the Universe is globally homogeneous and isotropic at scales greater than the large structures of the Universe.

This allows us to survey the deep Universe in search of the ancestors of current galaxies, just as an archeologist would attribute each layer of a core sample in the Earth's crust to a past age.

This approach assumes that we have access to deep images of the Universe that can detect galaxies at different epochs of the past, so that we can then relate these galaxies and deduce the history of present galaxies from them. We will see that, behind its apparent simplicity, the study of deep surveys of the Universe to unravel the cosmic history of the continuous formation of galaxies – the concepts of formation and evolution are no longer distinct in a context where galaxies are perpetually evolving – has many obstacles to overcome, which are discussed in the next section.

6.4.1. *Deep surveys of the Universe*

The quest to find distant galaxies in the Universe has several objectives. First, it aims to determine the epoch of the formation of the earliest stars and galaxies in the Universe. This is an observable quantity that should allow us to confirm or refute certain theoretical hypotheses about galaxy formation, or perhaps even our cosmological model, including the nature of dark matter. The intergalactic medium does not exhibit the Gunn–Peterson effect; in other words, light is not absorbed by hydrogen atoms, in the line of sight of quasars closer than a redshift of around z=6. This means that any light we receive from these sources passed through a globally ionized medium, not a medium containing neutral hydrogen. Given that the Universe became neutral and filled with hydrogen during recombination at around z~1,100, when protons and electrons combined to form hydrogen atoms 380,000 years after the Big Bang, the hydrogen atoms must have been reionized in intergalactic space before z = 6. The only known sources of ionizing radiation at these remote epochs are the earliest stars, galaxies and active galactic nuclei. Therefore, to answer the question of the physical origin of the reionization of the Universe we need to identify the earliest galaxies. This is one of the major goals of the Hubble satellite's successor, the James Webb Space Telescope (JWST). Another of its goals is to reconstruct the formation and evolution history of galaxies, in particular so that we may test it against our theoretical models and numerical or semi-analytical simulations of cosmic history.

The first searches for distant galaxies were notoriously unfruitful. They were based on a search for the Lyman α recombination line (Lyα, at 1,216 Å), which corresponds to the transition from the first excited state of the hydrogen atom to its ground state. Given that this is the most frequent transition of hydrogen, and hydrogen is the most abundant element, this line is the primary source of nebular emission from star-forming galaxies. In the 1980s, the prevailing idea (based on a suggestion in an article by Partridge and Peebles 1967a) was to search for distant Lyα emitters as candidates for "primordial galaxies", i.e. the earliest phases in the history of a galaxy. The correctness of this approach seemed obvious, and a technical trick was found to explore it: deep surveys using narrow-band filters that can effectively distinguish Lyα emitters from noise measured in the image. Yet repeated attempts were only able to detect a few distant galaxies, far fewer than the numerous primordial galaxies that had been expected. In their review on this series of failures, Djorgovski and Thompson (1992) concluded that there had been "no convincing detection of a population of normal galaxies in formation". Must we conclude that distant galaxies in fact form extremely few stars? In reality, the authors of these unfruitful studies had underestimated the diffusion of Lyα photons due to resonance with the neutral hydrogen atoms in the interstellar medium – a mechanism that causes the Lyα photons to be absorbed then reemitted in another direction (at the same frequency) – meaning that the proportion of photons propagating in the direction of the observer was actually almost negligible. Deeper and more extensive surveys in the 2000s, made possible by the class of 8- to 10-m telescopes, as well as more sensitive CCDs, later revealed the existence of these Lyα-emitting galaxies or LAEs (Lyman alpha emitters), but it was a wideband photometric technique – the Lyman break technique that led to the discovery of Lyman break galaxies (LBGs; Steidel *et al.* 1996) – together with the advent of ultradeep surveys, such as the HDF (Williams *et al.* 1996), that would ultimately profoundly revolutionize the field. With a field of view of 2.6 arcmin, observed for around 40 h for each position in the sky, the Hubble telescope was able to detect 3,000 galaxies up to a maximum magnitude of 28.2 that is 500 million times weaker than the limiting magnitude of visibility for the naked eye (6.5) (in the visible band, at 6,060 Å, in the AB system). Eight years later, it was succeeded by an even deeper field (Hubble eXtreme Deep Field [XDF], 2012). A window to the distant Universe had been opened and was being explored at every wavelength by every observatory on the ground and in space. The technique used to identify LBGs and distant galaxies by "photometric redshifts" is discussed in the next section. In just a few years, from 1995 to 2000, it took us from a Universe where only a few examples of extreme galaxies were known, at redshifts of $z > 1$, to a Universe where over a thousand such galaxies have been observed, up to redshifts of $z{\sim}6$–10.

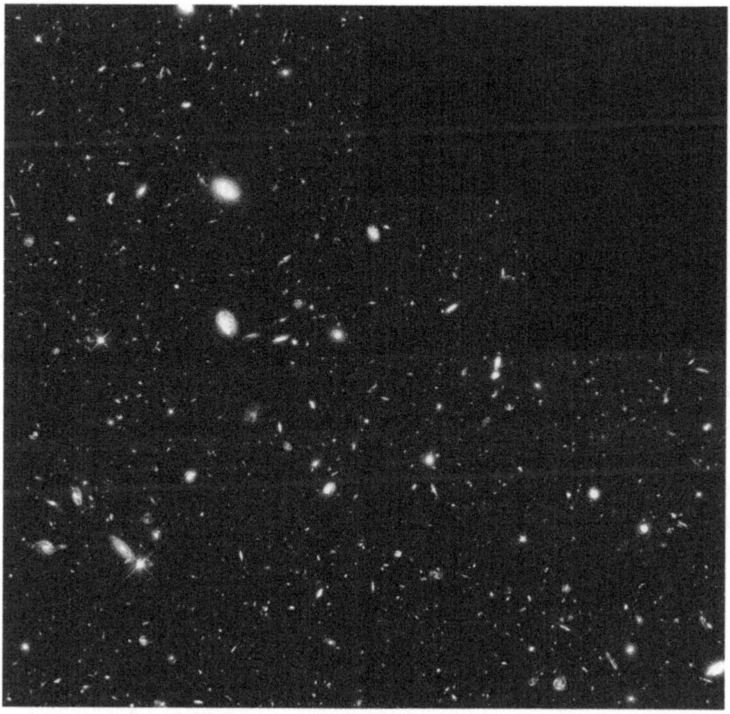

Figure 6.10. *Composite image of the Hubble Deep Field (HDF), constructed from images in the blue (4,500 Å), visible (6,060 Å) and red bands (8,140 Å) (figure reproduced from Williams* et al. *1996)*

6.4.2. *Large samples of distant galaxies*

6.4.2.1. *Lyman break technique*

Figure 6.11 illustrates the method used to identify an LBG. As discussed in section 6.2.1, the presence of neutral hydrogen along the line of sight of a galaxy triggers a break, or discontinuity, in the SED of a galaxy, with an abrupt drop in radiation at wavelengths lower than 912 Å. This break resembles a stairway; its effect is to make certain galaxies visible in green band (here, G filter, 5,100 Å) or red band images (R, 6,000 Å) and invisible in the U band (3,600 Å). Note that the break is reinforced by absorption from hydrogen encountered by light as it crosses clouds of neutral hydrogen gas in a largely ionized universe, the "Lyman alpha forest". The photons are absorbed at wavelengths where there are Lyman transitions from the ground state to excited states, i.e. at wavelengths below 1,216 Å, which adds an additional signature to the Lyman break.

In the case presented here, the break at 912 Å is shifted to an observed wavelength located between 3,600 Å and 5,100 Å, which indicates a redshift of between z = 3 and z = 4.5. The break does not occur abruptly at 912 Å, but continuously, due to the accumulation of absorption lines.

Clearly, this method can be used to detect large numbers of galaxies, since we can record their brightness on wide filters, but this comes at the cost of high uncertainty in their redshift. Nevertheless, by the late 1990s, it had already allowed hundreds of galaxies to be found at redshifts of 3 or 4, at a time where only a few such galaxies were currently known (see the review by Giavalisco 2002). A combination of many filters is used to refine the measurement of the redshift using the so-called "photometric redshift" method. Ultimately, the class of 8- to 10-m telescopes was the key that allowed us to measure the exact redshift of these galaxies by spectroscopy. Finally, note that the cosmological surveys performed in very recent years with the WFC3 camera of the Hubble telescope have successfully identified sources up to redshifts of $7 < z < 10$ by extending this "drop-out" technique to the near-infrared domain.

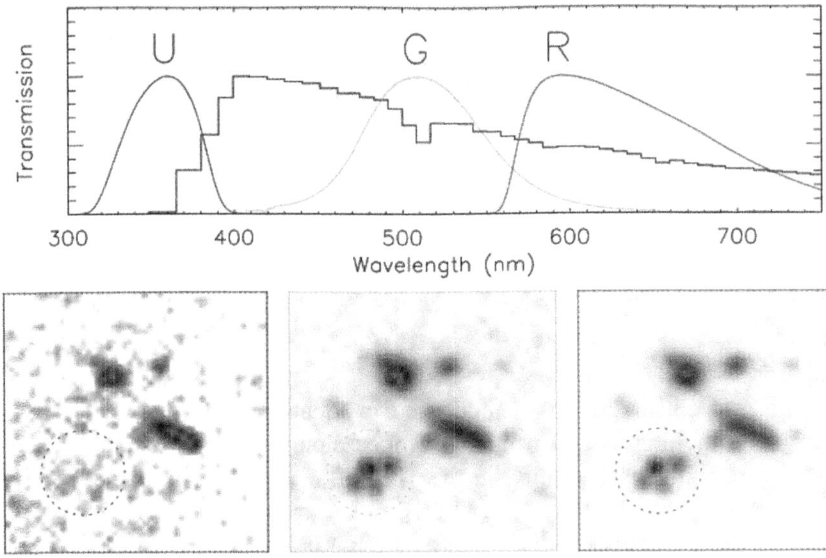

Figure 6.11. *Illustration of the method used to identify a Lyman break galaxy (LBG) (credit: P.U.Fynbo)*

6.4.2.2. *So-called "color–color" methods and photometric redshifts*

From the ground, the Lyman break technique can only be applied from the U band, i.e. for galaxies above a redshift of z~3. For sources with lower redshift (1 < z < 3), other techniques based on color–color diagrams are also very widely used to identify distant galaxies. These methods use the fact that these galaxies are characterized by apparent colors that change as a function of the redshift of their SED, which enables us to isolate source populations in certain specific redshift bands, depending on the combination of filters used. For example, some techniques are based on B, z and K filters ("BzK" method; Daddi *et al.* 2004); on the J-K color criterion for selecting "distant red galaxies" (Franx *et al.* 2003); on U, G and R filters to identify BM/BX sources (Steidel *et al.* 2004), or on the Spitzer satellite near-infrared bands to localize the stellar emission peak at around z~2 (Sawicki 2002). All of these "color–color" techniques are especially useful when, on a given cosmological field, imaging observations are only available through a limited number of filters.

However, today, many fields of distant galaxy observation have very rich multi-wavelength coverage with wideband photometry across the entire optical (U, B, V, g, R, I and z) and near-infrared (Y, J, H and K) domains, sometimes even with photometry obtained with narrower filters. These photometric data allow the SED of observed galaxies to be reconstructed relatively finely, at which point the "photometric redshift" technique can then be used to estimate the distance of these galaxies. This method involves determining the redshift that gives the best fit for the observed photometry based on a carefully selected library of spectral distribution models. Of course, the precision of the method depends on the number of photometry points available to sample the spectrum of each source, as well as the apparent magnitude of the galaxy (low-intensity galaxies have a lower signal-to-noise ratio, which means that models of their spectra are less accurate). The photometric redshift technique is very widely used in modern studies of cosmological surveys, as it can be systematically applied to every galaxy detected up to a certain flux threshold. It has proved to be an excellent complement to the approach of determining "spectroscopic" redshifts (by identifying spectral emission or absorption lines), which generally produces more accurate values but requires observations with much higher integration times.

6.4.2.3. *Multi-wavelength approaches*

Now that the same regions of the sky have been observed at every wavelength, we have access to additional information about measurements of the photometric redshifts of galaxies (e.g. Bolzonella *et al.* 2000) and the nature of sources: X-rays enable us to identify the presence of black holes in accretion phase, or active galactic nuclei (AGN), because of gas temperatures of a million degrees Kelvin in the close environment of the galaxy's black hole; UVs offer insight into massive and hence young stars; the near-infrared domain tells us about low-mass stars; the mid-to-far infrared bands give us data about interstellar dust, which is heated by young generations of stars, absorbing the UV radiation that they emit, then radiates in these bands after reaching temperatures of several dozen or hundreds of degrees Kelvin; and radio waves allow us to learn about synchrotron emissions in supernova remnants. But, for distant galaxies, we encounter a series of problems when attempting to assemble these various pieces of information.

To identify the same galaxy in different images of the same region of the sky, one obstacle is posed by differences in angular resolution between images. If the angular resolution is low, the diffraction spot will be large, and two different sources may be superimposed, which leads to a threshold for detecting individual sources called the confusion limit (see Dole *et al.* 2006). Additionally, even if the source is isolated, larger diffraction spots at higher wavelengths lead to uncertainty in determining the counterpart of this source at shorter wavelengths (see Figure 6.12).

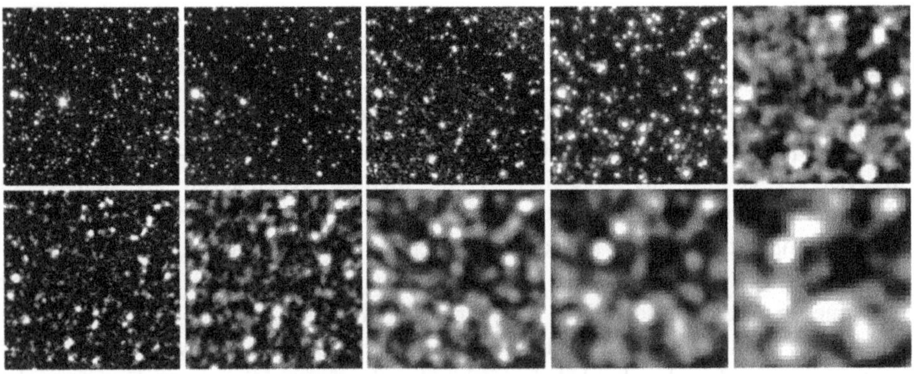

Figure 6.12. *Images of the same 5'x5' region in the sky in the GOODS-North field at 3.6, 8, 16, 24, 70, 100, 160, 250, 350, 500 µm (top-left to bottom-right; figure reproduced from Elbaz* et al. *2011)*

6.4.3. *K correction*

In a Euclidean space, the flux of a source located at a distance D can be calculated very easily from its luminosity ($F_\lambda = L_\lambda/4\pi D^2$). However, since our

Universe is expanding and frequencies are redshifted, the relation between the luminosity and the apparent magnitude of a distant galaxy in a specific band depends not only on the distance but also on a color term that accounts for the variation of the SED of the source between the wavelength of observation and the wavelength surveyed in the object's rest frame. As a concrete example, consider a galaxy at redshift z = 1, observed in the red band at 0.6 μm. The flux measured at this wavelength originates from the galaxy's emission in the blue band at 0.3 μm (=0.6 μm/(1 + z)), and the luminosity estimate in the red band therefore needs to account for the variation in the object's spectrum between the two bands.

This color term is called the "K correction". As discussed above, it depends on the SED of the observed source and can strongly affect the interpretation of the measured flux as a function of the redshift if the latter is not known. If so, we need to make a hypothesis based on previously established characteristics of the object, such as its morphological type or the intensity of its star formation activity (passive source? star-forming? starburst? and so on).

The K correction is generally expressed as follows:

$$K(z) = -2.5 * \log_{10}\left[\frac{\int F\left(\frac{\lambda}{1+z}\right)T(\lambda)\,d\lambda}{\int F(\lambda)\,T(\lambda)\,d\lambda}\,(1+z)\right],$$

where the function F describes the SED of the observed source and T is the bandpass transmission function. On some wavelength intervals, the shape of the SED requires a K correction term that can compensate most of the effect of the distance modulus. This is called a "negative" K correction. This effect is, for example, observed in the submillimeter domain, where galaxy emissions originate from the Rayleigh–Jeans tail of cold dust, which attains its maximum in the far infrared. At these frequencies, galaxies are surveyed on a portion of their SED that intrinsically becomes more luminous as the distance increases, which compensates for the decrease in the term $1/(4\pi D^2)$. At fixed bolometric luminosity, the flux measured at submillimeter wavelengths is then almost independent of the distance across the redshift interval ~2 < z < 10 (see Figure 6.13). This effect means that detecting very distant galaxies is just as easy as detecting galaxies at intermediate redshifts.

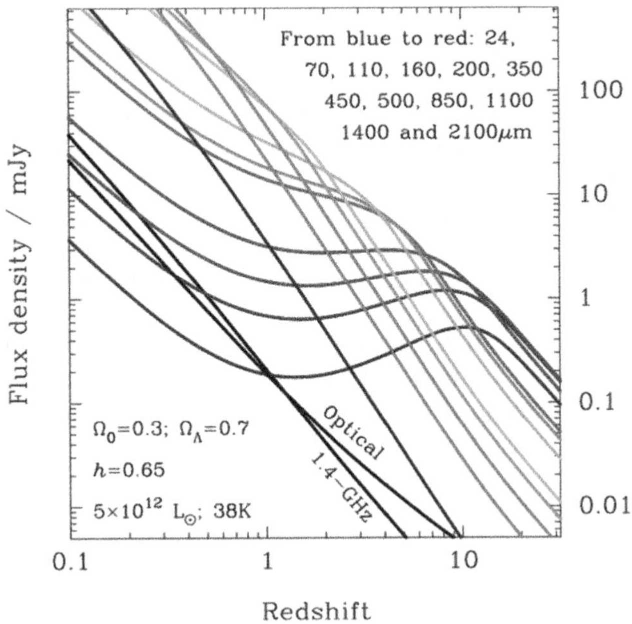

Figure 6.13. *Predicted flux at different wavelengths for an ultraluminous infrared galaxy as a function of the redshift (Blain* et al. *2002). Unlike in the visible, mid-infrared and radio bands (black and dark blue curves), where the observed flux decreases rapidly with the distance, the negative K correction in the submillimeter domain implies that the flux is almost constant at these wavelengths across a very wide interval of redshifts (red curves)*

6.5. Cosmic history of galaxies

6.5.1. *History of star formation*

In the 1990s, the first imaging surveys of the distant Universe revealed that galaxy counts at low fluxes had an excess of sources relative to the distribution that would be expected for a scenario without any evolution. This difference was notably stronger than tolerated by uncertainties in the cosmological parameters, which consequently implies that galaxies must evolve over time (either they were intrinsically more luminous in the past, or they must have been more numerous at a given luminosity). Subsequent spectroscopic follow-up studies then allowed distance of these galaxies to be determined and, for the first time, established more quantitative constraints on the evolution of their luminosity function and star formation activity, as well as on the evolution of the average star formation density in the Universe as a function of time. At cosmological scales, the latter quantity (often denoted $\rho_{SFR}(z)$) tells us about the total mass of new stars that form at any given moment in time per unit of volume. Its evolution as a function of the redshift, often called the "cosmic history of star formation", is directly linked to the mass growth of galaxies over time; it therefore represents a very strong constraint on structure formation models.

The very first estimate of this history of star formation was jointly published by Lilly *et al.* (1996) and Madau *et al.* (1996) for galaxies identified between z = 0 and z = 1 in the Canada France Redshift Survey, as well as for sources at z~3 selected from the HDF using the Lyman break technique. These observations show a very clear evolution in the star formation density over time, with a possible peak at around z~1. But the error bars were still very high at the time, due to various corrections that were proving difficult to master. First, recall that measurements of the SFR derived from visible wavelengths (which survey UVs in the rest frame of distant galaxies) depend very strongly on the dust extinction correction. In the 1990s, estimating this correction on a case-by-case basis was impossible because galaxies were generally only detected in a few bands, which meant that an average extinction value had to be used. Additionally, estimates were strongly affected by cosmic variance, due to the small size of early surveys. This variance manifests as density variations in galaxies across the cosmic web at small scales, depending on whether we encounter an empty zone or an overdensity. Its impact can be reduced by observing sufficiently large deep fields, or by combining several small fields in disjoint zones of the sky (like the approach used more recently by the CANDELS consortium in the 2010s; see Grogin *et al.* 2011). Finally, estimates of the star formation density need to consider *every* galaxy at a given epoch in cosmic history, not just the galaxies above a certain sensitivity threshold found by surveys. The completeness correction that must consequently be applied depends on the hypotheses about the shape of the luminosity function of the weakest galaxies that cannot be seen; it affects measurements more and more at larger distances, as more and more sources are "lost" below the detection threshold.

Over the last two decades, many deeper, broader and more comprehensive surveys have been conducted to complete the early surveys of the 1990s. They have allowed us to precisely constrain the evolution of the luminosity function of galaxies, both in the visible domain, where the sensitivity of our instruments enables us to sample their distribution right down to very low luminosities, and in the infrared and radio domains, where measurements are not skewed by extinction. The properties of galaxies have now been quantified much more reliably up to redshifts of z~3 (going back in time by ~11 billion years), and it has been shown that the density $\rho_{SFR}(z)$ at this epoch was ~10 times higher than today (see Figure 6.14). In fact, up to these cosmological distances, surveys performed in the infrared, millimeter and radio domains revealed that a non-negligible fraction of star formation occurred in environments that are heavily buried in dust, meaning that the extinction correction that needed to be applied to measurements at short wavelengths (UV, visible) was especially significant.

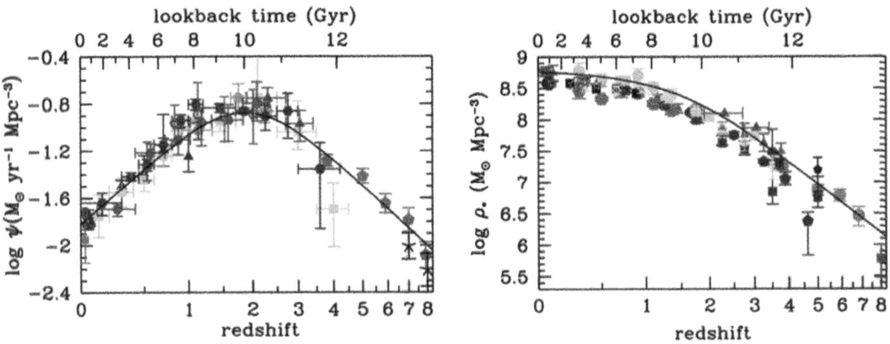

Figure 6.14. *Cosmic history of star formation (left) and mass growth (right) of galaxies, compiled from deep multiwavelength samples (Madau and Dickinson 2014). Star formation activity in the Universe experienced very rapid growth over the first few billion years after the Big Bang. It peaked at around a redshift of z~2–3, then gradually collapsed to a value 10 times lower than this peak today. On the left, the black curve shows the best fit to the data. The right-hand diagram shows its integral with respect to time*

At larger redshifts ($z > 4$), various estimates of $\rho_{SFR}(z)$ have also been established, primarily based on the identification of distant galaxies in deep observations of the Hubble telescope using the "drop-out" technique discussed in section 6.4.2. Even today, surveys at larger wavelengths are not sensitive enough to detect every galaxy at very high distances, so the extinction corrections that need to be made for these distant objects are still uncertain, and the exact shape of the evolution of $\rho_{SFR}(z)$ at these distances is still heavily debated. However, we do seem to have reached a consensus on the fact the star formation history of the Universe experienced very strong growth in the first 3 billion years after the Big Bang, reaching a maximum at around z~2–3, then decreasing gradually until today. Finally, it is worth noting that this star formation history, deduced from the evolution of the SFR of galaxies, measured as a function of the redshift, has been corroborated by measurements of the evolution of the mass growth of galaxies. The latter quantity is simply the integral of $\rho_{SFR}(z)$ over time and can be directly estimated in deep surveys via the stellar mass function and its evolution as a function of redshift (see Figure 6.14).

6.5.2. *Evolution of the stellar mass function and formation of the red galaxy sequence*

The evolution of the average star formation density in the Universe is a key element for properly understanding the mass growth of structures over time. However, it does not allow us to understand the formation of *every* galaxy, since, by definition, it does not include the population of passive galaxies where star formation has stopped. Another, complementary analysis is therefore to study the evolution of the stellar mass function of galaxies, for example by separating the evolution of red galaxies from that of blue, star-forming galaxies. Over the past decade, new ultra-high-performance cameras have made it possible to conduct deep and extensive surveys in the near-infrared bands, allowing us to survey the stellar mass of galaxies more reliably than with observations in the visible domain. This has allowed us to precisely constrain the evolution of the mass function of galaxies up to very large cosmological distances (z~5), covering almost the whole global formation history of structures. This work also succeeded in separating blue and red populations using the "color-color" techniques described above (see section 6.2.2) in order to specifically focus on the mass growth of passive galaxies over time (e.g. Muzzin *et al.* 2013; Grazzian *et al.* 2015; Davidzon *et al.* 2017).

As can be seen in Figure 6.15, the first passive galaxies appeared very early in the course of cosmic history. In particular, we can distinguish two distinct phases of their evolution. At every mass combined, their density initially experienced very rapid growth in the very first few billion years after the Big Bang (up to a redshift of z~2), with a characteristic peak at around $\sim 10^{10.5}$ $M_\odot$. Then, below z~2, the formation of massive red galaxies slowed down considerably, and only the red galaxies with low mass continued to increase in density. Globally, we therefore consider that the most massive red galaxies primarily formed over the first 3 billion years of the history of the Universe, whereas the formation of the majority of smaller red galaxies unfolded over much larger time scales; this scenario, which was demonstrated more than two decades ago, is widely known as the "downsizing of galaxies" (Cowie *et al.* 1996).

Unlike for red galaxies, the mass function of blue galaxies evolved relatively little over time; galaxies of intermediate mass (10^{10}–10^{11} $M_\odot$) are the ones whose density seems to have experienced the most growth. At very large redshifts, the mass function of star-forming galaxies seems relatively close to a simple power law that transformed into a "Schechter" function over time as a result of the increase in blue galaxies at these intermediate masses. At high stellar masses, we also observe that the density of blue galaxies has stayed globally constant at every redshift, and the mass function never exceeds a certain critical mass of around $10^{11.5}$ $M_{\odot}$. It therefore seems that galaxies can increase in mass by accreting gas from the intergalactic medium and converting it into new stars up to a certain threshold, beyond which star formation is abruptly interrupted. After reaching this critical mass, blue galaxies migrate toward the population of red galaxies, increasing the density of the latter at high masses, as observed above z~2. This formation "channel" of red and massive galaxies, where star formation being halted is especially effective at the highest masses, is known as "mass quenching". The origin of the phenomenon remains hotly debated. It perhaps occurs when the dark matter halo around the galaxy itself reaches a critical mass, inducing shocks and a temperature increase that prevent the gas in the interstellar medium from cooling and collapsing to form new stars (Dekel and Birnboim 2007). Mass quenching might also be caused by feedback from an active nucleus (triggered either by an internal process or after a major merger of two galaxies) that is sufficiently powerful to eject the gas supplies out of the disc (Hopkins *et al.* 2006).

Today, red galaxies outnumber blue galaxies above ~10^{11} $M_{\odot}$ (see Figure 6.15). But, as we noted earlier, their density has evolved very little over the past 8 billion years (since z~1), even though the density of blue galaxies with the same mass has remained almost constant. In these blue galaxies, the gas fraction and the SFR gradually dropped by a full order of magnitude over the same period, primarily because of the decrease in the rate of cosmic accretion of cold gas within their dark matter halos. This explains why the growth of their stellar mass proceeded at a much slower rate than in the first half of the history of the Universe. Conversely, the density of low-mass red galaxies has continuously increased until today, even though it is currently still lower than that of blue galaxies. This growth in the population of small passive galaxies seems to indicate the existence of mechanisms that terminate star formation regardless of mass and which are therefore distinct from the phenomenon of mass quenching. For example, environmental effects can halt star formation in high-density regions such as groups or galaxy clusters by systematically stripping away their gas reservoirs (such effects are described as ram pressure, galaxy harassment and strangulation). This is known as environmental quenching, as opposed to mass quenching.

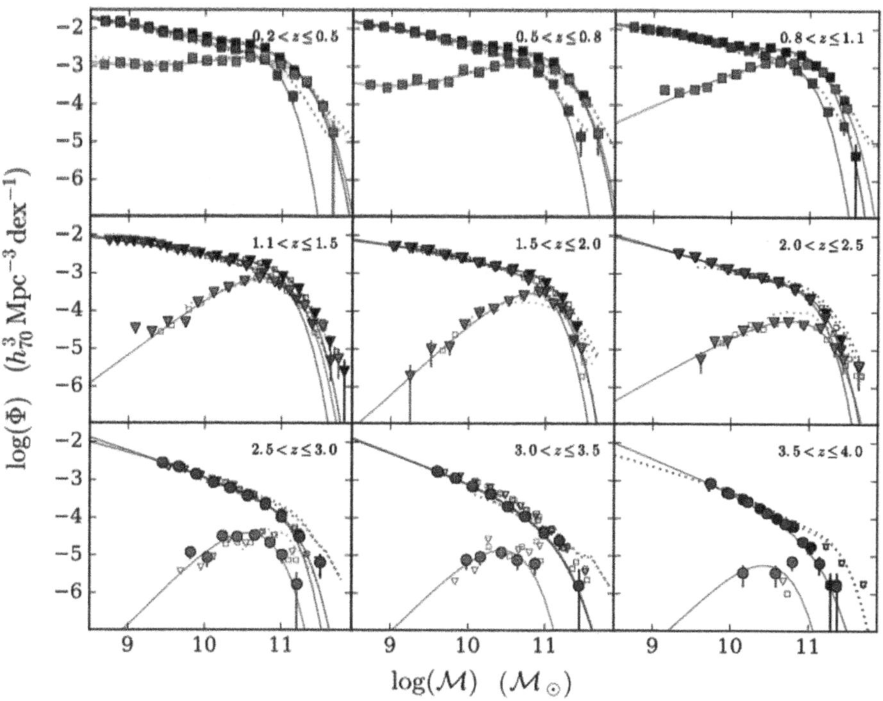

Figure 6.15. *Estimates of the stellar mass function of passive galaxies (red) and star-forming galaxies (blue) at different redshift intervals since z~4 (Davidzon* et al. *2017). The total mass function is shown in black*

6.5.3. *Evolution of the metallicity and of the average size of galaxies*

Other than the scaling relations between the stellar mass, the gas mass and the SFR within galaxies, correlations are also observed between the metallicity and average size. These relations are strongly dependent on the redshift, which imposes additional constraints on models of the evolution of galaxies.

Let us begin by considering the case of metallicity, which increases with the stellar mass. This correlation is observed up to redshifts of z~3 (Troncoso *et al.* 2014), but its normalization has evolved significantly; for a fixed stellar mass, the metallicity of galaxies was much less in the past (see Figure 6.16). At first glance, this evolution can simply be interpreted as an increase in the abundance of heavy elements, which is linked to the chemical evolution of stars and the interstellar medium over time. However, it has also been observed that low-mass galaxies experienced a much larger growth in metallicity than massive galaxies (Maiolino *et al.* 2008), which shows that the enrichment of galaxies in metals did not occur homogeneously at every stellar mass. This pattern is strongly reminiscent of the downsizing scenario of galaxies over the course of cosmic history; it reinforces the idea that the formation and chemical evolution of massive galaxies may have unfolded much more rapidly than for smaller galaxies. In today's Universe, the "mass-metallicity" relation is characterized by very low dispersion (~0.1 dex). It is generally interpreted as the signature of very powerful galactic winds, capable of ejecting heavy elements into the intergalactic medium, and whose action is much more efficient in shallow gravity potential wells and low-mass galaxies (Tremonti *et al.* 2004).

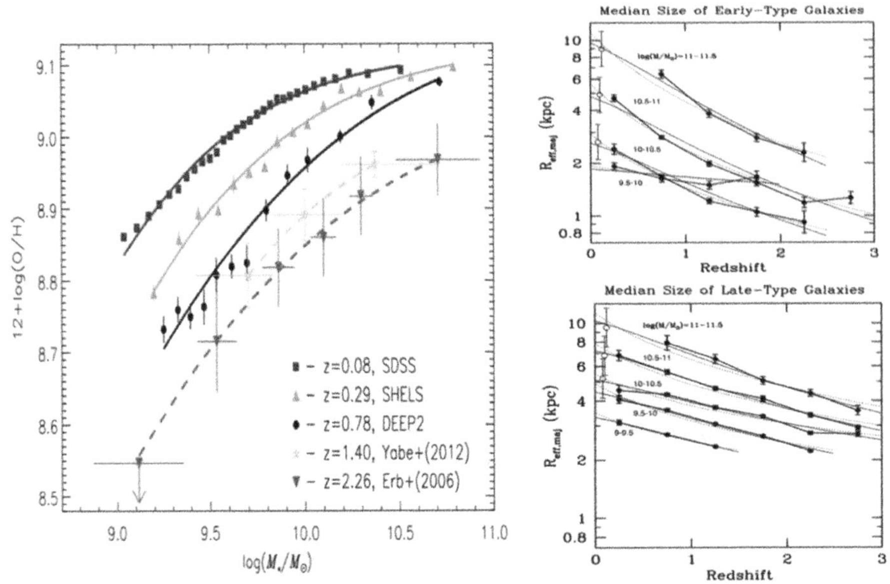

Figure 6.16. *Left: "stellar mass–metallicity" relation of galaxies, observed at different redshifts (Zahid* et al. *2013). Note the much more pronounced evolution of low-mass galaxies. Right: Evolution of the average size of passive galaxies (top) and star-forming galaxies (bottom) as a function of redshift for different stellar mass intervals (van der Wel* et al. *2014)*

The average size of galaxies also increases with their stellar mass; exactly as we observed for the metallicity parameter, their average size for a fixed stellar mass also gradually increased over the course of cosmic history (see Figure 6.16). Moreover, observations performed at high angular resolutions thanks to the Hubble telescope have shown, at least up to a redshift of z~3, that the relation between the mass and size of passive galaxies is not the same as the corresponding relation for blue star-forming galaxies. For a fixed redshift and stellar mass, blue galaxies are larger than red galaxies on average, and their dependence on the stellar mass is more pronounced at a given redshift. Over time, their size has also evolved more slowly than that of passive galaxies, which were conversely much more compact in the past than they are today. Therefore, the formation of these compact galaxies at large redshifts could only have taken place through some combination of particularly efficient compactification processes capable of compressing molecular gas into regions much denser than the star-forming regions observed in the Local Universe (Dekel & Burkert 2014).

The difference in average size between blue galaxies and red galaxies with the same mass (notably at low stellar mass, where the observed deviation is relatively large) raises questions about what physical mechanism could be capable not only of halting the star formation of a blue galaxy, but also of drastically reducing its size during the quenching phase (see section 6.6). Nevertheless, even though the "mass/size" relation of galaxies exhibits relatively little dispersion (~0.2 dex), the distribution of blue galaxies seems to have an asymmetric tail at small sizes, which matches the size distribution of passive galaxies (van der Wel *et al.* 2014). These blue galaxies might therefore provide the missing link of "transition" galaxies that are in the process of migrating toward the elliptical population.

In the even more distant Universe ($3 < z < 8$), the size of the blue galaxies selected by the "drop-out" technique has also been successfully constrained. The measured sources have smaller and smaller sizes as we go further back in the course of cosmic history, reaching effective radii as small as ~400 pc at redshifts of around z~8 (Shibuya *et al.* 2019).

6.5.4. *Evolution of the galaxy merger and starburst rates*

As we saw in section 6.3, some sources, known as starbursts, exhibit significantly higher SFRs than the average SFR of other galaxies with the same mass at the same redshift. In the Local Universe, the most intense starbursts are essentially triggered by interactions of galaxies, which create gravitational instabilities that can precipitate and compress the molecular gas reservoirs toward the center of the potential well to feed intense star formation activity. For example, "ultraluminous infrared" sources have an SFR exceeding 150 solar masses per year and are all associated with major galaxy mergers (Sanders and Mirabel 1996). These starburst episodes are generally short-lived (a few hundreds of millions of years; see Solomon and Vanden Bout 2005), but they have a drastic impact on the internal structure of galaxies (such as their morphology, their gas reservoirs and their star formation activity), and they may be the origin of the formation of the most massive spheroids. Similarly, minor galaxy mergers (i.e. mergers of galaxies whose mass ratio is 1:5 or higher) can also trigger a slight, temporary increase in the SFR, with more modest factors than major mergers, often too weak to knock these systems off the SFMS of galaxies. Thus, even though it can be difficult to quantify the increase in SFR that can be directly attributed to the interaction in the latter case, the phenomenon of starbursts in the Local Universe appears to be closely related to interactions between galaxies.

At larger cosmological distances, however, the relevance of this connection is more difficult to apprehend, in particular due to the increasing difficulty of properly discerning interacting systems among field galaxies. Indeed, in the past, galaxies were much richer in gas than they are today, which triggered larger gravitational instabilities within their discs. This could lead to morphologies that are naturally more perturbed and irregular, strongly resembling the morphologies that arise from major mergers. Today, several complementary approaches are used to evaluate the significance of galaxy mergers in the distant Universe with as little bias as possible. Non-parametric morphological analysis is one example; this approach automatically quantifies certain characteristics of galaxies, such as compactness, asymmetry, granularity and surface brightness distribution. By combining these properties, we can define criteria that allow us to separate interacting systems from isolated galaxies, while also reducing bias from irregular galaxies (Conselice *et al.* 2003; Lotz *et al.* 2007). Morphological classifications established with the human eye also remain widely used up to this day. Despite their apparent subjectivity, they escape certain systematic biases generated by automatic methods. In this type of approach, the impact of the "human" factor can be minimized by involving as many people as possible in the analysis and averaging the classifications produced by each of them (Kartaltepe *et al.* 2015). Finally, selecting "pairs" of galaxies by spectroscopy also allows certain interacting systems to be identified, especially ones that are still at an early stage (i.e. when the galaxies are still separated and there is not yet any visible perturbation of their morphology), which are consequently more difficult to find with the approaches mentioned above. In general, galaxies are considered to form a pair when their spectroscopic redshifts differ by less than 500 km/s, which usually implies the coalescence of two progenitors within less than a billion years (Kitzbichler and White 2008).

As observed in the Local Universe all of these methods for identifying merger phenomena between distant galaxies seem to indicate a very clear causal link between starburst episodes and interactions. Up to a redshift of at least $z{\sim}2$, we indeed observe that the fraction of major mergers increases very clearly with the extent of star formation activity relative to the average SFR of the main sequence.

For the most intense starbursts, with an SFR in excess by a factor of ~5 or more, this fraction even reaches nearly 100% (Cibinel *et al.* 2019; see the right-hand side in Figure 6.17). Similarly, it has been observed that pairs of galaxies, on the other hand, tend to remain within the main sequence. This is consistent with the idea that these systems are at an earlier stage where the interaction between the two progenitors has not yet triggered renewed star formation.

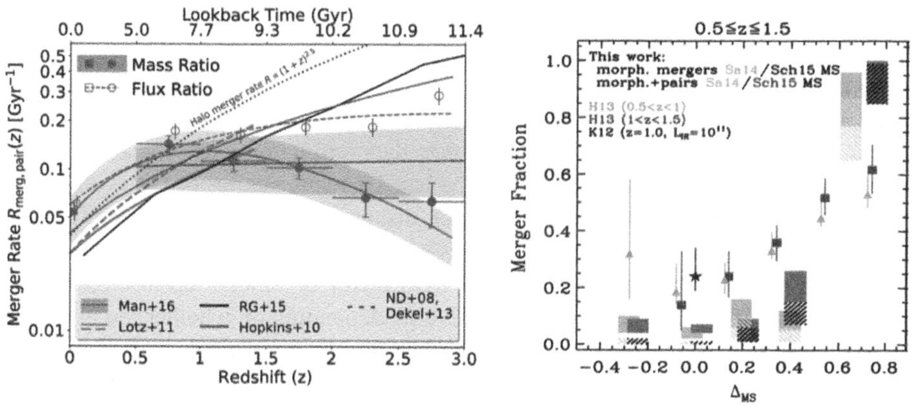

Figure 6.17. *Left: Evolution of the rate of major galaxy mergers (number of mergers per galaxy per billion years) compared to the predictions of various structure formation models (Mantha* et al. *2018). Note that there is a global consensus between observations (where mergers are identified by the presence of a neighboring galaxy at a distance ranging from 5 to 50 kpc, i.e. 50–500 thousand lightyears) and theory at 0 < z < 1, and significant divergence in the more distant Universe. The observations – represented by solid red dots fitted by a continuous red line – show growth between z = 3 (~11.4 billion years) and z = 1 (~8 billion years), followed by a drop until today (z = 0). The stellar masses of the two galaxies must be within a factor of four (the red circles fitted by the dashed red line contain more mergers at remote epochs because the technique used to identify major mergers is less reliable, as it is based on the flux rather than the stellar mass and consequently includes a greater number of "minor mergers"). Models do not explain the reduced rate of major mergers at z = 2–3 (from ~12 to ~10 billion years). Right: Fraction of galaxies at 0.5 < z < 1.5 classified as major mergers according to different techniques (morphological analysis, identification of spectroscopic pairs, etc.) as a function of the excess star formation relative to the average SFR of the main sequence (quantified by its logarithm Δ_{MS}). The measurements reveal a very pronounced link between "starburst phenomena" and "galaxy interactions" (Cibinel* et al. *2019)*

As we mentioned earlier, mergers between galaxies can have a considerable impact on their dynamics and internal structure. They often produce a major upheaval of the apparent morphology of galaxies; mergers might represent the origin of the formation of many elliptical galaxies (Hopkins *et al.* 2006). However, the contribution of these phenomena to the total star formation density seems to have been relatively negligible at the scale of galaxy formation. For instance, sources in starburst mode only account for a small fraction of field galaxies (around 5–10% depending on the definition chosen for the excess SFR relative to the main sequence), and this fraction seems to have stayed relatively constant over time (at least over the last 12 billion years; Schreiber *et al.* 2015). If we consider galaxies identified as pairs or classified as interacting systems based on their morphology, we again find fairly similar proportions, since they only represent 5% of the total population of the Local Universe. Over the last 10 years, studies of the evolution of the merger rate over time have led to much disagreement, due to the identification problems mentioned above. Nevertheless, we have successfully reached the consensus that this merger rate increased by a factor of ~2 or 3 up to a redshift of z~1. Beyond this point, recent estimates diverge depending on whether the impact of minor mergers is taken into consideration; the latter become much more difficult to detect as the distance increases (Mantha *et al.* 2018). Some studies seem to show that, although the global merger rate (minor + major) might have been ever higher beyond z~2, which agrees with the scenario of hierarchical galaxy formation, the significance of major mergers in fact diminished between z~1 and z~3 (Figure 6.17).

6.6. Origin of the cosmic history of galaxies

It might seem inappropriate to discuss a cosmic history of galaxies if we recall that, when considered individually, galaxies have (1) different shapes, (2) different proportions of stellar mass contained in bulges, thick or thin discs, bars or giant star formation regions (clumps of the order of the kpc), (3) different spectra and colors, indicating very different star formation histories. Nevertheless, when examining the star formation activity of a galaxy, we discover that it follows "scaling laws" that relate the galaxy's star formation intensity to its interstellar gas density (Schmidt–Kennicutt law; see Schmidt 1959, Kennicutt 1998) or stellar mass (SFMS). We can therefore use these two key parameters to examine the relevance of a cosmic approach to the cosmos, i.e. an approach that considers averages at the scale of the Universe, by studying the mass growth history of galaxies and hence their star formation history. This requires us to track the diffuse gas content (atomic and molecular) of galaxies over time, since this quantity is related to the star formation intensity; similarly, we need to know the progression of the stellar mass over time, since it describes the accumulation of all stars formed over time (and therefore the mathematical integral of the SFR, with a correction describing the loss of stellar mass over time). Finally, this cosmic history has a signature that can be measured from Earth, since the photons emitted over the 13.8 billion years since the Big Bang are still arriving at Earth today; they are simply spread out over the electromagnetic spectrum.

6.6.1. *The diffuse cosmic background as a signature of the cosmic history of galaxies*

Earlier, we cited the article "Are Young Galaxies Visible?" published in 1967, in which Partridge and Peebles presented their model of galaxy formation and recommended searching for the Lyα line as an ideal method to identify primordial galaxies. We saw that attempts to find these galaxies were initially predestined for failure due to the unexpected weakness of the Lyα line. Later the same year (1967), the same two scientists published a follow-up article, adding "II. The Integrated Background" to their original title. In this second article, the authors identify the signature of the cosmic history of galaxies for the first time among the background radiation that is constantly received on Earth. Their calculations were once again undermined by absorption effects, namely from interstellar dust. While they had predicted an emission peak between 5 and 15 μm, this peak was discovered 30 years later due to the COBE (Cosmic Background Explorer) satellite and its FIRAS (Far Infrared Absolute Spectrometer) instrument in the far-infrared band at a wavelength of 140 μm, 10 times higher than expected (Puget *et al.* 1996). With an intensity lower than that of the cosmic microwave background (CMB radiation, produced at recombination 380,000 years after the Big Bang, at a redshift of $z = 1{,}100$), the cosmic infrared background (CIB) holds a key element to the cosmic history of galaxies (see Figure 6.18). The CIB only contains around 2% of the energy of the CMB, but this energy does not come from the Big Bang; instead, it is radiated collectively by all of the star formation occurring throughout the entire history of galaxies. The black-body radiation underlying the CMB arises from the primordial plasma that preceded the formation of stars and galaxies, which was heated to 3,000 degrees Kelvin at the time of its emission, equivalent to a temperature of 2.7 K after accounting for the effect of the expansion of the Universe. The energy source of the CIB is twofold: nuclear fusion inside stars (*nucleosynthesis*) and the conversion of gravitational potential energy into light when matter collapses into a galactic black hole (*accretion*). This radiation is not emitted in the far-infrared band, even after taking into account the expansion of the Universe, but it does heat the interstellar dust, which radiates in turn with an emission peak corresponding to a black body heated to a temperature of around 30 K. This temperature is 100 times lower than the CMB, but the sources of this emission are found along the entire line of sight.

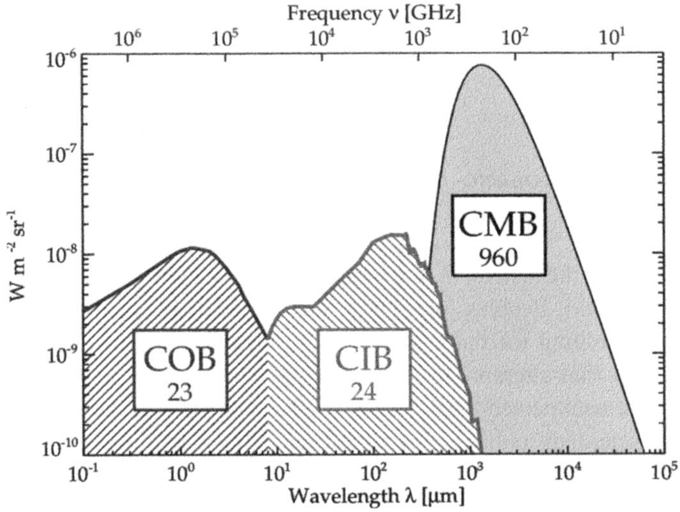

Figure 6.18. *Cosmic optical background (COB), infrared background (CIB) and microwave background (CMB) (figure reproduced from Dole* et al. *2006)*

The optical part of the cosmic diffuse background, the cosmic optical background (COB), is more difficult to determine, and there is currently no way to directly measure it. Instead, it is estimated from the emissions associated with individual galaxies along the line of sight. The intensity of the COB is approximately equal to that of the CIB, which allows us to deduce a global budget for the cosmic history of galaxies of around 47 nW m^{-2} sr^{-1}, or 47 billionths of a Watt received per meter square of sensor surface area from a celestial surface spanning one steradian (the entire celestial surface spans 4π steradians, and a steradian is equal to the square of a radian).

We can compare the intensity of the cosmic diffuse background, i.e. the integral of the radiation produced by galaxies over the course of their history, to the "ashes" of this cosmic fire by means of a simple calculation. Every photon emitted throughout the history of a galaxy originates either from the production of a heavy element, such as helium, carbon, oxygen or iron – the "ashes" of stellar nucleosynthesis – or from the production of a galactic black hole – the "ashes" of accretion. But we have a relatively good idea of the efficiency at which galaxies generate energy, and hence light, when producing these ashes. We therefore simply need to measure the mass of the ashes present in the Universe today to deduce the quantity of light generated over the 13.8 billion years since the Big Bang, then compare the amount of generated light to measurements of the cosmic diffuse background, in order to determine whether our picture of the history of galaxies is complete or incomplete.

Inside each star on the main sequence, the efficiency of light production during nucleosynthesis is determined by the binding energy of helium, which is equal to 0.74% of the mass energy, i.e. 0.0074 mc^2 (a proton of mass 0.94 GeV has a binding energy of 7 MeV). Accordingly, when a star converts one gram of hydrogen into helium, it radiates 0.74% of this mass in the form of light. The Sun radiates a luminous power of 3.9×10^{26} W; every second, it fuses 585 million tons of protons into helium nuclei, and 0.74% of this mass (4.3 million tons) is liberated in the form of light by the famous equation $E = mc^2$: $E = 4.3 \times 10^9$ kg $\times (3 \times 10^8$ m.s$^{-1})^2 = 3.9 \times 10^{26}$ W.

In 1 kg of matter in the Milky Way, there is 4% non-primordial helium (and 24% helium generated by the primordial nucleosynthesis of the Big Bang) and 2% heavy elements or metals, whose average binding energy is 0.84% mc^2. Thus, for 1 kg of galaxy matter, $0.0074 \times 0.04 + 0.0084 \times 0.02$=0.0464%, i.e. 46.4 g are converted into light, giving 4.2×10^{13} Joules. The Milky Way has a mass of around 7.8×10^{10} solar masses, which corresponds to around 6.5×10^{54} Joules produced. If we assume that this energy was radiated continuously over the 12 billion years of the Milky Way, we find a luminous power or luminosity of $6.5 \times 10^{54}/(12 \times 31.556 \times 10^6) = 1.72 \times 10^{37}$ W $= 4.55 \times 10^{10}$ solar luminosities ($L_\odot$), which is consistent with the luminosity of the Milky Way (4.55×10^{10} $L_\odot$, in the I band; Flynn *et al.* 2006).

Similarly, we can deduce the intensity of the cosmic diffuse background generated by stars, I_{stars}, from the quantity of helium synthesized in stars and the quantity of metals present in galaxies and intergalactic space, whether in diffuse gas or in stars. The density of light emitted by nucleosynthesis, ρ_{lum} ($\rho_{lum} = [0.0074 \times \rho_{He} + 0.0084 \times \rho_{metals}]$), is related to I_{stars} by the formula $I_{stars} = \rho_{lum} \times c^3/(4\pi \times (1 + z))$, where the factor of 4π means that the value is determined in sr^{-1}, and the factor of $(1 + z)$ accounts for the average redshift z of the galaxies that generated this light. The value of 3 in the exponent of c arises from multiplying the term c^2 in $E = mc^2$ by the velocity c at which the light is traveling toward us. Every second, over a collection surface spanning 1 m^2, we receive the light contained in a volume of c m^3. The values of the cosmic density of elements heavier than hydrogen and helium measured in the literature vary between ρ_{metals} = 4 and 27×10^{-31} kg m^{-3} = $1.5 \pm 1.1 \times 10^{-30}$ kg m^{-3} (Zepf and Silk 1996, Madau *et al.* 1996; Mushotzky and Lowenstein 1997; Calura and Matteucci 2004). Assuming that 4% of helium is not of primordial origin (corresponding to the difference between the cosmic primordial value of Yp = M_{He}/M_{total} = 0.24 and the solar value of $Y_{\odot}$ = 0.28), $\rho_{He} = 5 \times 10^{-30}$ kg m^{-3} (again with high uncertainty). Thus, for the order of magnitude of the cosmic diffuse background generated by stellar nucleosynthesis over the whole of cosmic history, we find $I_{stars} = 53 \pm 10$ nW m^{-2} sr^{-1} after assuming an average redshift of z = 1 (see Pozzetti and Madau 2001 for the COB and Lagache, Dole and Puget 2005 for the CIB).

To calculate the second component, the accretion by galactic black holes, we can use the cosmic density of galactic black holes, $\rho_{BH} = 4.6\ [-1.4,+1.9] \times 10^{-32}$ kg m^{-3} (Marconi and Hunt 2003) and the generally assumed value of around 10% for the conversion efficiency of gravity into light. This gives us a value of $I_{BH} = 5 \pm 3$ nW m^{-2} sr^{-1} at z = 1 for the contribution of accretion to the cosmic diffuse background.

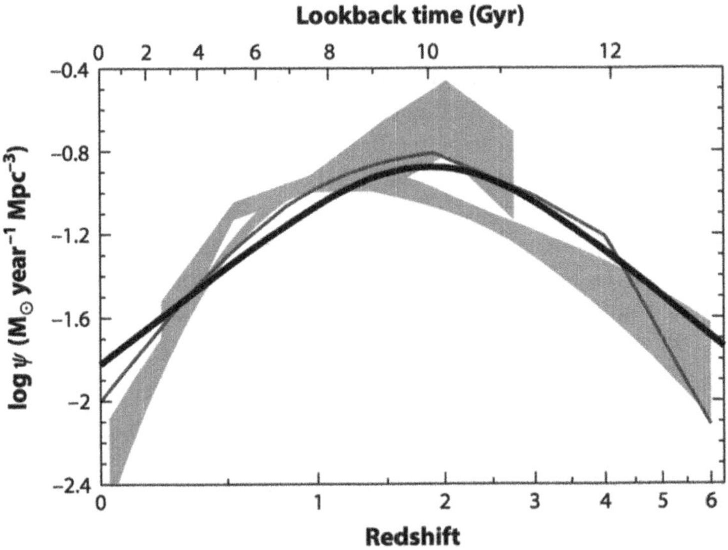

Figure 6.19. *Comparison of the cosmic history of star formation (black line, see Madau and Dickinson 2014) and the cosmic history of accretion by black holes based on X-rays (red line, see Shankar* et al. *2009; green zone, see Aird* et al. *2010) and infrared radiation (light-blue zone, see Delvecchio* et al. *2014) (figure reproduced from Madau and Dickinson 2014)*

In total, this gives $I_{galaxies} = I_{stars} + I_{BH} = 58 \pm 13$ nW m^{-2} sr^{-1}, a value that is consistent with the sum of the COB and the CIB, which is around 47 nW m^{-2} sr^{-1}, to within the large error bars. Thus, the amount of light that can be measured at any given time by collecting the light from the cosmic diffuse background does indeed coincide with the amount generated when the ashes of nucleosynthesis (helium and heavy elements) and accretion (galactic black holes) were produced over the course of cosmic history. Note that accretion is around 15 times more efficient at generating light per unit of mass than nuclear fusion, but the quantity of stars is much greater than that of galactic black holes, so the overall contribution of stars to the cosmic diffuse background is 10 times larger than that of accretion by black holes.

Since the global energy balance is of the correct order of magnitude, there does not seem to be any missing actors that we might have forgotten (to the first order of approximation); we can now attempt to unravel the origin of accretion and nucleosynthesis in the Universe, i.e. the cosmic history of the formation of stars and galactic black holes. To do this, we need to identify the individual sources responsible for the cosmic diffuse background. Although the galaxies responsible for the COB are relatively well known because of images from the Hubble Space Telescope, the galaxies responsible for the CIB have been the focus of many studies over the past couple of decades. It has gradually been established that the ISO (Elbaz *et al.* 2002), Spitzer (Dole *et al.* 2006) and Herschel (Oliver *et al.* 2010; Leiton *et al.* 2015) satellites have now identified the galaxies responsible for the majority of the CIB. Understanding the origin of star formation history in these galaxies will enable us to a give a physical interpretation of the cosmic history of galaxies. The IRAM and ALMA observatories have largely made this possible in recent years by measuring the reservoirs of interstellar matter at the origin of star formation in galaxies, and then, with ALMA, by establishing maps of regions where stars form within these galaxies.

6.6.2. *Unraveling the origin of the cosmic history of galaxies by studying their reservoirs of interstellar matter*

As we stated at the beginning of this section, the interstellar matter reservoirs of galaxies are one of the key parameters for understanding the origin of their evolution. Several methods are used to determine these interstellar matter reservoirs: the emission of the hyperfine structure line of atomic hydrogen (HI), which can be detected in the radio domain (21 cm) for atomic gas, the emission of the carbon monoxide line for molecular gas, and the emission of interstellar dust to deduce the total quantity of gas.

Due to its very weak dipole moment, the dominant molecule of dihydrogen (H_2) is very difficult to observe. However, its collisions with carbon monoxide (CO) molecules cause them to enter excited rotational states, emitting characteristic radiation that is measurable in distant galaxies. The CO molecule is said to be polar because it has a permanent electric dipole. Passing from its ground state to its first excited state generates light at a wavelength of 2.6 mm (115.27 GHz). By calibrating the relationship between the luminosity of the CO emission line and the mass of molecular gas, we can determine this mass by observing the CO line.

A complementary technique is to consider the emission of interstellar dust and deduce the mass of the dust responsible for this emission. The mass of the gas contained in a galaxy can then be calculated from the dust mass based on the ratio of dust mass to gas mass generally observed in galaxies. This ratio is higher in galaxies that are rich in heavy elements, so we also need to know this parameter to perform the calculation. To proceed, we use the strong correlation between the metallicity and the stellar mass of galaxies, the "mass–metallicity" relation discussed in section 6.5.3, since it is difficult to obtain a large number of spectra to measure the metallicity of a large number of galaxies, especially when they are distant.

The history of the gas fraction of galaxies as a function of redshift (Figure 6.20) shows that galaxies experienced a strong decrease in their interstellar gas fraction since the epoch where star formation peaked at the scale of the entire Universe, the "cosmic noon". This proves that the primary cause of the decrease in star formation comes from the reduction of the interstellar gas reservoirs of galaxies. In the past, the fraction of interstellar gas was larger by a factor of $(1+z)^{2.6}$, whereas the efficiency of the conversion of interstellar matter into stars has experienced very little variation, evolving by a factor of only $(1+z)^{0.6}$.

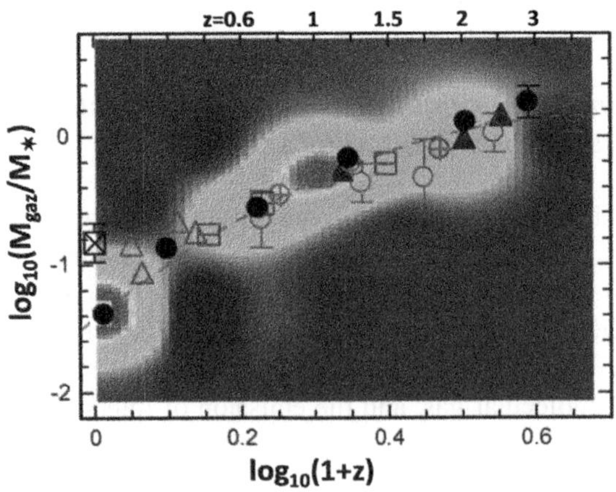

Figure 6.20. *Evolution of the interstellar gas fraction (M_{gaz}/M_*) as a function of the redshift z (figure adapted from Tacconi* et al. *2018)*

The drop of star formation in galaxies since the epoch of cosmic noon is therefore essentially explained by the exhaustion of gas reservoirs in galaxies. But the vast majority of baryons are found outside of galaxies, which raises the question of why they do not collapse into these galaxies. It is thought that the death of elliptical galaxies, i.e. the end of their star formation, initially occurs due to the heating of matter in the peripheral regions of the galaxy, known as circumgalactic matter, by shocks generated when this matter falls into the galaxy (Dekel and Birnboim 2008). These shocks destroy the filaments of intergalactic matter, which very heavily reduces the supply of matter to the galaxy. The jets of matter produced by active galaxy nuclei (galactic black holes surrounded by an accretion disk), then release enough energy into this circumgalactic gas to prevent it from cooling and keep it outside of the galaxies (maintenance mode; Dubois *et al.* 2016).

Thus, the cosmic history of galaxies is a testament to the paradoxical role of dark matter. At first, dark matter facilitates the formation of galaxies. Later, the strong gravitational potential of massive galaxies prevents them from growing, leading to the destruction of intergalactic filaments and heating from shocks as the intergalactic matter collapses.

6.7. Conclusion

The study of the cosmic evolution of galaxies has made remarkable progress over the past few decades. We now have a much better understanding of the physical mechanisms involved in the history of a galaxy. Nevertheless, the remaining questions are still crucial. Why is the Universe globally so inefficient at forming stars (10% of baryons)? Why are around two-thirds of all stars in the Local Universe found in regions that no longer have any star formation (galactic bulges and elliptical galaxies)? How do the thin, spread-out discs of galaxies form while resisting encounters with other galaxies? Which galaxies are responsible for the reionization of intergalactic medium? How do galaxies change shape over the course of their history? And, finally, why do galaxies evolve according to universal scaling laws that cause the SFR to depend on the stellar mass as well as on the gas mass or gas density, while also creating a relation between the color and the mass? As long as these mysteries continue to obscure part of the cosmic history of galaxies, it will be difficult to use these galaxies to test our cosmological model. One or several of these open questions might conceal some major conflict with our theoretical models. This perspective alone justifies the efforts undertaken today to advance our knowledge of the structuration of the Universe, as well as those planned for the future.

6.8. References

Aird, J., Nandra, K., Laird, E.S., *et al.* (2010). *MNRAS*, 401, 2531.

Baldry, I.K., Glazebrook, K., Brinkmann, J., *et al.* (2004). *Astrophysical Journal*, 600, 681.

Barro, G., Faber, S., Koo, D.C., *et al.* (2017). *Astrophysical Journal*, 840, 47.

Blain, A.W., Smail, I., Ivison, R.J., *et al.* (2002). *Physics Reports*, 369, 111.

Bolatto, A., Wolfire, M.G., Leroy, A. (2013). *Annual Review of Astronomy & Astrophysics*, 51, 207.

Bolzonella, M., Miralles, J.-M., Pelló, R. (2000). *Astronomy and Astrophysics*, 363, 476.

Bruzual, G., Charlot, S. (2003). *Monthly Notices of the Royal Astronomical Society*, 344, 1000.

Butcher, H., Oemler, A. (1978). *Astrophysical Journal*, 219, 18.

Calura, F., Matteucci, F. (2004). *Monthly Notices of the Royal Astronomical Society*, 350, 351.

Carilli, C., Walter, F. (2013). *Annual Review of Astronomy & Astrophysics*, 51, 105.

Charlot, S., Worthey, G., Bressan, A. (1996). *Astrophysical Journal*, 457, 625.

Cibinel, A., Daddi, E., Sargent, M.T., *et al.* (2019). *Monthly Notices of the Royal Astronomical Society*, 485, 5631.

Condon, J. (1992). *Annual Review of Astronomy & Astrophysics*, 30, 575.

Conselice, C., Bershady, M.A., Dickinson, M., *et al.* (2003). *The Astronomical Journal* 126, 1183.

Cowie, L.L., Songaila, A., Hu, E.M., *et al.* (1996). *The Astronomical Journal*, 112, 839.

Daddi, E., Cimatti, A., Renzini, A., *et al.* (2004). *Astrophysical Journal,* 617, 746.

Davidzon, I., Ilbert, O., Laigle, C., *et al.* (2017). *Astronomy & Astrophysics*, 605, 70.

Dekel, A., Birnboim, Y. (2008). *Monthly Notices of the Royal Astronomical Society*, 383, 119.

Dekel, A., Sari, R., Ceverino, D. (2009). *Astrophysical Journal,* 703, 785.

Dekel, A., Burkert, A. (2014). *MNRAS*, 438, 1870.

Désert F.X., Boulanger, F., Puget, J.L. (1990). *Astronomy & Astrophysics,* 237, 215.

Désert, F.-X., Boulanger, F., Puget, J.-L. (1990). *A&A,* 237, 215.

Delvecchio, I., Gruppioni, C., Pozzi, F., *et al.* (2014). *MNRAS* 439, 2736.

Dole, H., Lagache, G., Puget, J.-L., *et al.* (2006). *Astronomy & Astrophysics*, 451, 417.

Draine, B., Li, A. (2001). *Astrophysical Journal*, 551, 807.

Dubois, Y., Peirani, S., Pichon, C., *et al.* (2016). *MNRAS*, 463, 3948.

Elbaz, D., Cesarsky, C. J., Chanial, P., *et al.* (2002). *Astronomy and Astrophysics*, 384, 848.

Elbaz, D., Daddi, E., Le Borgne, D., *et al.* (2007). *Astronomy and Astrophysics*, 468, 33.

Elbaz, D., Jahnke, K., Pantin, E., *et al.* (2009). *Astronomy and Astrophysics*, 507, 1359.

Elbaz, D., Dickinson, M., Hwang, H.S., *et al.* (2011). *Astronomy and Astrophysics*, 533, A119.

Ellis, R. (1998), Nature, 385, 6701.

Flynn, C., Holmberg, J., Portinari, L., *et al.* (2006). *MNRAS*, 372, 1149.

Franx, M., Labbé, I., Rudnick, G., *et al.* (2003). *Astrophysical Journal*, 587, 79.

Galliano, F., Galametz, M., Jones, A.P. (2018). *Annual Review of Astronomy & Astrophysics*, 56, 673.

Giavalisco, M. (2002). *Annual Review of Astronomy and Astrophysics*, 40, 579.

Grazian, A., Fontana, A., Santini, P., *et al.* (2015). *Astronomy & Astrophysics*, 574, 96.

Grogin, N.A., Kocevski, D.D., Faber, S.M., *et al.* (2011). *Astrophysical Journal Supplement Series*, 197, 35.

Harrison, C.M. (2017). *Nature Astronomy*, 1, 165.

Heavens, A., Panter, B., Jimenez, R., *et al.* (2004). *Nature*, 428, 625.

Hogg, D.W. (2005). *NewAR*, 49, 379.

Hopkins, P.F., Hernquist, L., Cox, T.J., *et al.* (2006). *Astrophysical Journal Supplement Series*, 163, 1.

Kartaltepe, J.S., Mozena, M., Kocevski, D., *et al.* (2015). *Astrophysical Journal Supplement Series,* 221, 11.

Kauffmann, G., Nelson, D., Ménard, B., *et al.* (2017). *MNRAS*, 468, 3737.

Kitzbichler, M.G., White, S.D.M. (2008). *Monthly Notices of the Royal Astronomical Society* 391, 1489.

Kennicutt, R. (1998). *Annual Review of Astronomy & Astrophysics*, 36, 189.

Lagache, G., Dole, H., Puget, J-L. (2005). *Annual Review of Astronomy & Astrophysics*, 43, 727.

Leiton, R., Elbaz, D., Okumura, K., *et al.* (2015). *Astronomy & Astrophysics*, 579, A93.

Lilly, S., Le Fevre, O., Hammer, F., *et al.* (1996). *Astrophysical Journal,* 460, 1.

Lotz, J., Davis, M., Faber, S.M., *et al.* (2008). *Astrophysical Journal*, 672, 177.

Madau, P., Ferguson, H.C., Dickinson, M.E. *et al.* (1996). *Monthly Notices of the Royal Astronomical Society,* 283, 1388.

Madau, P., Dickinson, M. (2014). *Annual Review of Astronomy & Astrophysics,* 52, 415.

Maiolino, R., Nagao, T., Grazian, A., *et al.* (2008). *Astronomy & Astrophysics*, 488, 463.

Mantha, K.B., McIntosh, D.H., Brennan, R., *et al.* (2018). *Monthly Notices of the Royal Astronomical Society,* 475, 1549.

Marconi, A., Hunt, L.K. (2003). *Astrophysical Journal Letters*, 589, L21.

Molnár, D.Cs., Sargent, M.T., Elbaz, D., *et al.* (2017). *Monthly Notices of the Royal Astronomical Society*, 467, 586.

Montero-Dorta, A.D., Prada, F. (2009). *Monthly Notices of the Royal Astronomical Society*, 399, 1106.

Mushotzky, R.F., Loewenstein, M. (1997). *Astrophysical Journal Letters*, 481, L63.

Muzzin, A., Marchesini, D., Stefanon, M., *et al.* (2013). *Astrophysical Journal*, 777, 18.

Oliver, S.J., Wang, L., Smith, A.J., *et al.* (2010). *Astronomy & Astrophysics Letters*, 518, L21.

Partridge, R.B., Peebles, P.J.E. (1967a). *Astrophysical Journal*, 147, 868.

Partridge, R.B., Peebles, P.J.E. (1967b). *Astrophysical Journal*, 148, 377.

Pozzetti, L., Madau, P. (2001). *IAU Symposium*, 204, 71.

Puget, J.L., Abergel, A., Bernard, J.-P., *et al.* (1996). *Astronomy and Astrophysics Letters*, 308, L5.

Sanders, D., Mirabel, I.F. (1996). *Annual Review of Astronomy & Astrophysics*, 34, 749.

Sawicki, M. (2002). *Astronomical Journal,* 124, 3050.

Schreiber, C. Pannella, M., Elbaz, D., *et al.* (2015). *Astronomy & Astrophysics*, 575, 74.

Schmidt, M. (1959). *Astrophysical Journal,* 129, 243.

Shankar, F., Weinberg, D.H., Miralda-Escudé, J. (2009). *ApJ*, 690, 20.

Shibuya, T., Ouchi, M., Harikane, Y., *et al.* (2019). *Astrophysical Journal*, 871, 164.

Silk, J. (2013). *Astrophysical Journal*, 772, 112.

Solomon, P., Vanden Bout, P. (2005). *Annual Review of Astronomy & Astrophysics*, 43, 677.

Steidel, C., Shapley, A.E., Pettini, M., *et al.* (2004). *Astrophysical Journal*, 604, 534.

Tacconi, L.J., Genzel, R., Neri, R., *et al.* (2010). *Nature*, 463, 781.

Tacconi, L.J., Genzel, R., Saintonge, A., *et al.* (2018). *ApJ*, 853, 179.

Tremonti, C., Heckman, T. M., Kauffmann, G., *et al.* (2004). *Astrophysical Journal*, 613, 898.

Troncoso, P., Maiolino, R., Sommariva, V., *et al.* (2014). *Astronomy & Astrophysics*, 563, 58.

Tyson, J.A., Jarvis, J.F. (1979). *Astrophysical Journal Letters*, 230, L153.

van der Wel, A., Franx, M., van Dokkum, P.G., *et al.* (2014). *Astrophysical Journal*, 788, 28.

Williams, R.E., Blacker, B., Dickinson, M., *et al.* (1996). *Astrophysical Journal*, 112, 1335.

Zahid, H.J., Geller, M. J., Kewley, L.J., *et al.* (2013). *Astrophysical Journal Letters*, 771, 19.

Zepf, S.E., Silk, J. (1996). *Astrophysical Journal*, 466, 114.

List of Authors

Ronald BUTA
Astrophysics
University of Alabama
Tuscaloosa
USA

Françoise COMBES
Observatoire de Paris (LERMA)
and
Collège de France
Paris
France

Christopher J. CONSELICE
Astrophysics
The University of Nottingham
UK

Paola DI MATTEO
Observatoire de Paris (GEPI)
Meudon
France

David ELBAZ
CEA
Saclay
France

Eric EMSELLEM
European Southern Observatory
(ESO)
Garching
Germany

Emeric LE FLOC'H
CEA
Saclay
France

Author Biographies

Ronald BUTA is Professor Emeritus at the University of Alabama, USA. He is a world expert in galaxy classification. He produced a catalog dedicated to galaxy rings and participated in the Third Reference Catalogue of Bright Galaxies.

Françoise COMBES is a Professor at Collège de France and Chair of Galaxies and Cosmology. She is a specialist of galaxy evolution through bars and spiral waves, and has uncovered dynamical mechanisms to fuel central black holes in active nuclei.

Christopher J. CONSELICE is Professor of Astrophysics at the University of Nottingham, UK. He has worked on modern automatic galaxy classification (based on concentration, asymmetry, and clumpiness), and has quantified the fraction of interacting galaxies across the Universe.

Paola DI MATTEO is Adjunct Astronomer at Paris Observatory and the Principal Investigator of the GalMer project in France. She is an expert in numerical simulations of galaxies and studies the dynamical evolution of the Milky Way, using the fine results of the GAIA satellite.

David ELBAZ is researcher at the Commissariat à l'Energie Atomique (CEA-Saclay) in France. His expertise is in the evolution of galaxies across the Universe, using their infrared dust emission. He has discovered totally obscured galaxies in the early Universe.

Eric EMSELLEM is an astronomer, Head of the Office for Science at ESO (European Southern Observatory) in Germany. He specializes in red spheroidal galaxies, and quantified the degree of rotation of their stellar component, in fast and slow rotators.

Emeric LE FLOC'H is researcher at the Commissariat à l'Energie Atomique (CEA-Saclay) in France. His main interests are galaxy formation and evolution, in particular luminous starbursting galaxies and obscured active nuclei. He also studies gamma-ray bursts in the early Universe.

Index

T

V, W